Local Area Network Interconnection

Local Area Network Interconnection

Edited by

Raif O. Onvural

IBM
Research Triangle Park, North Carolina

and

Arne Nilsson

North Carolina State University
Raleigh, North Carolina

Springer Science+Business Media, LLC

Library of Congress Cataloging-in-Publication Data

Local area network interconnection / edited by Raif O. Onvural, Arne
Nilsson.
    p.   cm.
    "Proceedings of the First International Conference on Local Area
Network Interconnection, held October 20-22, 1993, in Research
Triangle Park, North Carolina"--
    Includes bibliographical references and index.
    ISBN 978-1-4613-6282-1   ISBN 978-1-4615-2950-7 (eBook)
    DOI 10.1007/978-1-4615-2950-7
    1. Local area networks (Computer networks)--Congresses.
I. Onvural, Raif O., 1959-   . II. Nilsson, Arne.
III. International Conference on Local Area Network Interconnection
(1st : 1993 : Research Triangle Park, N.C.)
TK5105.7.L582   1993
004.6'8--dc20   93-6370
                                                        CIP

Proceedings of the First International Conference on Local Area Network
Interconnection, held October 20–22, 1993, in Research Triangle Park, North
Carolina

ISBN 978-1-4613-6282-1

© 1993 Springer Science+Business Media New York
Originally published by Plenum Press, New York in 1993
Softcover reprint of the hardcover 1st edition 1993

COMMITTEES

Conference Co-Chairs

R. O. Onvural, IBM, Research Triangle Park
A. A. Nilsson, NCSU

Organizing Committee

J. Fjeld, IBM
J. Hart, 3Com
L. Kleinrock, UCLA
F. Marshall, CISCO
H. Salwen, Proteon
M. Schwartz, Columbia University
G. C. Stone, Network Systems

Program Committee

P. Baker, England
S. Bradner, USA
W. Bux, Switzerland
A. Danthine, Belgium
S. Fdida, France
L. Fratta, Italy
B. Gavish, USA
J. Gray, USA
U. Korner, Sweden
J-Y. LeBoudec, Switzerland
P. Martini, Germany

H. Perros, USA
G. Pujolle, France
D. Roffinella, Italy
O. Spaniol, Germany
D. Stevenson, USA
H. Takagi, Japan
K. Tolly, USA
K. Trivedi, USA
Y. Viniotis, USA
H. Yamashita, Japan

Conference Secreteriat

Y.C. Liu, IBM

Local Arrangements

M. Hudacko, NCSU
B. Sampair, IBM

Sponsored by
IEEE Comm. Soc. Eastern NC Chapter
IBM-Research Triangle Park
The Center for Communications and Signal Processing

In cooperation with
Interlab
IFIP TC6 Task Force on the Performance of Computer Networks
IFIP Working Group 7.3
IFIP Working Group 6.4

PREFACE

There are many exciting trends and developments in the communications industry, several of which are related to advances in fast packet switching, multi-media services, asynchronous transfer mode (ATM) and high-speed protocols. It seems fair to say that the face of networking has been rapidly changing and the distinction between LANs, MANs, and WANs is becoming more and more blurred. It is commonly believed in the industry that ATM represents the next generation in networking. The adoption of ATM standards by the research and development community as a unifying technology for communications that scales from local to wide area has been met with great enthusiasm from the business community and end users.

Reflecting these trends, the technical program of the First International Conference on LAN Interconnection consists of papers addressing a wide range of technical challenges and state of the art reviews. We are fortunate to have assembled a strong program committee, expert speakers, and panelists. We would like to thank Professor Schwartz for his keynote speech. We would like to thank Professor Yannis Viniotis and his students for the preparation of the index. We gratefully acknowledge the generous financial support of Dr. Jon Fjeld, Mr. Rick McGee, and Mr. David Witt, all of IBM-Research Triangle Park. We also would like to thank Ms. Mary Safford, our editor, and Mr. John Matzka, both at Plenum Press, for the publication of the proceedings.

Raif O. Onvural
Arne A. Nilsson

CONTENTS

QoS ENHANCEMENTS AND THE NEW TRANSPORT SERVICES

André Danthine, Olivier Bonaventure[1], Yves Baguette[2],
Guy Leduc[3] and Luc Léonard[2]

[1] Research Assistant of the Université de Liège
[2] Research Assistant of the F.N.R.S. (National Fund for Scientific
    Research, Belgium)
[3] Research Associate of the F.N.R.S.
    Institut d'Electricité Montefiore, B28
    Université de Liège,
    B-4000, LIEGE, Belgium
    E-mail: danthine@vm1.ulg.ac.be

INTRODUCTION

The transport protocols TCP and TP4 have been designed in the late seventies at a time when the network environment was essentially based on the switched telephone network and on leased-lines and when the packet switching was an emerging concept.

In ISO, the transport service was seen at the beginning as a unique service based on connection mode and, in this framework, one protocol class, TP4, was designed for the poor network service environment [Dan 90]. TP4, as well as TCP, were designed to be able to recover from the worst situation in terms of packet losses and packet disorders.

In the last ten years, we have seen a tremendous change in the communication environment. The LANs have drastically changed the scene and the high-speed LANs have accelerated the trend. More recently, the MANs have opened new possibilities not to mention the broadband ISDN which is expected to be deployed starting in the middle of this decade.

Not only the communication environment has drastically changed during the eighties, but we have also faced a drastic change in application requirements of which we will mention only the client-server paradigm and the multimedia-based applications.

In this paper we will first justify the need of new transport services and protocols based on the analysis of the consequences of the changes we have been facing

- in network performance
- in network services
- in application requirements.

We will, in the second part, present an enhancement of the QoS semantics in order to be able to specify the new transport service in a way compatible to the requirements of the transport service users. In our proposed semantics, a QoS parameter is seen as a structure of three values, respectively called "compulsory", "threshold" and "maximal quality", all these values being optional. Each one expresses a contract between the service users and the service provider. This means, and this is the most fundamental difference with the present situation, that the service provider is now submitted to some well defined duties.

The last part of the paper will be a brief presentation of examples coming from the OSI 95 transport services.

CHANGE IN NETWORK PERFORMANCE

Let us first point out the robustness of the OSI Reference Model which has been able to integrate the LANs in its layered architecture [Dan 89] and is very likely to integrate the B-ISDN.

The basic contributions of the LANs have been a drastic improvement of the network performance in terms of data rate as well as in terms of bit error rate (BER) and packet error rate. The error recovery was not any more a basic goal of the DL layer and this explained the tremendous success of the connectionless-mode.

The LANs have not only introduced an increase of the global communication capability due to the high value of their data transmission rate but, more important, they have offered the user an access data rate several orders of magnitude greater than what was available before.

If, as usual, the first implementations of the LAN controllers were not always able to offer the maximum access rate theoretically available at the MAC level [DHH 88a], the market, in about five years, was able to offer well-designed and well-implemented LAN interfaces.

It did not take long to realise that an access rate available at the MAC or LLC levels was far from being available above the transport layer [DHH 88a][Svo 89a, 89b][HeS 89][CCC 91]. This is not surprising, because opening a highway does not by itself change drastically the speed of the bicycles.

How to Improve ?

From the observations made in the second half of the eighties about the transport performance on top of LANs, detailed studies of mechanisms and implementations have been done [CJR 89], [GKW 89], [Mei 91], [JSB 90], [Zitt 91] and ended up in two different schools of thought.

The first school advocated the design of new transport protocols such as NETBLT [CLZ 87], VMTP [Che 88],[ChW 89], SNR [NRS 90] and XTP [Ches 89], [Wha 89], [PE 91].

The second school is putting the burden of the poor transport performance to the bad implementation quality and is claiming the capability of existing or slightly modified transport protocols to better handle the performance available at the MAC level [CJR 89].

Both schools have arguments and we would like here to discuss them, keeping in mind that these LANs do not only offer a better access rate, they are also characterised by a very low probability of corrupted packets by transmission errors.

Mechanisms and Implementations

We claim that there is a basic interest to adapt or to modify the protocol mechanisms when the network service is drastically changing. The following example will support our claim.

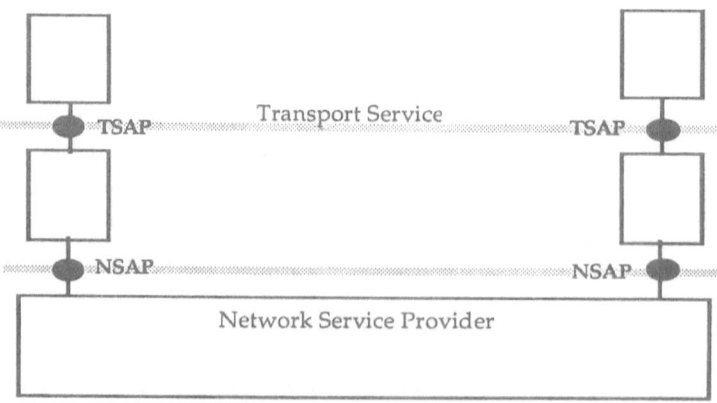

Figure 1. The basic OSI reference model

The figure 1 represents a simplified view of the basic OSI reference model and we assume, on the NSAP-to-NSAP association, a guaranteed throughput and a fixed round trip time. On top of such an well-defined N-service, we assume an ideal implementation of the transport entities which does not introduce any latency to process TSDUs into NSDUs. This does not mean however that the throughput and the round-trip delay on the TSAP-to-TSAP association, say a transport connection, will be equivalent to the values at the network level.

The product of "throughput x round-trip delay" - also called "bandwidth x delay" - is becoming, with high speed networks, an important characteristic to be dealt with. The value of such a product has evolved from 1 Kb (2 Kbps x 500 msec) to 600 Kb (10 Mbps x 60 msec), not to mention the very high speed environment. If the window associated with a transport connection does not allow the pipelining of enough megabits, the transport entity will be obliged to stop transmitting by lack of acknowledgements and of flow control credits. Therefore the performance at the transport level will be much lower than the performance at the network level **even with an ideal implementation of the transport layer.**

On the other hand, if protocol mechanisms may have such an influence on the performance even with an ideal implementation, it is also true that some mechanisms may induce very poor implementations. For instance a time-out per packet is not the best mechanism from an implementation point of view.

With a transport entity trying to send as many packets as possible, a lot of ACKs to handle represents a burden and the classical ARQ mechanism has therefore to be reviewed to reduce the penalties on performance.

Last but not least, there remain implementation problems which are independent of any protocol mechanism such as buffer handling and interaction with the operating system [CJR 89], [MaB 91].

There exist of course many choices of implementation from VLSI to pure software, more or less integrated in the operating system, using additional dedicated computing power or not [GKW 89], [JSB 90], [Zitt 91] .

The syntaxes of the PDU are not neutral with respect to the implementation results. Variable fields may save bandwidth but are more difficult to "siliconize". The place of the CRC field in the PDU is also important if hardware implementation with insertion and removal "on the fly" is envisaged.

With the availability of microprocessors of more than 10 Mips, with the possibility to use parallelism in the implementation of the transport entities, with the possible use of dedicated chips, we have a wide variety of possible implementations of the transport entities, all of them attempting to offer a transport service with a minimum drain on the host capabilities.

The conclusion to this section on the link between mechanisms and implementations is that even if it is true that the implementation of the protocol mechanisms is not always the most time consuming part of the transport activity, it will not be bad to have mechanisms which are well-aligned with the objectives and services of the protocol and which will ease the implementation problems. This does not prevent of course a careful choice of the syntaxes of the PDUs and an in-depth analysis of the memory management.

The Congestion Problem

During the eighties, we have not only seen the widespread usage of the LANs and recently of the high-speed LANs. We have also seen the building of very big internet-based networks following the efforts supported by the US DoD.

Such an internet network may give rise to very difficult problems of congestion which are almost impossible to tackle with the error recovery mechanism which has been introduced in TCP and TP4. The correcting action of the transport protocol in order to recover the lost packets due to the congestion is based on retransmission on time-out. These retransmissions contribute to an increase of the offered load and therefore to increase the congestion problems.

This is why congestion avoidance mechanisms which try to operate the network at the knee of the response time curve [RaJ 88], [ChJ 89] have to be preferred to the congestion control approach. Time-out retransmission must be associated with a reduction of the load on the network [Jai 86].

In the congestion problems, we have also two possible viewpoints. Either the mechanisms are aiming at the protection of the provider of the service, i.e. the network itself, in order to avoid a complete collapse of communication capability. Either, they are aiming at offering, to a service user, the best service possible at any given time. We do not have today a clear trend between these two views.

Congestion control may be exercised by the network service user, and the slow-start method in TCP [Jac 88] is the best known way to do it, or it may be exercised by the network service itself through some kind of rate control.

In any case, we have to deal with some reduction of the rate exercised either by the network service user or by the network service provider. It is very difficult today to have a clear view of the best solution. There exist a lot of proposals based on different ways to collect the congestion information. The congestion control may indeed be exercised on the basis of the information collected by the user through the observations of the behaviour of the service provider. As measurement bases already proposed, we have the rate of packet loss, the increase of round-trip delay, the occurrence of a timeout on an unacknowledged packet. Another proposal intends to base congestion control on information provided by the network in the form of loss-load curves [WiC 91]. The provider of service may also be involved by setting a special bit in packets going through a congested area. This piece of information is used to report internal congestion to the transport recipient which is able to act on the flow control mechanism in order to ease the congestion problem [RaJ 88], [RaJ 90].

The congestion problem is much more difficult to handle with a connectionless network service. In [Zha 87a,87b], it has been proposed to introduce, on a network association, the idea of a flow which will be used to handle a stream of packets. Setting up a flow, transmitting on a flow and terminating a flow appear similar to what we do for a connection but this flow is used only to handle resources on the way and is not associated with any service property. The protocol ST-II [Top 90] is based on the same idea and try to handle the congestion problem by a resource reservation scheme allowing to offer well defined performance. This is also the congestion problem which push XTP to handle the transport and network layers in an integrated transfer layer.

The deployment of corporate communication networks give rise to corporate internet networks where we face the same problem due to congestion of bridges or routers. The high-speed environment will just make things a little more difficult. For instance, if the high throughput goal requires not to use time-out on data packets, it will be impossible to base congestion control on time-out occurrences.

Increasing the resources to avoid packet loss due to congestion may be done by increasing the buffering capability of the intermediate nodes, but this will drastically increase the round-trip delay and may not be considered as a good solution [Jai 90].

The QoS (Quality of Service) in OSI as well as the ToS (Type of Service) in IP [Pos81], [Alm92] are specified in qualitative terms and are completely useless for the congestion control or the congestion avoidance.

For LANs, the probability that a frame be corrupted by a transmission error is very, very low, but the probability of packet loss due to congestion in the access device to a host, is not to be neglected. With integrated LANs through bridges and routers, the problem may get more complex and the congestion may be responsible for packet losses which will jeopardise the quality of the service.

To tackle this congestion problem, in the new high speed environment, we believe it is essential to have first a better specification of the service user requirements through more adequate QoS parameters associated with a well defined semantics and second a better way to specify the responsibilities of the service users on one side and of the service provider on the other side.

CHANGE IN NETWORK SERVICE CAPABILITY

If it has been possible to integrate the basic functionalities of the LANs in the OSI Reference Model, there exist subnetwork services which are available neither at the network level nor at the transport level.

Multicast Service

One of these services is related to the capability of addressing not only a given service access point, but a group of SAPs through multicast or all SAPs through broadcast. As we will see in the next section, many applications have a direct interest to see this multicast service not limited to the subnetwork level but offered at the network level as well as at the transport level. Some of the applications will require more than the connection-less based multicast and this will again require an enhancement of the QoS.

Synchronous Service

In FDDI, we have two types of services : the asynchronous service and the synchronous service. The synchronous service offered by FDDI I is a service where the average throughput is guaranteed as soon as it has been reserved.

The stream of bits coming from a constant rate source or from a variable rate source may be packetised and sent through an association offering a jitter on latency which does not compromise the re-synchronization at the acceptor [DRB 87]. This approach was applied in 1988 with the BWN, a 140-Mbps FDDI-like network developed in the framework of the ESPRIT project 73 [DHH 88b]. Videoconferencing with 2-Mbps codecs has been demonstrated on the connectionless BWN and this system was able to operate with a network loaded with more than 100 Mbps of background traffic.

With FDDI I and its synchronous service such an approach is possible even with a higher rate and with a guaranteed average throughput. However such a synchronous service is not available today neither at the network service level nor at the transport service level. Such an extension of the service will require at the transport service, a drastic change in the QoS semantics

Broadband Service

For the B-ISDN, even if the pace of the standardisation of the AAL (ATM Adaptation Layer) is slow, the AAL Type 3-4 is offering a service identical or comparable to SMDS or DQDB. It is not excluded that, through the signalling protocol still under development, the connectionless service will be associated with a guaranteed average rate.

For AAL Type 1, an isochronous service on SDU of more than 48 bytes may not be excluded and, if it happens, it is very likely to be associated with a guaranteed average or peak rate or to something in between.

For AAL Type 5, the data orientation is clear but the extension of the QoS is not yet in its final stage.

A Clear Trend

From the discussion of the new network services, it is obvious that we need to be able to introduce in all the lower layers of the OSI Reference Model, better addressing capabilities such as multicast, as well as to find a way to link the performance such as throughput into the QoS not only in qualitative terms but in quantitative terms.

CHANGES IN APPLICATION REQUIREMENTS

Beside the classical applications such as file transfers and conversational remote access to databases, the eighties have seen new orientations in the applications and these new applications have introduced new requirements for communication.

Client-Server Paradigm

The Client-Server paradigm is one of the important changes which took place during the last decade. From a communication point of view, it implies a low latency and an extended addressing capability.

The search of a server may take place by broadcasting or multicasting a request and the blocking character of the request implies the requirement for a low latency. This low latency is difficult to achieve when using the complete OSI stack. A system such as ANSA [Her 88] has chosen a collapsed OSI architecture based on a convergence sublayer, located in the lower part of the layer 7, acting as a transport service and giving access to the SAP of the subnetwork at the layer 2 interface. Others systems have developed better focused transport service and protocol [ChW 89],[MaB 91].

Today, most of the client-servers are implemented on a local basis, but the clear trend to use this new paradigm at the corporate network level will require a re-evaluation of the architecture presently used. If we hope to see the distributed computing systems integrated in the OSI architecture, it is mandatory for the transport service to offer the needed addressing capabilities as well as a transaction-oriented service with low response time as requests are most often of the blocking type.

Graphics and Colours

We have seen a constant increase in the power of the microprocessors used for the workstations and PCs but an analysis of the evolution puts into evidence that quite an important part of the power has been invested in the user interface and in a more advanced form of presentation of the results to the users. Multiple windowing, graphics usage as well as use of colours have fundamentally changed the friendliness of the workstations and paved the way for the widespread usage of them.

Of course, the volume of data which has to be exchanged in interactions involving windows, graphics and colours, is much more important than for the classical ASCII pages. The user, however, is not ready to accept to have long delays associated with the dynamic of the data presentation. Therefore the higher volume of data to be exchanged through the communication system on one side and the need to achieve what may be called the "client expected latency" on the other side mean that we will have a request for a high data throughput with a still very important bursty character. It is only if the global network service is offering the high value of the throughput associated with an acceptable latency that this kind of system will be usable on a wide area.

Multimedia

By multimedia we mean, as usual, a mixing of text, voice, graphics, image, audio and video. It is however important to make the distinction between two types of multimedia communications.

The videophone is a multimedia terminal as it is able to handle voice and video. But here the terminal is a source of a stream of bits which is very likely to be carried out on a physical or on a virtual circuit from one equipment to another equipment of the same type. The complete OSI architecture is very unlikely to be used as the information exchanged is not represented in machine-processable form, able to be processed, stored, retrieved and disseminated at any time.

The workstation where we will have the text in one window, a fixed image in another window and a video in a third one completed by some audio is an example of equipment requiring the second type of multimedia communication. Here all information which may be processed, stored, retrieved and disseminated are in machine-processable form. Here the video part of such an equipment will be seen as a sequence of data structures which have to be submitted to the presentation device at a rate compatible with the type of presentation we are looking for. As indicated earlier, such a video may be transmitted as a sequence of video frames, each frame being divided into packets at the transport level to be reconstructed at the remote end. The correct reconstruction requires from the transport service a sufficient throughput and a jitter on packet latency which is below some fixed value. Unfortunately, nothing today allows a user of TP4 or of TCP to request a quantitative value for the throughput on its connection, nor to express a limit to the jitter associated with the TSDU transport delay.

With video, the error recovery by retransmission does not work properly as it conflicts with the low jitter requirement. Such a data stream is not flow-controllable and if the transport service is not able to provide the required throughput, it will be reasonable to cancel the connection.

Most of today's multimedia applications are implemented on a stand-alone workstation. The trend to expand on network in the local area is already there and if some existing systems are working properly, it is very often because the transport service is able, through best effort, to offer a throughput sufficient for the application. This is true for an application implemented on a lightly loaded LAN, but may become less evident when the LAN will be more heavily loaded, when the LAN will be built with bridges and routers and, last but nor least, when we will try to have this application running on the wide area internet network. Here, we will face, as we go along these different environments, an increase of the latency and a possible reduction of the throughput.

All applications mentioned in this section require the possibility to associate, to a transport connection, quantitative compulsory QoSs with regard to the throughput, transit delay and delay jitter, and the service provider may have to intervene on the connection if the performance is not fulfilled.

NEED OF A NEW TRANSPORT SERVICE

In the previous sections, we tried to stress that the improvement of the transport service and protocol does not lie only on the performance, but that it will be necessary to extend the characteristics of the offered services. This opinion was already expressed in [ChW,89] to advocate a new generation of communication systems : *"By 'new generation' we mean that the design of protocols, networks and network interfaces should be rethought from the first principles in the context of the new environment. Refining existing designs and tuning conventional approaches in the area are inadequate, given the magnitude of changes that have taken place. At the same time, we need to be extremely careful to identify the key problems of the current standard protocols and come up with a convincing solution that is significantly better than these protocols. That is, we clearly recognise that an incremental improvement over existing protocols, such as the stream-oriented TCP or ISO TP4, would not warrant a new protocol or modifications to the existing protocols, given the enormous investment already in place."*

It is in complete agreement with this position that started, in October 1990, the ESPRIT II Project, OSI 95¹. The design of a new transport service and a new transport protocol, was the basic goal of the ESPRIT Project OSI 95 which involved Alcatel Bell, BULL, Alcatel Austria, INRIA, Institut National des Télécommunications, Intracom, Olivetti Research, as well as the universities of Madrid, Lancaster and Liège.

THE OSI 95 APPROACH

In the OSI 95 project, the University of Liège, which assumed the technical direction of the project, was responsible for the design and the formal specification of a new transport service and a new transport protocol. Two other tasks were done in parallel in order to better assess the new environment. The first one defined the specific requirements coming from the distributed multimedia applications and ODP systems which may have an impact on the new transport service. The second one defined the data link layer services provided by ATM networks.

The first step of the design of a new transport service has been the enhancement of the QoS (Quality of Service).The term Quality of Service (QoS) is the collective name given to certain characteristics associated with the invocations of (N)-service facilities as observed between (N)-SAPs. The QoS is described in terms of a set of parameters.

Taking into account the previous discussion about network performance, network services and application requirements, we believed in this context that it was essential to define, for the lower layers (4 to 2), a new model of QoS involving a new semantics of the QoS parameters and the definition of new parameters.

The strong requirements of some applications have driven us to define, through the semantics of the QoS, the characteristics of the contract between the service users and the service provider with a well defined behaviour if the requirements cannot be met and it is on the basis of this new semantics of the QoS that the new transport services have been designed [Dan 92a], [Dan 92b], [DBL 92c].

But before presenting our proposal, we will review the situation of the QoS since their introduction in the OSI RM in the beginning of the eighties. We will address mainly the connection-mode service, the problem of the QoS in the connectionless-mode being a subset of the first one.

TYPES OF QoS NEGOTIATIONS

We will restrict our discussion to the peer-to-peer case where the three actors of the negotiation are the calling (service) user, the called (service) user and the service provider. It is however possible to extend the discussion to the 1-to-N, N-to-1 and N-to-N multicast cases.

All negotiations are based on the classical 4-primitives exchange; request, indication, response and confirmation.

---

¹ OSI 95 is the acronym of ESPRIT II Project 5341 "High Performance OSI Protocols with Multimedia Support on HSLANs and B-ISDN".

For some performance parameter such as the throughput, the higher the better. For some other performance parameter such as the transit delay jitter, the smaller the better. For the following discussion we will use the terms "weakening" and "strengthening" a performance parameter to indicate the trend of the modification. Weakening a throughput parameter means reducing its value but weakening a transit delay jitter means increasing its value.

Triangular Negotiation for Information Exchange

In this type of negotiation, the calling user introduces in the request primitive the value of a QoS parameter. This value may be considered as a suggested value because the service provider is free to weaken it as much as it wants before presenting the new value to the called user through an indication primitive. The called user may also weaken the value of the parameter before introducing it in the response primitive. This final value will be included without change by the service provider in the confirm primitive. At the end of the negotiation, the three actors have the same value of this QoS parameter.

Taking into account the freedom for the service provider to weaken the value suggested by the calling user, the service provider will reject directly the request only if it is unable to offer the service whatever the value of the QoS.

The calling user has always the possibility to request a disconnection if it is unsatisfied by the value resulting from the negotiation .

The goal of such triangular negotiation is essentially to exchange information between the three actors.

The ISO Transport Service is using this type of negotiation for the performance-oriented QoS [ISO DIS 8072]. The classes 1 and 3 of the ISO Network Service [ISO 8348] is also based on the same scheme.

Triangular Negotiation for a Bounded Target

In this type of negotiation, the calling user introduces, in the request primitive, two values of a QoS parameter, the target and the lowest quality acceptable. The service provider is not allowed to change the value of the lowest quality acceptable. Here, the service provider is free, as long as it does not weaken it below the lowest quality acceptable, to weaken the target value before presenting the new value of the target and the unchanged value of the lowest quality acceptable to the called user, through an indication primitive. It will be the privilege of the called user to take the final decision concerning the selected value of the target. This selected value of the QoS will be returned by the called user in the response primitive. This selected value will be included without change by the service provider in the confirm primitive. At the end of the negotiation, the three actors have agreed on the value of this QoS parameter.

Here the service provider does not have anymore the possibility to weaken without limit the target value proposed by the calling user, due to the existence of the lowest quality acceptable value. The service provider may have to reject the request if it does not agree to provide a QoS in the requested range.

The called user may also reject the connection attempt if it is not satisfied by the range of values proposed in the indication primitive.

With respect to the target value introduced by the calling user, the only possible modification introduced by the negotiation is the weakening of the target but limited by the lowest quality acceptable value.

The class 2 of the ISO Network Service [ISO 8348] is also based on this scheme.

Triangular Negotiation for a Contractual Value

In this type of negotiation the goal is to obtain a contractual value of a QoS parameter which will bind both the service provider and the service users. Here the calling user introduces, in the request primitive, two values of a QoS parameter, the minimal requested value and the bound for strengthening it. The service provider is not allowed to change the value of the minimal requested value. Here, the service provider is free, as long as it does not weaken it below the minimal requested value, to weaken the bound for strengthening value before presenting the new value of the bound for strengthening and the unchanged value of the minimal requested value to the called user, through an indication primitive. It will be the privilege of the called user to take the final decision concerning the selected requested value. This selected value of the QoS will be returned by the called user in the response primitive.

8

This selected value will be included without change by the service provider in the confirm primitive. At the end of the negotiation, the three actors have agreed on the value of this QoS parameter.

The service provider may have to reject the request if it does not agree to provide a QoS in the requested range.

The called user may also reject the connection attempt if it is not satisfied by the range of values proposed in the indication primitive.

With respect to the minimal requested value introduced by the calling user, the only possible modification introduced by the negotiation is the strengthening of the minimal requested value but limited by the bound for strengthening value. The service provider may weaken the bound for strengthening and the called user may strengthen the minimal requested value but to the limit accepted by the service provider.

This scheme of negotiation has been used in OSI 95 for two types of requested values, the compulsory QoS value and the threshold QoS value which will defined later on.

Bilateral Negotiation versus Triangular Negotiation

For some QoS parameters, the negotiation takes place between the two service users, the service provider being not allowed to modify the value proposed by the service user. However, this bilateral character is more formal than real, the service provider having always the possibility to reject the request.

When not only the service provider but the calling service user is not allowed to modify the value proposed by the calling user, the negotiation is reduced to a "take it or leave it" approach. This type of negotiation may have it merits in special situation.

BEST EFFORT QoS VALUE

Coming back to the result of the negotiation for information exchange, it is clear that it is difficult to attach a strong semantics to the resulting value of the QoS parameter. It is only a value agreeable to all three actors. The QoS parameter does not require a permanent monitoring by the service provider because it is not possible to specify a particular behaviour of the service provider if the real value of the QoS parameter is weaker than the agreed value. The service users do not expect such particular behaviour in such case. They are just expecting the *"best effort"* from the service provider.

In such loosely defined environment, if a service user introduces, in a request primitive, the value of a QoS parameter, it is not always clear if this suggested value is related to a boundary or to an average value. In the latter case, the measurement sample or the number of SDUs to be considered is often far from obvious. This is not a great problem if no monitoring has to be done.

For negotiation where the service provider is not allow to weaken the value suggested by the calling user, it is possible for the service provider to reject the request due to the requested QoS value but, in case of acceptance, nothing will be done if the service provider does not reach the QoS value. In this case, the service user is not even informed about the situation by the service provider as no REPORT indication primitive has been defined.

It is therefore not surprising that in many cases, the QoS is expressed in qualitative terms without any specification of a given value. This confirms the lack of relationship between the QoS parameter and a real performance parameter. The only way for the service users to assess the value of a QoS is to monitor it.

If, to operate in a correct way, an application requires a well defined set of performance parameters, the present approach will not be suitable.

The Monitoring and the Protocol Functions.

Monitoring is not the only way to assess the performance of a service provider. In TP4, the error control function based on time-out and retransmission achieves a very low value of the residual error which will be impossible to assess by monitoring on the life time of most connections.

For the data communication, the ISO transport service has been considered as adequate because it was offering a reliable service. The throughput, the transit delay and the transit delay jitter were not considered as critical factors of performance.

GUARANTEED QoS VALUE

In some situation, it would be nice to be able to have the concept of a guaranteed QoS, especially when the guarantee value result from a negotiation for a bounded target or from a negotiation for a contractual value. The possibility for the service provider of giving such a guarantee is related to the existence of resources associated with some protocol function for allocating the resources and managing it during the connection.

A residual error rate on a connection may be guaranteed by the service provider if enough buffers and numbering space can be allocated to the connection and if it operates an error control function with known property.

A minimum throughput on a connection may be guaranteed by the service provider if each protocol entities can be allocated enough processing resources and if the underlying service provides the corresponding throughput guarantee.

If the allocation of resources is only partial, leaving room for uncertaintity, or if the resource allocation is done on a statistical basis or on an overbooking scheme, it is possible to associate some probability measure with the QoS value. The same is true if the resource allocated to a connection may be recover by another connection of higher priority as in ST-II [Top 90] by the introduction of the precedence mechanism.

We may in these cases either speak of guaranteed QoS with a low probability or we may qualify differently the QoS value when it has to be associated with a probability far from 1. It is was has been done in [Par 92] where a distinction is made between a guaranteed service and a predicted service.

WHAT IF ?

If one associates the semantics of guaranteed QoS to the value resulting from a negotiation for a bounded target or from the negotiation for a contractual value, it is clear that the service provider may have to reject the request if it does not agree to provide a QoS in the requested range.

However if the request is accepted, it less clear what will have to done if during the connection the value of the QoS parameter is weaker that the guaranteed value.

If the service provider do nothing in such case, we are back to the best effort.

The service provider may abort the connection as it is not able to maintain the guaranteed value.

The service provider may also indicate to the service user(s) that it cannot maintain the selected value and leave to the service user(s) the reponsibility of aborting.

A NEW SEMANTICS FOR THE QoS

In OSI 95, a new semantics for the performance QoS parameters has been introduced. In this semantics, a parameter is seen as a structure of three values, respectively called "compulsory", "threshold" and "maximal quality", all these values being optional. Each one has its own well defined meaning, and expresses a contract between the service users and the service provider.

This means, and this is the most fundamental difference with the best effort, that the service provider is now submitted to some well defined duties, known by each side. In other words, the rules of the game are clear.

The existence of a contract between the service users and the service provider implies that, in some cases, the service users are also submitted to well defined duties also derived from the application environment.

THE COMPULSORY QoS VALUE

The idea behind the introduction of a "compulsory" QoS value is the following one: *when a compulsory value has been selected for a QoS parameter of a service facility, the service provider will monitor this parameter and abort the service facility when it notices that it cannot achieve the requested service.*

It must be clear that no obligation of results is linked to the idea of compulsory value. The service provider **tries** to respond to the requested service facility and, by monitoring its execution, it will

- either execute it completely without violating the selected compulsory value of the performance parameter;
- or abort it if the selected compulsory value of the performance parameter is not fulfilled.

The compulsory concept reflects the fact that, in some environments (e.g. a lightly loaded LAN), the compulsory QoS value may be achieved without resource reservation. Of course, the same LAN, which does not provide any reservation mechanism or any priority mechanism, may, when heavily loaded, prevent the service provider from reaching the compulsory QoS value and oblige it to abort the execution of the requested service facility.

THE THRESHOLD QoS VALUE

Some service users may find that the solution of aborting the requested service facility, when one of the compulsory QoS values is not reached, is a little too radical. They may prefer to get information about the degradation of the QoS value.

To achieve that we propose to introduce a "threshold" QoS value with the following semantics: *when a threshold value has been selected for a QoS parameter of a service facility, the service provider will monitor this parameter and indicate to the service user(s) when it notices that it cannot achieve the selected value.*

This threshold QoS value may be used without an associated compulsory value. In this case, the behaviour of the service provider is very similar to the one it has to adopt with a compulsory value. The main difference is that, instead of aborting the service facility, it warns, when it notices it is unable to provide the specified value, either or both users depending of the service definition. If the service provider is able to provide a QoS value better than the threshold value, everything is fine.

Threshold QoS versus Best Effort QoS

If the threshold QoS is used without any compulsory QoS, the main difference between the threshold and the best effort is that in the former case, the service provider has the obligation to monitor the parameter and to indicate if the threshold value is not reached.

Threshold and Compulsory QoS values

It is possible to associate, to the same QoS parameter, a threshold and a compulsory QOS values with, of course, the threshold value "stronger" than the compulsory one.

If the performance parameter degrades slowly and continuously, an indication will be delivered to the service users before the abortion of the service facility. Until such a threshold indication occurs, the service user knows that the service facility is not endangered by the current parameter value.

THE MAXIMAL QUALITY QoS VALUE

In most cases, if the service provider is able to offer a "stronger" value of the QoS parameter than the threshold, the service user will not complain about it. But it could happen that the service user wants to put a limit to a "richer" service facility.

A called entity, for instance, may want to put a limit to the data arrival rate or a calling entity may want, for reasons of cost, to prevent the use of too many resources by the service provider.

Such a parameter may be useful to smooth the behaviour of the service provider. Introducing a maximal quality QoS value on a transit delay, i.e. fixing a lower bound to the transit delay values will reduce the transit delay jitter and facilitate the resynchronization at the receiving side.

To achieve that we propose to introduce a "maximal quality" QoS value with the following semantics: *when a maximal quality value has been selected for a QoS parameter of a service facility, the service provider will monitor this parameter and avoid occurrence of interactions with the service users that would give rise to a violation of the selected value.*

It is possible to associate, to the same QoS parameter, a maximal quality, a threshold and a compulsory QOS values with, of course, the maximal quality "stronger" than the threshold value, itself "stronger" than the compulsory value.

QoS PARAMETERS AND INFORMATION PARAMETERS

The introduction of compulsory QoS values implies that the service provider will have a more difficult task to fulfil. It is therefore not surprising that the service user may have to provide the service provider with more information about the characteristics of the elements associated with the request in order to facilitate the decision of rejection or acceptance of the request. Hence, the introduction of the concept of compulsory QoS requires the introduction, in the primitives associated with a request, of additional parameters. These additional parameters may be designated as information parameters to distinguish them from the QoS parameters proper.

Information values of QoS parameters may also have to be introduced to control the negotiation process in the case of compulsory or threshold QoS to preserve the semantics associated with the negotiated value.

THE NEGOTIATED QoS PARAMETERS

Depending upon the type of facilities, these QoS parameters will be submitted to a negotiation or not. In case of a negotiation, some rules related to the rights of strengthening or weakening (lowering) the QoS values are defined for each type of value and for each participant in the negotiation. This has to be done to keep the final result coherent with the meaning of the parameter value.

Negotiation of a Compulsory QoS value

When a negotiation occurs on a compulsory value, the selected compulsory value of a performance parameter may be different from the value proposed first by the calling service user. However, in order to remain coherent with the semantics attached to a compulsory value, some rules on the negotiation procedure must be imposed.

First, it must be clear that if a service user introduces a compulsory QoS value for a performance parameter to be negotiated, the only possible modification is the strengthening of this compulsory value. In particular, it is absolutely excluded for the service provider to modify this value in order to relax the requirement

However the calling service user may not be interested in an unlimited strengthening of the proposed compulsory QoS value. It introduces therefore in the request primitive, a second value which fixes a bound indicating to what extend the proposed compulsory QoS value may be strengthened.

When the service provider analyses the request of the calling service user, it has to decide whether it rejects it or not (it can already do so as it knows that the request could only be strengthened).

This reject decision may be based:

- on the rejection by the underlying service provider of the underlying request with related compulsory values;
- on the overall knowledge of the capability of the underlying service;
- on the past history of other associations of the same type involving the same set of service users. An interesting way to be explored is the feasibility for the service provider to rely on the information collected by the management entities of the service.

In the latter case, it has to examine the bound of strengthening. This bound may be made weaker (brought closer to the compulsory value) by the service provider, before issuing the indication primitive to the called service user, in such a way to give, to the called service user, the range of compulsory values acceptable by both the calling service user and the service provider.

The service provider does not have to strengthen the compulsory QoS value which must be seen as the expression of the requirements of the service users.

After receiving the indication primitive, the called service user may accept or reject the request. If it accepts it, it may modify (strengthen) the compulsory QoS value up to the value of the bound and return it in its response. In this case the negotiation is completed and the service provider may confirm the acceptance of the request and provide the final selected compulsory QoS value to the calling service user.

If the negotiation is successful, the bound is of no interest anymore (the bound is an example of information values mentioned earlier) and the selected compulsory QoS value reflects now the final and global request to the service provider from both service users.

Negotiation of a Threshold QoS value

The negotiation procedure of a threshold value is similar to the negotiation procedure of a compulsory value. Here also the only possible modification is the strengthening of the threshold value. Here also the calling service user introduces, in the request primitive, an information parameter which fixes a bound indicating to what extend the proposed threshold QoS value may be strengthened.

If a compulsory and a threshold values are associated to the same QoS parameter, there exist a set of order relationship between the compulsory, the threshold and their bounds values which must be verified in the request primitive and maintained during the negotiation.

Negotiation of a Maximal Quality QoS value

If a service user introduces a maximal quality QoS value for a performance parameter, the only possible modification is the weakening of this maximal quality QoS value. This value can be weakened during the negotiation by the service provider that indicates by this way the limit of the service it may provide and by the called service user.

If the maximal quality is the only value associated with a given QoS parameter, no bound will be introduced in the request primitive and the negotiation will result in the selection of the weakest of the maximal quality values of the service users and the service provider. This is an example of negotiation for information exchange.

If the maximal quality value and a compulsory or/and a threshold values are associated to the same QoS parameter, there exist an order relationship between the maximal quality value and the bound value on the threshold (or the bound value on the compulsory value if no threshold value is specified) which must be verified in the request primitive and preserved during the negotiation.

THE NON-NEGOTIATED QoS PARAMETERS

The Non-negotiated Compulsory QoS

When a service user asks for a service facility with a performance parameter at a given compulsory value, the service provider will reject the request, without even trying to provide the requested service, when the service provider knows that it will be unable to succeed.

This decision may be based:

- on the rejection by the underlying service provider of the underlying request with related compulsory values;
- on the overall knowledge of the capability of the underlying service;
- on the past history of other associations of the same type involving the same set of service users. An interesting way to be explored is the feasibility for the service provider to rely on the information collected by the management entities of the service.

The Non-negotiated Threshold and Maximal Quality QoSs

When a service user asks for a service facility with a performance parameter at a given threshold value, the service provider will accept the request. The same is true for a service facility with a performance parameter at a given maximal quality value.

THE DEFINITION OF THE PERFORMANCE QoS PARAMETERS

Remember that the QoS refers collectively to certain characteristics associated with the use of the available service facilities, as observed between the SAPs. The only aspects, regarding the use of the service facilities, that are observable at the SAPs are the interactions which take place by means of service primitives.

The QoS is specified in terms of QoS parameters. The performance QoS parameters, in particular, are those QoS parameters whose definition is based on a performance criterion to be assessed.

It is very important to have a clear view about the way to define a performance QoS parameter. Of course the assessment of a performance criterion requires the introduction of timing considerations. Since the only events observable by a service user, when it is using a service facility, are the primitives that occur at its SAP, the only notion of time which can be relied on to introduce timing considerations is the notion of time of occurrence of a service primitive at a SAP. Such a time is an absolute time which is not usable as such.

Time has to appear in the definition of the performance QoS parameters in the form of time intervals between occurrences of service primitives. Thus, any performance QoS parameter has to be defined in relation with the occurrences of two or more related service primitives. These primitives may be related in very different ways. They may be related just by the fact that they pertain to a same connection, or by the fact that they are occurring successively at a same SAP, or by the fact that one or several parameters of one of the primitives occurring at a given SAP have been replicated in the other primitive(s) occurring at peer SAP(s), etc. The figure 2 give example of time intervals usable in relation with the performance parameter definition. More complex relationships resulting from combinations of several basic relationships are also possible.

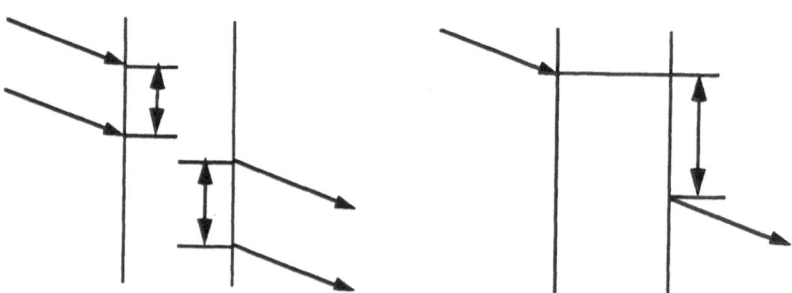

Figure 2. Time intervals usable in relation with the performance parameter definition

Transit Delay Definition

For example, the **transit delay** associated with an invocation of a peer-to-peer SDU transfer facility may be defined as *the time interval between the occurrence, at the SAP of the sending user, of the DATA request primitive that conveys the SDU and the occurrence, at the peer SAP, of the corresponding DATA indication primitive*. This definition may be used in the connectionless-mode case as well as in the connection-mode case. Here, the DATA request and indication primitives are related by the fact that the SDU parameter of the DATA indication has been replicated from the SDU parameter of the DATA request.

Transit Delay Jitter Definition

In connection mode, it is conceivable that a performance parameter calculated at each invocation of a service facility intervenes in the definition of a more complex performance parameter whose calculation is based on a certain number of invocations of this service facility. A **transit delay jitter** may be defined for one direction of data transfer of a connection as *the difference between the longest and the shortest transit delays observed on this direction since the connection establishment*. Here, each pair formed by a DATA request and the corresponding DATA indication primitives is related to the other pairs of primitives by the fact that they pertain to a same direction of data transfer on a connection.

Throughput Definition

An example of a performance parameter whose definition is based on the time of occurrence of service primitives related by the fact that they occur at the same SAP is the throughput of a direction of data transfer on a connection. Unlike the definitions of the previous two performance parameters, the definition of the throughput does not only use time intervals between occurrences of service primitives, but uses an additional quantity, namely the length of the transferred SDUs.

In the specifications of the ISO services [ISO 8072], the throughput for one direction of transfer of a connection is defined in terms of a sequence of n (with $n \geq 2$) successfully transferred SDUs.

When particularised to the case $n = 2$, this definition becomes:

"the throughput for one direction of transfer is defined to be the smaller of:
a) the number of SDU octets contained in the last transferred SDU divided by the time interval between the previous and the last T-DATA requests; and
b) the number of SDU octets contained in the last transferred SDU divided by the time interval between the previous and the last T-DATA indications",

where the throughput measured in a) and b) may be referred to respectively as the sending user's throughput and the receiving user's throughput for the direction of transfer considered. When $n = 2$, the definition corresponds to an instantaneous throughput.

The figure 3 represent the ISO definition of the sending user's throughput when $n = 2$.

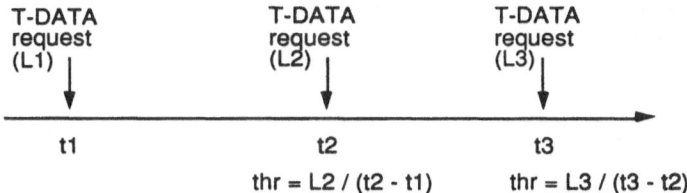

Figure 3. ISO definition of the sending user's throughput

An alternative way to define the throughput associated with an invocation of the SDU transfer facility on one direction of transfer of a connection is the following: the **global throughput** for one direction of transfer is still defined to be the smaller of the **sending user's throughput** and the **receiving user's throughput** but these two throughputs are now defined differently from the ISO's view:

- the **sending user's throughput** is now defined as *the number of SDU octets contained in the last transferred SDU divided by the time interval between the last and the next T-DATA requests.*
- the **receiving user's throughput** is now defined as the *number of SDU octets contained in the last transferred SDU divided by the time interval between the corresponding last and next T-DATA indications.*

This definition corresponds also to an instantaneous throughput. It is this definition of the throughput that has been adopted in the framework of the OSI 95 Enhanced Transport Service Definition [DBL 92b, DBL 92c, BLL 92b]. The figure 4 represent the OSI 95 definition of the sending user's throughput

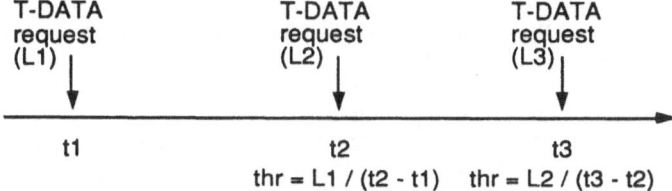

Figure 4. OSI 95 definition of the sending user's throughput

In comparison with the ISO's one, the latter definition of the global throughput has a great advantage. Indeed, this definition is much more convenient for the service provider that has to check its ability to correctly fulfil its contract, in particular as regards the selected compulsory and threshold throughput values, during the data transmission phase.

With the latter definition, the behaviour of the service provider between two occurrences of DATA request primitives on the sending side is influenced only by the time that is elapsing. This is due to the fact that the length of the SDU that is used for the calculation of the current throughput value (i.e. the value associated with the last invocation of the SDU transfer facility) is known. Only the time interval between the last DATA request and the next expected DATA request is unknown until the occurrence of this next primitive. This means that the constraints which the service provider has to obey on the sending side will be expressed only in terms of the time already elapsed since the last DATA request, regardless of the possible lengths of SDU in the next DATA request. Of course a similar conclusion is true on the receiving side.

With the ISO's definition, the monitoring of a compulsory or a threshold throughput value is more intricate. In fact, the behaviour of the service provider between two occurrences of DATA request primitives on the sending side depends not only upon the time that is elapsing but also upon the possible lengths of SDU in the next expected DATA request. This results from the fact that the length of the SDU that is to be used for the calculation of the current throughput value (i.e. the value associated with the next invocation of the SDU transfer facility) is obviously unknown, as well as the time interval between the last DATA request and the next expected DATA request, until the occurrence of this next primitive. This means that the constraints which the service provider has to obey on the sending side will be expressed in terms of the time already elapsed since the last DATA request but also in terms of the possible lengths of SDU in the next DATA request. Of course a similar conclusion is true on the receiving side.

EXAMPLES

The purpose of the following examples is to illustrate the use of the various values of a performance QoS parameter.

Data Transfer in a Connectionless-mode Service

If we introduce, in an UNITDATA request, a compulsory value for the transit delay , the time between an UNITDATA request and the corresponding UNITDATA indication will never exceed this value: either it will be shorter, or no UNITDATA indication will ever occur. In the latter case, no indication of failure will be given to the sending user, which is coherent with the semantics of a connectionless transmission.

Introducing a threshold value for the transit delay in an UNITDATA request may appears meaningless as the semantics of this transmission service implies that the sending user does not receive any further information about its request. However, it is possible to envisage the signalling of the threshold violation in the UNITDATA indication on the receiving side.

Introducing a maximal quality value for the transit delay in an UNITDATA request seems difficult to justify as there is very little rationale for a sending user in a connectionless-mode to prevent the transit delay to be lower than a given value.

Data Transfer in an Acknowledged Connectionless-mode Service

In an acknowledged connectionless-mode, such as the one introduced in [Dan 92a, DBL 92a], it is possible to introduce a threshold value for the transit delay. A violation of the threshold value (actual transit delay value above the threshold value) could be signalled to the sending user by a special parameter of the confirm primitive as well as to the receiving user by a special parameter of the indication primitive.

In an acknowledged connectionless-mode it is also possible to introduce a compulsory value for the transit delay. A violation of the compulsory value will prevent the ACKDATA indication to be issued and the sending user will receive the notification of non-delivery via the REJECT- ACKDATA.indication primitive.

Introducing a maximal quality value for the transit delay appears difficult to justify as there is very little rationale for a sending user in an acknowledged connectionless-mode to prevent the transit delay to be lower than a given value.

Data Transfer in a Connection-mode Service

If a compulsory value for the transit delay has been negotiated during the connection establishment, the time between a DATA request and the corresponding DATA indication will never exceed this value: either it will be shorter, or no DATA indication will ever occur. In the latter case, the service provider will abort the connection by issuing a DISCONNECT indication with the reason of the abortion.

It is possible to negotiate, during the connection establishment, a threshold value for the transit delay. If the time between a DATA request and the corresponding DATA indication is above the threshold value, the service provider will signal it to the service user(s) by a REPORT indication primitive.

Introducing a maximal quality value is not anymore meaningless in a connection-mode transmission. Preventing the transit delay to be lower than a given value implies that the DATA indication may be delayed at the receiving user SAP and this may be an efficient means to reduce the transit delay jitter. With a maximal quality value corresponding to a minimum value of the transit delay and with a compulsory value corresponding to a maximum value of the transit delay, it is possible to introduce an upper bound on the transit delay jitter.

A More Complex Example of Negotiated QoS Parameter

As a more complex example of a negotiated QoS parameter, let us consider the throughput on a direction of transfer of a connection with the associated compulsory, threshold and maximal quality values.

The compulsory is the smallest value and represents the throughput that the service provider must be able to offer to the sending and to the receiving users. If at anytime during the whole life of the connection, the service provider happens to be unable to maintain this throughput a DISCONNECT indication will occur.

If the traffic on the connection is dedicated to a video channel, the compulsory value may represent the throughput which allows to maintain the synchronisation between the sending and the receiving users.

However, in the most general case, two particular situations have to be considered. Both are dealing with the responsibilities of the users on the one hand and of the service provider on the other hand when the selected values are not matched.

At the sending SAP, the compulsory value may be violated because no DATA request occurred in due time owing to the behaviour of the sending user only (i.e. the service provider was ready to accept a DATA request in due time). This does not imply a failure of the service provider but only results from the behaviour of the sending user. In this case, it may be reasonable to believe that the service provider does not have to issue a DISCONNECT indication.

If at the receiving SAP, no DATA indication occurs due to the receiving user, the compulsory QoS value will also be violated. Here however it seems reasonable to recognise that the compulsory value does not only constrain the service provider but also constrains the receiving user. Its acceptance of the compulsory value implies that it believes to be able to support the compulsory value. Therefore a DISCONNECT indication seems appropriate. This asymmetrical behaviour may not be acceptable in every application environment.

It is possible, for the threshold value to apply the same rules as for the compulsory value, issuing, of course, a REPORT indication instead of a DISCONNECT indication. However, taking into account the warning characteristic associated with the threshold value, the service provider may also issue a REPORT indication to the service users as soon as the threshold value is violated.

The maximal quality QoS value is the highest value of the three. The service provider, by controlling the occurrence of the interactions with the service users, prevents the sending user's throughput and the receiving user's throughput from crossing the limiting value.

An interesting point to mention is that the combination of the compulsory and maximal quality values helps to make more precise the model of the service provider. The internal queues may build up at a maximum pace linked to the difference between the maximal quality and the compulsory values of the throughput.

The Connection-mode Transport Service of OSI 95

For the connection-mode transport service of OSI 95, we used the structure of three values, compulsory, threshold and maximal quality.

The introduction of compulsory QoS values in the QoS negotiation implies that the TS provider will have a more difficult task to fulfil. It is therefore not surprising that the TS user may have to provide the TS provider with more information about the characteristics of the sequence of TSDUs it intends to submit. Requesting a throughput of 2 Mb/s with TSDUs of 10 Kbytes is different from requesting a throughput of 2 Mb/s with TSDUs of 40 bytes. Hence, the introduction of the concept of compulsory QoS requires the introduction, in the primitives associated with the opening of a TC, of additional parameters such as the maximum and the minimum size of TSDU. As indicated earlier, these additional parameters will be designated here as information parameters to distinguish them from the QoS parameters proper.

Assume that through a successful negotiation we have the three values for the throughput defined as indicated at the figure 4. It may be of interest to see how the service provider will operate.

The minimum compulsory value for the throughput is the value that the TS provider must be able to maintain during the whole TC lifetime. Otherwise it must immediately shut down the TC. So, if a T-DATA.request interaction occurs at time T_0, with a TSDU of size L, the TS provider must be ready to offer another T-DATA.request interaction at the latest at time $T_0 + \Delta t_{max}$, where Δt_{max} is given by the ratio of L to the minimum compulsory value for the throughput. Besides, the TS provider must be able to offer the first T-DATA.request interaction immediately after the receipt of the T-CONNECT.response on the called side and the issuance of the T-CONNECT.confirm on the calling side.

The maximum value for the throughput is a value that may never be exceeded for the duration of the TC. So, if a T-DATA.request interaction occurs at time T_0, with a TSDU of size L, the TS provider shall offer another T-DATA.request interaction at the earliest at time $T_0 + \Delta t_{min}$, where Δt_{min} is given by the ratio of L to the maximum value for the throughput. The maximum value for the throughput can thus be used for rate control purposes.

Figure 5 shows the relation between the TSDU size L and the Δt_{min} and Δt_{max}.

If the TS provider is not able to offer another T-DATA.request interaction at the latest at time $T_0 + \Delta t_{max}$, the minimum compulsory throughput cannot be achieved and the TS provider must shut down the TC. This does not mean that the throughput associated with a

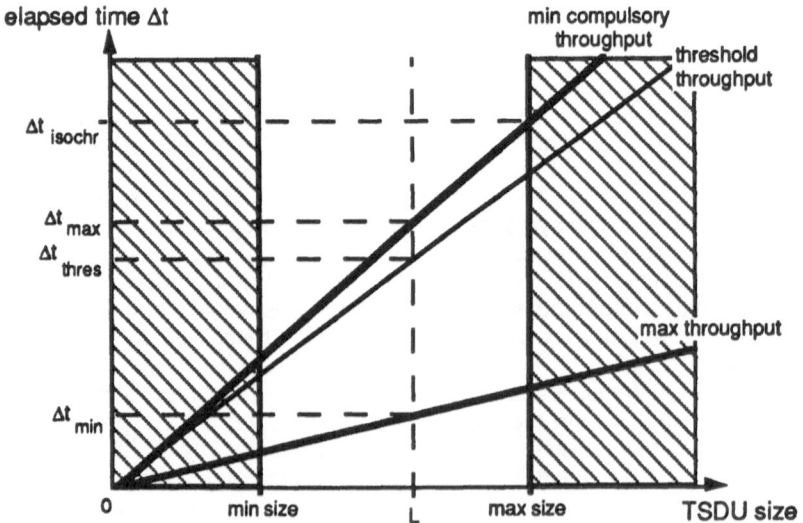

Figure 5. Relation between the TSDU size and the Δt_{min} and Δt_{max}

particular invocation of the data transfer facility may never go under the minimum compulsory value. This may happen if the offer of the TS provider for another T-DATA.request interaction is not matched in time by the sending TS user. In such a case however, the inability of maintaining the throughput at or above the minimum compulsory value is entirely the responsibility of the sending TS user. Our choice of an "instantaneous" throughput measured at each invocation of the data transfer facility has the advantage of putting a clear separation between the responsibilities of the TS user and those of the TS provider. The TS provider does not have to disconnect in case of a late submission of TSDU from the part of the TS user.

If a sending TS user generates an isochronous traffic, that is if it produces TSDUs (possibly of variable length) at a fixed rate, this sending TS user may want the TS provider to be ready to accept the submission of TSDUs at the same rate characterised by the constant time interval Δt_{isochr} between the submission of successive TSDUs whatever their length. For this purpose, the TS user may negotiate with the TS provider a minimum compulsory throughput given by the ratio of the maximum size authorised for the TSDUs to the constant time interval Δt_{isochr}. With such a minimum compulsory throughput, if T_0 is the time at which the last T-DATA.request occurred, the sending TS user is assured that the TS provider will be able to offer another T-DATA.request interaction at the latest at time $T_0 + \Delta t_{isochr}$ in case the TSDU submitted in the interaction at time T_0 is of maximum size or even earlier in case the TSDU is of smaller size, and will disconnect otherwise.

However, by forcing the TS provider to be able to accept a new T-DATA.request before $T_0 + \Delta t_{isochr}$ when the last TSDU is not of maximum size, we impose an unnecessary constraint on it. In fact, for a really isochronous traffic, forcing the TS provider to be able to offer a T-DATA.request interaction every Δt_{isochr} units of time, whatever the size of the TSDU submitted in the previous interaction, is a sufficient constraint. That is why we have decided to add a traffic type indicator to the QoS components pertaining to the throughput parameter. If this traffic type indicator is "non-isochronous", then the Δt_{max} is calculated as explained above, i.e. by taking the size L of the previous submitted TSDU into account. By contrast, if the traffic type indicator is "isochronous", then the Δt_{max} is simply taken equal to Δt_{isochr} as if all the TSDUs were of maximum size (Figure 5).

Obviously, if the negotiated minimum compulsory throughput is equal to zero, Δt_{max} is becoming infinite and we come back to the concept of best effort QoS.

To respond to certain requirements, we have also introduced, in our list of QoS components pertaining to the throughput parameter, the idea of a threshold value for the throughput that would be slightly above the minimum compulsory value (Figure 5). If the TS provider was not able to maintain the threshold throughput, i.e. if it was not ready to offer another T-DATA.request interaction at time $T_0 + \Delta t_{thres}$, then it would have to indicate its inability to the TS users but the TC would remain open.

In summary, the behaviour of the TS provider will be time-dependent according to the following table where T_0 is the time at which the last T-DATA.request occurred:

In $]T_0, T_0 + \Delta t_{min}[$, the provider will not accept a T-DATA.request.

In $[T_0 + \Delta t_{min}, T_0 + \Delta t_{thres}[$, the provider may accept or not a T-DATA.request.

In $[T_0 + \Delta t_{thres}, T_0 + \Delta t_{max}[$, the provider must accept a T-DATA.request, or otherwise must indicate it to the TS user.

In $[T_0 + \Delta t_{max}, \infty[$, the provider must accept a T-DATA.request or otherwise must start the release of the TC.

In this table, $\Delta t_{max} = L \,/\, minimum\ throughput$,

where L is either the size of the last TSDU (in the non-isochronous case) or the maximum TSDU size (in the isochronous case).

CONCLUSION

We present in this paper an enhancement of the QoS model able to match the communication requirements of the new application environment.

In our proposed semantics, a QoS parameter is seen as a structure of three values, respectively called "compulsory", "threshold" and "maximal quality", all these values being optional. Each one has its own well defined meaning, and expresses a contract between the service users and the service provider.

This means, and this is the most fundamental difference with the present situation, that the service provider is now submitted to some well defined duties, known by each side. In other words, the rules of the game are clear. These duties correspond to services that we found useful for a service user in the above-mentioned current environment.

The existence of a contract between the service users and the service provider implies that, in some cases, the service users are also submitted to well defined duties also derived from the application environment.

We have applied the ideas of this paper in the design of the OSI 95 Enhanced Transport Service [Dan 92b, BLL 92a, DBL 92b, BLL 92b] and we are currently working in the framework of the CIO[2] project on the integration of the new QoS semantics into the protocol developed to provide this enhanced service

The goal of the OSI 95 project was not only to specify new transport service and protocol. It is also to use LOTOS as formal description technique, from the very beginning of the design process. The methodology for the design of the service and the protocol has been be based on the constraint-oriented style in order to allow an incremental development of the protocol. This method allows the addition of, the removal of or a modification in a mechanism without jeopardising the other part of the work. This methodology has been tested successfully on TP4 [Led 92] and has been applied for the service and the protocol [BLL 92b].

The work done in OSI 95 and in CIO has been introduced in the ISO in the framework of the ECFF (Enhanced Communication Functions and Facilities) new working group of ISO/SC6 [BLL 92a], [DBL 92a], [DBL 92b], [BLL 93], [Bag 93], [DBL 93]. The document [DBL 93] is presently circulated in the National Bodies and Liaison Organisation of SC6 for review and comments.

REFERENCES

[Alm 92]   P. ALMQUIST, **Type of Service in the Internet Protocol Suite**, Network Working Group RFC 1349, July 1992

[Bag 93]   Y. BAGUETTE, **Enhanced Transport Service Definition (Informal specification in English - Version 1)**, ISO/IEC JTC1/SC6/WG4/N822, 19 Apr. 1993, 85 p. (R2060/ULg/CIO/DS/P/003/b1, Feb. 1993, SART 93/03/15)

[BLL 92a] Y. BAGUETTE, L. LEONARD, G. LEDUC, A. DANTHINE, **Belgian National Body Contribution - Four Types of Enhanced Transport Services and their LOTOS Specifications**, *SC6 plenary meeting, San Diego, July 8-22, 1992*, ISO/IEC JTC1/SC6 N7323, May 1992, 58 p. (OSI 95/ULg/A/22/TR/P, May 1992, SART 92/12/09).

[BLL 92b] Y. BAGUETTE, L. LEONARD, G. LEDUC, A. DANTHINE, O. BONAVENTURE, **OSI 95 Enhanced Transport Facilities and Functions**, OSI 95/Deliverable ULg-A/P, Dec.1992, 277 p. (SART 92/25/05)

[BLL 93]   Y. BAGUETTE, L. LEONARD, G. LEDUC, A. DANTHINE, **OSI 95 Enhanced Transport Services**, ISO/IEC JTC1/SC6/WG4/N821, 19 Apr. 1993, 107 p. (R2060/ULg/CIO/DS/S/002/b1, Jan. 1993, SART 93/02/15)

[CCC 91]   COMER M.H, CONDRY M.W., CATTANACH S., CAMPBELL R., **Getting the Most for Your Megabit**, *Computer Communication Review*, 1991, Vol. 21, nr 3, pp. 5-12

[Che 88]   D. CHERITON, **VMTP : Versatile Message Transaction Protocol - Protocol Specification**, Network Working Group RFC 1045, Feb. 1988

[Ches 89] CHESSON G., **XTP/PE Design Considerations**, *Protocols for High-Speed Networks*, Zurich, May 9-11, 1989, Rudin H. & Williamson R. Ed., Elsevier (North Holland), pp. 27-32

[ChJ 89]   CHIU D-M, JAIN R., **Analysis of the Increase and Decrease Algorithms for Congestion Avoidance in Computer Networks**, *Computer Networks and ISDN Systems*, 1989, Vol. 17, nr 1, pp. 1-14

[ChW 89]   CHERITON D.R., WILLIAMSON C.L., **VMTP as the Transport Layer for High-Performance Distributed Systems**, *IEEE Communications Magazine*, 1989, Vol. 27, n° 6, pp. 37-44

---

[2] CIO is the acronym of RACE Project 2060 "Coordination, Implementation and Operation of Multimedia Services".

[CJR 89]   CLARK D.D., JACOBSON V., ROMKEY J., SALWEN H., **An Analysis of TCP Processing Overhead**, *IEEE Communications Magazine*, 1989, Vol. 27, nr 6, pp. 23-29

[CLZ 87]   CLARK D.D., LAMBERT M.L., ZHANG L., **NETBELT : A Bulk Data Transfer Protocol** Network Working Group RFC 998, March 1987

[Dan 89]   DANTHINE A., **Communication Support for Distributed Systems - OSI versus Special Protocols**, *Proceedings of the 11th World Computer Congress on Information Processing*, San Francisco, 28 August-1 September 1989, pp. 181-190

[Dan 90]   DANTHINE A., **10 Years with OSI**, *INDC-90*, Lillehammer, Norway, 26-29 March 1990,(Ed. : D. KHAKHAR, F. ELIASEN), Elsevier (North Holland), 1990, pp. 1-16

[Dan 92a]   A. DANTHINE, **A New Transport Protocol for the Broadband Environment**, *IFIP Workshop on Broadband Communications*, Estoril, January 20-22, 1992, A. Casaca, ed., Elsevier (North-Holland), 1992, pp. 337-360.

[Dan 92b]   A. DANTHINE, **Esprit Project OSI 95 - New Transport Services for High-Speed Networking**, *3rd Joint European Networking Conference*, Innsbruck, May 11-14 1992, also in: *Computer Networks and ISDN Systems 25 (4-5)*, Elsevier Science Publishers (North-Holland), Amsterdam, November 1992, pp. 384-399.

[DBL 92a]   A. DANTHINE, Y. BAGUETTE, G. LEDUC, **Belgian National Body Contribution - Issues Surrounding the Specification of High-Speed Transport Service and Protocol**, *SC6 interim meeting, Paris, February 10-13, 1992*, ISO/IEC JTC1/SC6 N7312, January 1992 58 p.(OSI 95/ULg/A/15/TR/P/V2, 46 p., January 1992, SART SART 92/04/05).

[DBL 92b]   A. DANTHINE, Y. BAGUETTE, G. LEDUC, L. LEONARD, **Belgian National Body Contribution - The Enhanced Connection-Mode Transport Service of OSI 95**, *SC6 plenary meeting, San Diego, July 8-22, 1992*, ISO/IEC JTC1/SC6 N7759, June 1992 16 p. (OSI 95, OSI 95/ULg/A/24/TR/P, 16 p., June 1992, SART 92/14/05).

[DBL 92c]   A. DANTHINE, Y. BAGUETTE, G. LEDUC, L. LEONARD, **The OSI 95 Connection-mode Transport Service - The Enhanced QoS**, *IFIP Conference on High Performance Networking*, Liège, December 16-18, 1992, to appear in: A. Danthine, O. Spaniol, eds., High Performance Networking, Elsevier Science Publishers (North-Holland), Amsterdam, 1993.

[DBL 93]   A. DANTHINE, Y. BAGUETTE, L. LEONARD, G.LEDUC, **An Enhancement of the QoS Concept**, ISO/IEC JTC1/SC6/N8010, Jan. 1993, 16 p. (OSI 95/ULg/A/28/TR/P, Jan. 1993)

[DHH 88a]   DANTHINE A., HENQUET P., HAUZEUR B., CONSTANTINIDIS C., FAGNOULE D., CORNETTE V., **Access Rate Measurement in the BWN Environment**, Proceedings of High Speed Local Area Networks II, Liège, 14-15 April 1988, Elsevier Science Publishers BV(North Holland), 1990, pp. 89-115

[DHH 88b]   DANTHINE A., HAUZEUR B., HENQUET P., CONSTANTINIDIS C., FAGNOULE D., CORNETTE V., **Corporate Communication System by LAN Interconnection**, *EURINFO 88 - Information Technology for Organisational Systems*, Athènes, 16-20 May 1988 Bullinger H.J. & al., Ed., Elsevier (North Holland), 1988, pp. 315-326

[DRB 87]   DE PRYCKER M., RYCKEBUSCH M., BARRY P., **Terminal Synchronization in Asynchronous Networks**, *IEEE Conference on Communications*, Seattle, 1987

[GKW 89]   GIARRIZZO D.,KAISERWERTH M., WICKI T., WILLIAMSON R.C., **High-Speed Parallel Protocol Implementation**, *Protocols for High-Speed Networks*, Zurich, May 9-11, 1989, Rudin H. & Williamson R. Ed., Elsevier (North Holland), pp. 165-180

[Her 88]   HERBERT A., **Communications Aspects in ANSA**, *Computer Standards & Interfaces*, 8 (1988) pp 49-56

[Hes 89]   HEATLEY S., STOKESBERRY D., **Analysis of Transport Measurements Over a Local Area Network**, *IEEE Communications Magazine*, 1989, Vol. 27, nr 6, pp. 16-22, also in *Protocols for High-Speed Networks*, Zurich, May 9-11, 1989

[ISO 8348] ISO/IEC JTC1, **Information Technology - Telecommunications and Information Exchange between Systems - Network Service Definition for Open Systems Interconnection**, ISO/IEC JTC1/SC6 N7558, 21 Sep. 1992

[ISO DIS 8072] ISO/IEC JTC1, **Information technology - Transport service definition for Open Systems Interconnection**, ISO/IEC DIS 8072 (DIS ballot terminates on April 1993).

[Jac 88]   JACOBSON V., **Congestion Avoidance and Control**, *SIGCOMM 88*, Stanford, Aug. 16-19, 1988, pp. 314-329

[Jai 86]   JAIN R., **A Timeout-Based Congestion Control Scheme for Window Flow-Controlled Networks**, *IEEE Journal on Selected Areas in Communications*, Vol. SAC-4, nr 7, Oct. 1986, pp. 1162-1167

[Jai 90]   JAIN R., **Congestion Control in Computer Networks : Issues and Trends**, *IEEE Network Magazine*, 1990, Vol. 4, nr 3, pp. 24-30

[JSB 90]   JAIN N., SCHWARTZ M., BASHKOW T.R., **Transport Protocol Processing at GBPS**

Rates, *SIGCOMM 90,* Philadelphia, Sep. 24-27, 1990, *Computer Communications Review,* 1990, Vol. 20, nr 4, pp. 188-199

[Led 92]   LEDUC G., **A Methodology for the Design of Large LOTOS Specifications and its Application to ISO 8073,** Report from the ESPRIT II Project OSI 95, OSI 95/ULG/A/16/TR/R/V2, April 1992.

[Mab 91]   MAFLA E., BHARGAVA B., **Communication Facilities for Distributed Transaction-Processing Systems,** *Computer,* 1991, Vol. 24, nr 8, p. 61

[Mei 91]   MEISTER B.W., **A Performance Study of the ISO Transport Protocol,** *IEEE Transactions on Computers,* 1991, Vol. 40, nr 3, pp. 253-262

[NRS 90]   NETRAVALI A.N., ROOME W.D., SABNANI K., **Design and Implementation of a High-Speed Transport Protocol,** *IEEE Transactions on Communications,* 1990, Vol. 38, nr 11, pp. 2010-2024

[Par 92]   C. PARTRIDGE, **A proposed Flow Specification,** Network Working Group RFC 1363, Sep. 1992

[PE 91]   PROTOCOL ENGINES, **XTP Protocol Definition — Revision 3.6 — Editor's Second Draft,** November 4, 1991.

[Pos 81]   J. POSTEL, Ed., **Internet Protocol - Darpa Internet Protocol Specification,** Network Working Group RFC 791, Sep. 1981

[RaJ 88]   RAMAKRISHNAN K.K., JAIN R., **A Binary Feedback Scheme for Congestion Avoidance in Computer Networks with a Connectionless Network Layer,** *SIGCOMM 88,* Stanford, Aug. 16-19, 1988, pp. 303-313

[RaJ 90]   RAMAKRISHNAN K.K., JAIN R., **A Binary Feedback Scheme for Congestion Avoidance in Computer Networks,** *ACM Transactions on Computer Systems,* 1990, Vol. 8, nr 2, pp. 158-181

[Svo 89a]   SVOBODOVA L., **Measured Performance of Transport Service in LANs,** *Computer Networks and ISDN Systems,* 1989, Vol. 18, nr 1, pp. 31-45

[Svo 89b]   SVOBODOVA L., **Implementing OSI Systems,** *IEEE Journal on Selected Areas in Communications,* 1989, Vol. 7, nr 7, pp. 1115-1130

[Top 90]   C. TOPOLCIC, Ed., **Experimental Internet Stream Protocol, Version 2 (ST-II),** Network Working Group RFC 1190, Oct. 1990

[Wha 89]   WHALEY A.D., **The XPRESS Transfer Protocol,** *14th Conference on Local Computer Networks,* Minneapolis, Oct. 10-12, 1989, pp. 408-414

[WiC 91]   WILLIAMSON C.L., CHERITON D.R., **Loss-Load Curves : Support for Rate-Based Congestion Control in High-Speed Datagram Networks,** *SIGCOMM 91,* Sept. 1991, pp. 17-23

[Zha 87a]   ZHANG L., **Some Thoughts on the Packet Network Architecture,** *ACM Computer Communication Review,* 1987, Vol. 17, nr 1-2, pp. 3-17

[Zha 87b]   ZHANG L., **Designing a New Architecture for Packet Switching Communication Networks,** *IEEE Communications Magazine,* Sept. 1987, Vol. 25, nr 9, pp. 5-12

[Zitt 91]   ZITTERBART M., **High-Speed Transport Components,** *IEEE Network Magazine,* 1991, Vol. 5, nr 1, pp. 5

PERFORMANCE EVALUATION AND MONITORING OF HETEROGENEOUS NETWORKS

José Neuman[*], Serge Lalanne, Fouad Georges, and J.P. Claudé

Laboratoire MASI
45 avenue des Etats Unis
78000 VERSAILLES
FRANCE

ABSTRACT

This paper presents the different perspectives of performance management for heterogeneous networks. On top of different interconnected networks, transport connections may be monitored for an overall QOS evaluation and management. The presented platform is a special environment spread on top of several chosen processing nodes. This construction augments the different nodes facilities to provide the overlying management applications with a uniform support across the 'peer nodes'. The platform is composed of several modules, each performing a given functionality. In the present context, the platform's functionalities are repartitioned amongst its modules in an atomic way i.e. each module regroups all functionalities belonging to a given aspect (e.g. storage, notifications, communications, etc.) which permits it to achieve its role in an integral, but not independant, manner. Obviously, all modules must "cooperate", in order for the platform to provide its intended services correctly. The potential heterogeneity of the underlying network was evaluated in the experimental phase of the PEMMON european project by the use of prototypes integrating two different network technologies, ethernet (TCP/IP based) and X25, and an ATM simulator.

Key Words: Network Management, TMN, MIB, Supporting Platform

1. INTRODUCTION

Network management specifications address a group of designers, each of which, perceiving management from a particular viewpoint. TMN [1] providers take specifications to build a "user-generic" system, Network Management Center (NMC) Administrators will then take the provided system and install it, by the introduction of knowledge, to their user-specific network. Finally NMC Customers who maybe network operators or simple clients use the provided network administration functions to develop their desired applications. In order to reduce the complexity of the system's specifications and to separate the different

---

[*] Supported by CNPq and BNB S/A (Brazil)

envolved issues, a Performance Management System (PMS) is described from several perspectives [2], each highlighting a certain facet of the system while abstracting away the un-related details.

The PEMMON project is an european ESPRIT II project which addresses the area of heterogeneous network management like MAN (ATM), WAN (X.25) and LAN (Ethernet) focusing on performance management. The main objective of the project was to explore, analyse and specify all the implications, aspects and solutions concerned with the design and the future implementation of a Network Performance Management System. The scope of this project was intentionally restricted to networks' performance management in a heterogeneous environment. The reason for this choice was that this category of network management is the least advanced, as far as standardisation is concerned, and is currently a subject for Research and Development [3].

The role of TMN supporting platforms is mainly to provide for the engineering mechanisms of network management systems. In fact, for a given system, we distinguish between two categories of functionalities, namely computational and engineering ones, where as the first concerns applications implementing the intended management activities, the second is only about infrastructure mechanisms (i.e. those recognized and offered as basic, common or public to all management applications). This distinction leads to regarding the same management system from two perspectives (or viewpoints). Thus, from an engineering perspective, a management system view (fig. 1) is that of a set of cooperating applications running on top of a supporting platform and controlling underlying network elements. Management applications depend on the platform for supporting their access to the network and their interactions. This view gives the functional requirements detailed in the following section and which form the basis for the platform's specifications. The design of the platform structure is based on an architecture proposed by the IDEA [4] methodology which was developed in the MASI laboratory. Validation of the current specifications has resulted several prototypes in the experimental stage in the same laboratory.

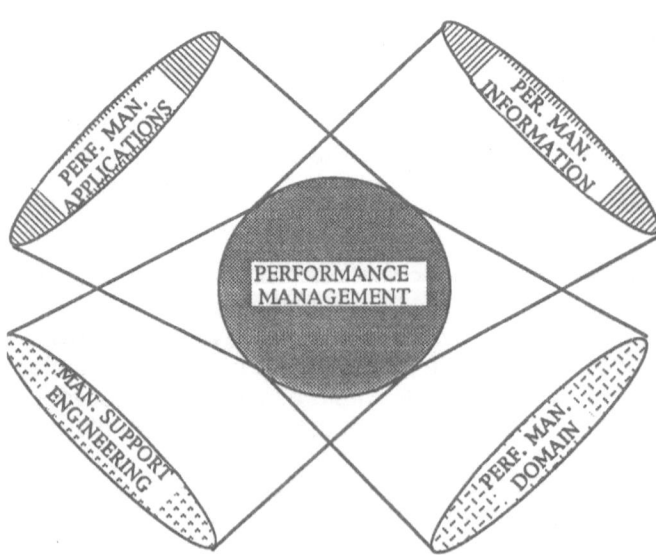

fig 1 : PEMMON Perspectives

2. PERFORMANCE DOMAIN PERSPECTIVE

It is premordial for any design to identify the system's context and the role it is taking as part of this context, which is commonly called the enterprise viewpoint. In the present case, this viewpoint describes performance management as a domain of the overall integrated network management. It should be noted that at this level no distinction is made between those tasks that will be automated and the human interventions, both are valid resources for the system to achieve its role. In fact, PEMMON recognises that a performance management system will coexist with other management centers (e.g. configuration, fault, etc.) where it will maintain Manager/Agent relationship on top of a Q interface for information exchange. It also shows that the management system will have to cooperate with other TMNs across a X-interface. With respect to the familiar management loop, namely: Awareness creation, Decision making and Implementation (commonly known as ADI [1]), PEMMON activities will cover the first two phases. More precisely, the role of PEMMON is to monitor the network performance (Awarness creation) and to participate in the Decision making by diagnostics and solution proposition. The human intervention (that of the network administrator) is expected to complete the management cycle, still with the system's support, by giving orders of the required actions.

2.1 PEMMON DOMAIN

Identifying a management system's domain is important as it implies the layout of logical (authority) boundaries for the system. Accordingly, the system's behaviour, especially that of the support platform, depends on whether events are internal to the domain or traversing the domain's boundaries. By drawing the PEMMON domain boundaries, both managed and managing entities belonging to its system (collectively called domain members) are identified. Indeed, other domains may comprise network elements controlled by PEMMON, either for simply using their communications facilities (in which case the elements control belong entirely to PEMMON) or to participate in their management, (e.g. case of configuration management, where a higher level management is needed to coordinate the activities of both domains). It should be noted that users will dialogue with the system through specialized Human Machine Interfaces controlling each user access rights and view of the underlying network.

In general, the management system distinguishes between three types of activities.

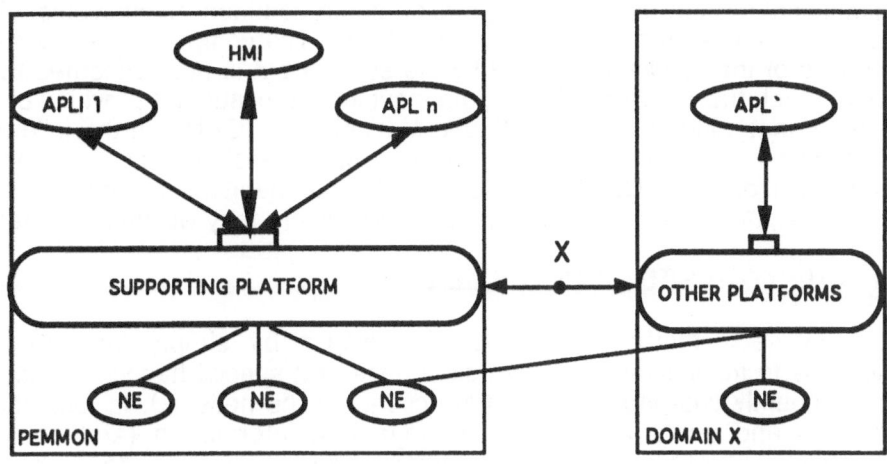

fig 2 : PEMMON Domain

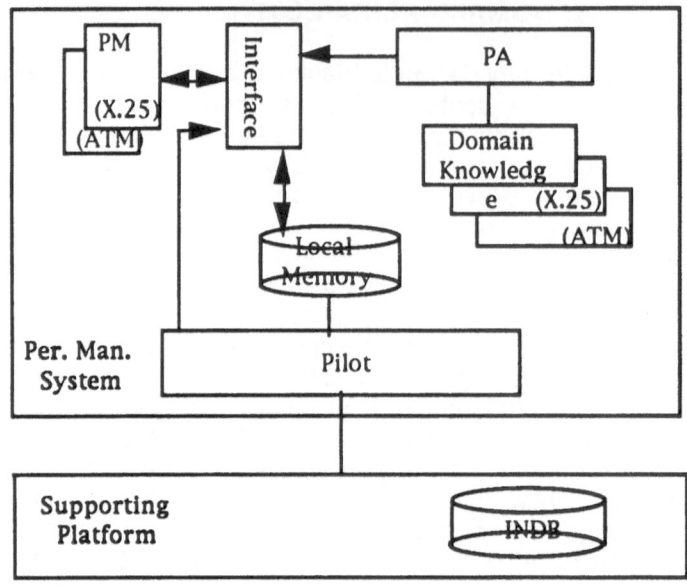

fig 3 : Performance System Perspective

Activities happening entirely within the domain (interactions between PEMMON members) are identified as intra-domain interactions. Activities traversing the domain boundaries, will happen when one or more PEMMON members interact with an authorised entity (taking a Manager/Agent role) of the external domain. In this case the external domain architecture is determined either to be conformant (i.e. another PEMMON system), where activities are called inter-domain or non-conformant where they are called extra-domain.

2.2 PERFORMANCE SYSTEM PERSPECTIVE

Once the system's role is specified with respect to the overall management enterprise, its internal management functions are described. Also, the automated management activities implementing the necessary computations are detailed. Figure 3 depicts the performance management system architecture.

The system's function is to exercise real time QOS (Quality of Service) control for end-to-end connections and to provide medium and long term statistical analysis of the network's performance [5]. It should however be noted that the computational system uses for its operation some common mechanisms which constitute what is known as the system's support. The computational system is thus aware of the services provided by the support but is not concerned by its structure or internal operation. In fact, the most important support provided by the platform is that concerning Query-Based Management (QBM). In the following, this approach is presented as it directly influences the design of the performance management domain. Details concerning the mechanisms performing such support take part of the engineering perspective, where they will be presented.

2.2.1 QUERY-BASED MANAGEMENT

QBM is a method for developing NMCs by simply implementing applications to manipulate the information base of a special handler (which will be consequently charged to reflect the actions on the network). In fact, NMCs developers and users will access management information (and not only managed information) by means of "data-base queries" and thus, do not have to tackle with the network themselves. It is very important not to confuse that

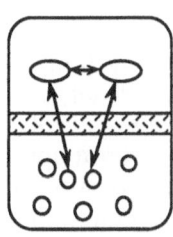
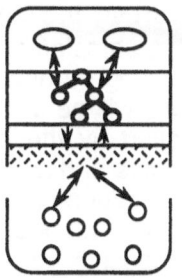

a) Classical Approach   b) MBM Approach

fig 4 : Query Based Management

approach with an Object-Oriented based system where applications should be programmed in OO languages. In contrast, in QBM, the engineering system presents an image (fig. 4) of the environment's information accessible by a predefined set of methods. Although this implies message-based interactions the interface is functionality based (a super set of CMIS [6] operations) as opposed to the object-based messages.

This kind of buffering allows for complexity hiding but also for a better control of the management activities integrity. Moreover, the network image may be constructed to directly reflect the applications perception of the network (e.g. an image of the network performance behaviour maybe built based on a queuing theory model). This approach allows performance management applications to query the necessary indicators directly i.e. instead of having to calculate them from more elementary variables.

2.3 PERFORMANCE INFORMATION PERSPECTIVE

Performance evaluation is the first step of awarness creation of the network's status. This is based on calculation of performance indicators to model the network's behaviour. The details of the network management information to be exchanged across the system are given in this perspective. The schema for the PEMMON Performance Model specifies the available performance information represented as objects in the INDB. Once more, it is necessary to dessociate the QBM and OO paradigms. The choice of Object Oriented Paradigm for the sake of information representation (modelling) was merely done on the basis of its suitability to the architecture requirements (abstraction, encapsulation, etc..).

The INDB schema describes (and restricts) the Performance Management Applications view of the underlying network i.e. the information semantics across the system. Therefore, in spite the fact that other perspectives treat only schemas from the meta level (i.e. as variable objects, classes, attributes, etc.), it is essential for any implementation undertaken in these perspectives to comply to the schema defined in the system's information perspective. NMC customers are provided with views of this information (depending on their profile) including the management information to which they may have access.

2.4 PERFORMANCE ENGINEERING PERSPECTIVE

In contrast to the performance system perspective, only mechanisms recognised as infrastructure are taken into the system's engineering account. These mechanisms are identified as those providing common basic services which are not directly involved in the management activities. They should not be confused with certain management applications implemented as servers to others[1]. In fact, a complementary criteria applies to distinguish between these two cases, depending on how these applications are controlled. Thus, whereas server applications will be controlled by the managing system, the infrastructure ones are controlled by the engineering system itself, which views them as processing resources and not as clients (case of server applications).

PEMMON engineering system is based on a platform approach where the previously mentioned computational entities are regarded as potential service users (clients) imposing certain requirements on its communications, storage and processing capabilities. In order to satifsy its requirements, the support platform exploits the available resources of its hosting environment (one or more processing nodes). Although minimal nodes facilities will be required, a platform construction will have to cope with variable existing capabilities. Therefore, the needed mechanisms and their operation to support performance management systems are described in this perspective at a conceptual (implementation-free) level. Mapping the given specification to a physical platform will be done (per node) by the TMN providers. However, by following the given framework, the resulting platform is guaranteed a correct operation with respect to the other nodes as well as to the management system's applications.

2.4.1 MANAGEMENT SUPPORT FUNCTIONS

The PEMMON architecture, from the engineering perspective, highlights the support functionalities (fig 5) of its support platform. This perspective is detailed in the next chapter.

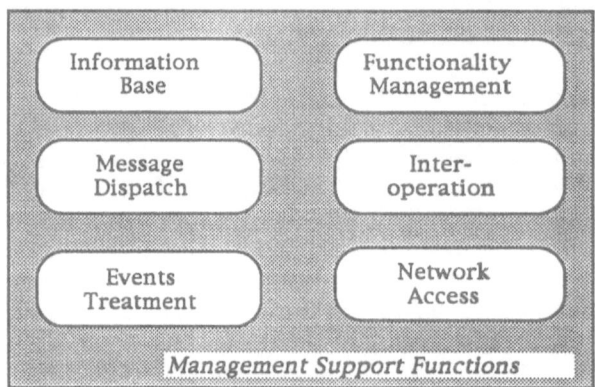

fig 5 : The Support Platform Functional Diagram

Message Dispatch: The support and control of message exchange between the platform's modules as well as between applications and the platform.
Network Access: The functionality of interfacing to the underlying networks via their manufactures' access points and representing the network parameters as elements of a virtual data base.
Events Treatment: This interprets the network status changes and decides the occurance of events. Accordingly, events are treated either by sending notifications or taking other defined actions.
Information Base: The management of a repository for the platform's management (and not only managed) information. The integrity of the system's information is assured by this function.

---

[1] Those management applications implementing the "User Generic" mechanisms of an OSF.

Interoperation: The function providing the possibility for the platform's applications to cooperate with external management centers.
Functionality Management: Control and coordination of the management services available on the platform by their representation as objects into a repository of services. This allows applications to fulfill their needed activities by invoking actions on shared (public) or imported functional elements.

3. THE SUPPORTING PLATFORM STRUCTURE

The platform (fig. 6) is composed of several modules generated on top of a group of TMN nodes. These modules interact by exchanging messages via the DISPATCHER [7]. There is a conceptually unique access point to this platform which is realised by the visible interface. In fact the interface shown in the figure as well as the other modules might be replicated for availability or reliability reasons. During the platform installation, one or more instances of the different modules are generated and configured across the system e.g. the shown INDB maybe actually implemented using distributed or federated databases. Moreover, the whole support is not intended to be implemented on one node. This implementation freedom is needed for the optimization of the development efforts by maximally exploiting the capabilities of the underlying systems (the INDB might profit from a specialised object oriented data base machine). The module responsible for keeping the virtual image of one platform together is the DISPATCHER, which is considered as the kernel. The description of the support's main components is as follows:

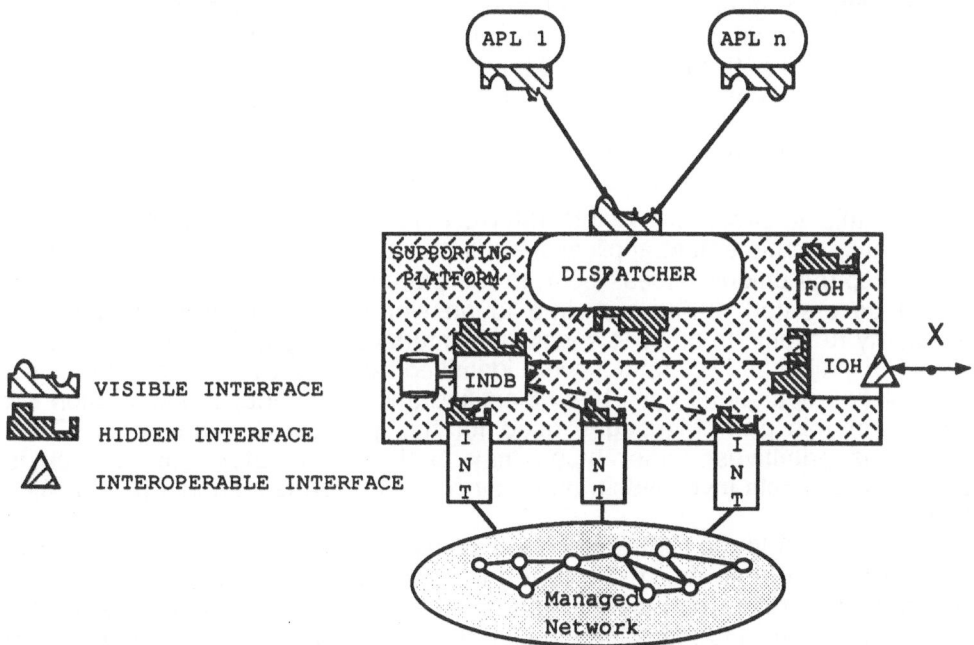

fig 6 : The Platform Structure

DISPATCHER: It represents the platform kernel. It controls the communications and the exchange of messages between the platform's modules as well as between applications and the platform.
INTEGRATOR: This component handles the functionality of interfacing to the manufactures access points and representing them as a common virtual access

point. In order to do this, it provides a syntactic unification of the access methods (e.g. the translation of an incoming request to the respective manufactures management commands) and a semantic unification of the manufacturers management operations.

INDB: This component supports the integration and the manipulation of the platform's information (those of the existing INTEGRATORs as well as the data bases). Any interogation concerning management information must be directed to this module.

IOH: The Inter-Operability Handler is the component providing the possibility for the platform's applications to cooperate with external management centers. In contrast to the other components, it must support one or more interoperable interfaces (e.g. CMIP for ISO-based and CMOT for internet-based management centers). It also controls the visiblity of management information and maintains their correct semantics on top of the different interfaces.

FOH: This is the Functional Object Handler. It provides a repository of the management services available on the platform. This allows applications to fulfill their needed activities by invoking actions on shared (public) or imported functional elements [8].

Visible Interface: This is a set of primitives (offering 3 communication modes) used by the platform's users (e.g. performance management applications) to communicate with it. Management messages (which is CMIS [9] based) are encapsulated within a given structure before sending it to the platform. Messages are first collected by the DISPATCHER and then delivered to the proper module (or to another application).

It should be noted that such an architecture is suitable for network administration in general. The orientation of the platform towards performance management concerns the management information i.e. during the specification of the models of both the INDB and the INTEGRATOR and also by the generic services that are represented in the FOH model.

3.1 THE PLATFORM'S INTERFACE SPECIFICATION

In this section, the visible interface provided by the platform to the performance management applications will be specified. This interface (fig. 6) is the conceptual unique access point to the platform. Instances of this interface should be created during run time to answer the applications demand for availability reasons. In order to specify this interface the possible interactions i.e. the set of exchanged operations were identified. At the present stage, the object manipulation operations have been identified and the interface specification will only cover this part. Messages used for this purpose are based on CMIS operations primitives. Other types may be identified later, e.g. to include functional elements invocations, and their message formats should be specified then.

In contrast to the CMI Protocol which imposes an association between two CMI Service Elements, the platform's interface [10], controled by the DISPATCHER, offers both connection-oriented and non-connected possiblities. For the non-connected message transfer, the platform allows for three modes for module interaction, namely, synchronous, confirmed asynchronous and non-confirmed asynchronous modes. The synchronous mode is a blocking mechanism i.e. the calling procedure will be disabled until receiving a return message from the DISPATCHER. The confirmed asynchronous is a non-blocking mechanism where the procedure waits only for the message delivery acknowledgement before continuing execution i.e. it does not wait the answer to its invocation. Finally, in the non-confirmed asynchronous mode, the procedure does not even wait for delivery acknowledgement. The three primitives to send messages in one of the above three modes are respectively: **CALLtoIDEA**, **CCASTtoIDEA** and **CASTtoIDEA**.

Besides the above non-connected modes, a connection-oriented one is provided, mainly for applications having heavy interaction with a given destination (or that require frequent access to the network across the platform, the case of statistical gathering applications). A **BIND** primitive allows such applications to work in association with a designated block of the platform. In this case, the DISPATCHER will initiate a special instance of the visible interface, notifies the destinator of the new session creation and returns the interface reference to the application. As from this time, the DISPATCHER no longer intervenes until the initiation of an **UNBIND** primitive to demand the closure of the association. In this way, the application enters in direct communication with the destination (normally a platform module) and will no longer need to address via the DISPATCHER i.e. it will not need to pack its messages in the IDEA_Frame. This also permits the platform to answer a request with multiple responses as soon as they arrive (case of notification) or at determined intervals (case of data collect).

This diversity of modes, permits a developer to choose the appropriate one according to the application's activities needs. This should not be confused with the operations confirmations to invocations (discussed later) which can be collected with or without blocking. It is evident that the actual inter-process synchronisation is up to the client applications i.e. these applications must be able to handle the proper protocol (a connection or a rendez-vous) with another application, the platform's role being restricted to the transfer of the proper protocol primitives.

3.2 The DISPATCHER

The DISPATCHER may be regarded as the kernel of the proposed platform. It offers the access point to the platform i.e. it provides the visible interface via which the platform users (e.g. performance management applications) intercommunicate with each other or with the platform's components. In addition to this interface, the DISPATCHER manages an internal or hidden interface used essentially by the platform's components which are expected to be implemented in a distributed manner (and which might even migrate from one machine to another during the execution of the platform). In fact, a given module is developed with no prior knowledge of the whereabouts of its partner ones. The ensemble of components base their interactions on exchanging messages (via the hidden interface) irrespective of their actual organization (local/remote). Location resolving is one of the main roles of the DISPATCHER. There are four supported communications modes, of which three are connectionless and one is connection oriented. Connectionless modes allow for synchronous, confirmed asynchronous and non-confirmed asynchronous communications, while the connection mode implies session establishment between two modules. Exchanged messages are CMIS-based requests together with the indication about the desired component and communication mode. For the DISPATCHER to undertake its role, it interfaces with the local operations supporting systems and exploits their offered facilities (e.g. queues, sockets, rpc, etc.).

It should be noted that the information needed for the DISPATCHER operation is added as a header to the sent request. To facilitate the packing of the request into an IDEA message, a compilation-time library is provided. The applications have to fill a data-structure then call the library function that takes care of communicating the request to the DISPATCHER (a process commonly called marshalling/unmarshalling).

3.3 The INTEGRATOR

The INTEGRATOR is responsible for interfacing to the manufactures acces points and representing them as a common virtual access point. In order to do so, the INTEGRATOR groups whatever underlying local schemas into one set of objects defining the view of the network elements for the management functions. In fact, the unification procedure is started by partitioning the overall network into homogeneous domains, each having an access point (provided by a proprietary system) through which management of the internal objects may be achieved. The mapping between the INTEGRATOR model and the different domains is done individually for each domain. That is, for each object's property (attribute or method) in the INTEGRATOR's model there exists a per domain syntax and semantic interpretation functions. The above partitioning is logical and is done with respect to a certain management aspect (geographic, functional or other). That is several organizations for the same network may arise, each consisting of domains to which an INTEGRATOR instance is assigned. As such it is expected that more than one INTEGRATOR are needed to provide access to the network information through one reference point. This reference point is realized by an interface that the INTEGRATOR should offer to its clients and that defines the set of permissible operations on the objects of its model.

In this section a brief description of the INTEGRATOR's internal structure is given. Figure (7) shows the different building blocks needed by the INTEGRATOR to fulfill its role. The INTEGRATOR has two kinds of interfaces :

3.3.1 THE "ADMINISTRATION" INTERFACE

This interface is provided for the network expert that will feed the INTEGRATOR with its knowledge of the underlying network. It is used for the instanciation of a specific INTEGRATOR by creating its objects. The knowledge also includes all the operational information such as the mapping procedures of

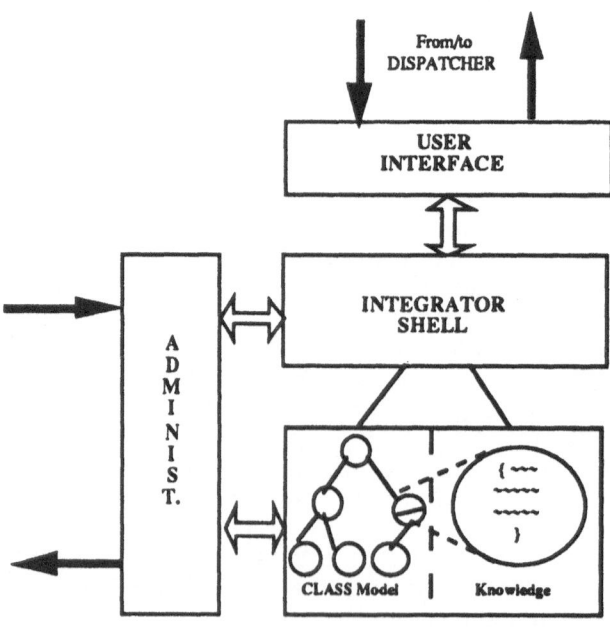

Fig. 7 : Internal Structure of the INTEGRATOR

each attribute in the model to the specific instances in the different domains. This block should undertake the necessary checks on its input (e.g. lexical and syntax checks, model integrity checks, etc..) before representing it in the form usable by the INTEGRATOR's shell. Presenting the knowledge includes stocking the schema of the model, its population (instances) as well as the encapsulation of the mapping procedures for the different attributes.

3.3.2 THE "USER" INTERFACE

It is to this interface that the INTEGRATOR's clients should import and use in order to manipulate its objects. Its type is that of the hidden interfaces and the operations invoked via this interface are CMIS based.

3.3.3 THE INTEGRATOR'S SHELL

This is the main building block where the INTEGRATOR functionalities are supported i.e. on reception of an external request to undertake the invoked operation on behalf of the Manager. To achieve this some subfunctions are identified:

- Object Management
- Object Persistance Handling
- Syntaxic Interpretation and mapping according to the domain.
- Semantic Interpretation and mapping.
- Notification of abnormal events

3.3.4 THE INTEGRATOR INFORMATION BASE

The remaining two-part block represents the information base of the INTEGRATOR. This should not be confused with the INDB which is the repository to all the system's shared information. In contrast to the INDB, this information base stocks the internal knowledge of the INTEGRATOR i.e. its object descriptions, along with the different instances, their static attributes as well as the code for the procedures that permits mapping the attributes (syntaxically and semantically) to their corresponding real resources.

3.4 INTEGRATED NETWORK DATA BASE (INDB)

The purpose of the INDB is to provide a common view of all management networks in order to permit users or management applications to interact with the elements of these networks via a unique semantic ([11], [12]) . The INDB module is the set of mechanisms providing a unique view of the real system which is composed of several networks to be managed. It is used by all management applications to interact with the real Network Elements (NEs) and to store their management informations (e.g. network and system configuration, customers and services, current and historic performance and trouble logs, security parameters and accounting information).

Management applications use the CMIS-based services to formulate their requests and send them to the INDB module via the DISPATCHER. In case of reading or modifying the parameters of real resources, the INDB must interogate one or several appropriate INTEGRATORs. The INDB also handles requests concerning the system informations which may be held in one or several databases or in the management model itself. On the reception of the responses to its interogations, the INDB processes the collected informations and returns the results to the management applications. Return messages are either values of attributes for a consultation or an operation confirmation for an action.

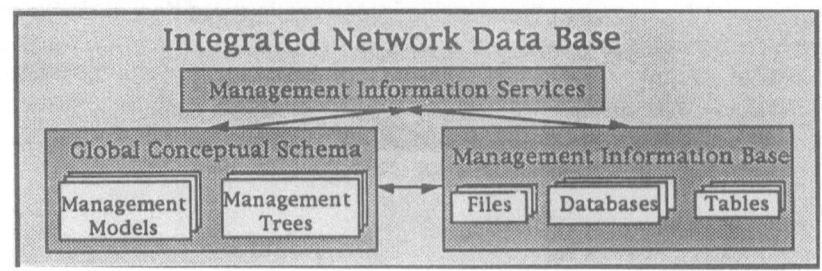

Fig 8 : INDB Functionalities

3.4.1 FUNCTIONALITIES

To ensure these roles, the INDB is concerned by three management functionalities (fig. 8) :

- to provide an INDB Global Conceptual Schema (GCS), i.e. a unique point of view of the managed system, which allow management applications to have the knowledge about the management informations of all the different NEs via Management Models (described in OSI NM/Forum Templates [13], [14]), and their interactions between them via Management Trees (Inheritance, Naming, Containment and Registration Trees [15]). The GCS gives an unique homogeneous view of several heterogeneous environments.

- to provide an INDB Management Information Base (MIB), i.e. a conceptual repository which contains all the management informations of the Network Elements,

- to provide INDB Management Information Services (MIS) based on CMIS, i.e. a user interface on which management applications can manage (creation, deletion, manipulation) NEs.

3.4.2 LOGICAL ARCHITECTURE

To ensure its functionalities, the INDB module has to manage three phases of treatment when an INDB request sended by users is received (fig. 9).

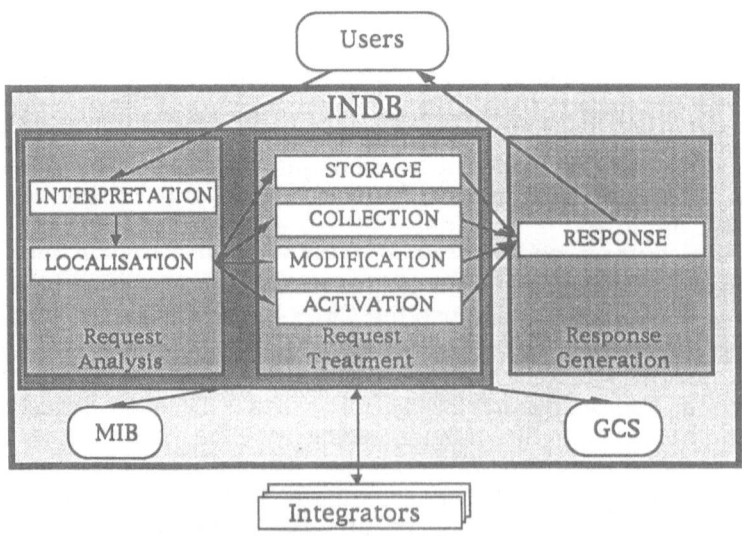

Fig 9 : INDB Logical Architecture

These phases of treatment are request analysis, request treatment and reply generation phase.

3.4.2.1 REQUEST ANALYSIS PHASE

This first phase of analysis of the INDB request consists on its understanding by the INDB module, that means, its control, its interpretation and the fashion to resolve it. A first subphase of interpretation (research of selected instances in applying request scoping and filtering) followed by the subphase of localisation (GCS Manager, INTEGRATORS, MIB Manager) are realised to analyse the request.

3.4.2.2 REQUEST TREATMENT PHASE

The second phase of treatment of the INDB request consists on its execution on the Network Elements via INTEGRATORS. Several kinds of execution are considered which are Storage, Collection, Modification and Activation of managed objects.

3.4.2.3 RESPONSE GENERATION PHASE

The last phase of generation of the response to the user consists on the collection of the results generated by the treatment phase and on the design of the INDB response.

3.4.3 PHYSICAL ARCHITECTURE

This section presents a brief description of the internal structure of the INDB which is composed of the following modules (fig. 10).
The INDB supports the organisation of the Network Elements. Organisation means the description and the relationships of these NEs

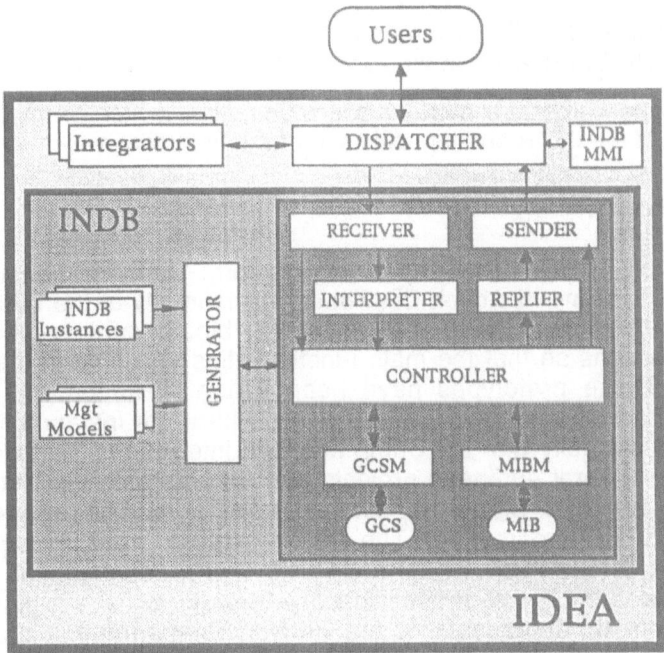

Fig 10 : INDB Physical Architecture

throughout a Global Conceptual Schema (GCS) plus the manipulation of their management informations throughout either a Management Information Base (MIB) for static informations or existing INTEGRATORS for dynamic informations. Any management application request concerning management information must be directed to the INDB module. The INDB Physical Architecture is composed of the modules :

Receiver: The Receiver module assumes the role of reception of messages from the DISPATCHER module and realises the unmarshalling of a message structure into an exploitable structure either for the Interpreter (Requests of management applications) module or for the Controller (Replies of INTEGRATORS).

Sender: The sender module receives either the INDB response from the Replier to the requested management application or requests from the Controller to a specific integrator and realises the marshalling of the reply or request structure into a message structure to send to the DISPATCHER.

Interpreter: The Interpreter module realises the functionality of the INDB requests analyser received from management applications.

Replier: The role of the Replier module is to format the response of the INDB request from the replies sended by the Controller which interrogates either GCSM or MIBM or INTEGRATORS.

Controller: The controller ensures the role of localisation of module (GCSM or MIBM or INTEGRATORS) to treat requests and send replies to the Replier.

GCSM: The Global Conceptual Schema Manager (GCSM) module has to manage a Global Conceptual Schema (GCS) i.e. the knowledge of the managed networks in terms of management trees and management models.

MIBM: The Management Information Base Manager (MIBM) module manages the Management Information Base (MIB) which is the repository of the management informations of the Network Elements.

INDB MMI: The INDB MMI is a module which communicates with the INDB via the DISPATCHER and ensures the role of GCS and MIB Browser.

Generator: The Generator module is a mechanism which permits to generate one instance of the INDB on a specific site of the network.

4. CONCLUSION

The architecture of the supporting platform for heterogeneous networks management systems has been developed continuously at the MASI laboratory. By now, the Integrator, INDB and Dispatcher blocks can be found in its first prototypes versions so that the main functionalities specified in the architecture are shown. These prototypes have been validated in the real world using environment like ethernet and X25 networks, where, for instance, the integrator block interacts actually with the Snmp and X25 integrators. This work has been undertaken in several europeen projects that are interested in investigating the construction of TMN platforms, like ADVANCE (RACE programme) and PEMMON (ESPRIT programme). The results obtained from these projects have been used as a feed-back on the support platform development. As a further work, we think to improve our specifications in order to have a generic platform and investigate other aspects of the network management information like security.

5. REFERENCES

[1]   Working group 2 of RACE project "GUIDELINE" (R1003) ,"*TMN Implementation Architecture Main Event 8"*, April 1992.

[2]   F. Georges et al, *"An Engineering Perspective for Performance Management"*, COMCON 4, June 1993, Greece.

[3]   PEMMON project, Final Report, University College Dublin's, November 1992.

[4]   J.P. Claudé et al., *"Unification of Heterogeneous Management by a Generic Object Oriented Agent"*, TMN Conf, Nov. 1990.

[5]   PEMMON project (Esprit II 5371), *"Detailed Functional Specifications of the Proposed Application for Performance Management"*, deliverable 22, Nov. 1992.

[6]   ISO/IS 9595, Information Technology - Open Systems Interconnection, Common Management Information Service Definition, July 1991.

[7]   PEMMON project (Esprit II 5371), *"Global Specification of a TMN Platform"*, deliverable 6, Dec. 1992.

[8]   M. TAG, *"Developpement D'Applications Reparties Sur le Systeme Distribué Orienté Objet CSA"*, Doctorat of the University Paris VI, June 1989.

[9]   ISO/IS 9595-2, Information Processing Systems, Open System Interconnection, Management Information Service Definition-Part 2, Common Management Information Service, July 1991.

[10]   PEMMON project (Esprit II 5371), *"Specification of a TMN Platform"*, deliverable 4, Nov. 1991.

[11]   PEMMON project (Esprit II 5371),*"Description of the INDB"*, Deliverable 17, May 1992.

[12]   PEMMON project (Esprit II 5371), *"Documentation of the INDB"*, Deliverable 18, Nov. 1992.

[13]   OSI/Network Management Forum : FORUM 003, *"Object Specification Framework"*, Issue 1.0, September 1989.

[14]   OSI/Network Management Forum : FORUM TR 102, *"Modelling Principles for Managed Objects Technical Report"*, Issue 1.0, January 1991.

[15]   OSI/Network Management Forum : FORUM 007, *"Managed Object Naming and Addressing"*, Issue 1.0, May 1990.

[16]   F. Georges et al, *"A Supporting Platform for Heterogeneous Networks Management Systems"*, 11th SBRC, May 1993, Bresil.

[17]   *A System Architecture for Updating Management Information in Heterogenous Networks*, J.N. De Souza et al., IEEE GLOBECOM' 92, Dec. 1992, U.S.A.

[18]   *Managing Heterogeneous Networks : Integrator-Based Approach*, J.N. De Souza et al., IFIP'93, April 1993, France.

[19]   *Unification de Contextes Hétérogènes par un Agent Générique Orienté Objet,* M. Claudé, Phd. Thesis from Paris VI University, Feb. 1990, France.

Application of High Speed Networks in Hospital Environment

Jai R. Rao[*]
Information Transfer Solutions, Inc.
St. Louis, MO 63131

ABSTRACT

This is a case study based on author's experience working with the Washington University medical center. The center includes three major hospitals with a total of 2000 beds and a major research oriented school of medicine. Medical research and hospitals are seeing the potential benefits of using high speed communications networks for sharing the patient care information. There are two facets to the increased demand for high speed networking: trends in tele-medicine and consolidation of patient records. This paper presents a networking scenario that has been evolving at Barnes Hospital/Washington University Medical Center. It consists of high speed internetworking among application driven sub-networks which enhance the strategic and research value of the patient data in the form of image, text, and annotated voice. The discussion leads into high speed networks based on star topologies.

1. Hospital Environment

Application environment is the mold for the network infrastructure. Hospital application environment considerably differs from other business environments: they are open and family oriented. That is, hospitals need to provide easy access to the patients and their families. This sort of *open environment* also leads to a less physically secure layout to the communications as well as information processing equipment unless special measures are incorporated. In other business environments, for instance, network wiring centers may not have to be locked, because chances are minimal that a stranger walks into the facilities and vandalizes the communication equipment.

The Washington University Medical Center (WUMC) located in St. Louis, Missouri consists of business oriented Barnes Hospital (BH), Children's and Jewish Hospitals and research/teaching oriented Washington University Medical School. With a total patient beds of around 2000 makes this one of the major centers in U.S. Hospitals have a centralized decision process, whereas medical school departments are autonomous. The departmental decisions vary depending on the research funding, grants, etc. A patient and his medical records are the common denominators.

[*]Adjunct Faculty, School of Technology and Information Management, Washington University, St Louis

The operational differences between the hospital and the university work environment make it difficult to have a compatible common network layout and support services. It is a challenge to identify a unified platform to use as a basis for procurement decisions.

Hospitals are phasing through a bottom up revamping of the patient care services. In a traditional setting, hospital MIS department would be providing all the patient billing, financial and administrative services through a centralized processing center. However, this process has been reversed over the past several years. Service providers such as surgery, internal medicine, pediatrics, etc. have taken up the cost control issues by internalizing the services. The question of whether if that is the proper approach is not pertinent to this paper. This has lead to the formation of many sub-networks at department or organization level. Sub-networks snake through the hospital complex. The sub-networks have either separate cabling plans, or in certain cases physical cable is the same, but the logical networks are managed separately.

Business environments in general use per unit product cost as a barometer for comparison purposes. In hospitals patient is the single most important product entity. In the present operating mode, each patient service creates an independent record; examples of which are ophthalmology, radiology, LAB tests, strategic services, etc.. These services maintain separate database repository. Except for certain specifics related to the service being provided, the other information contained in the patient record such as history, age, insurance, financial, etc. are common information. A strong effort is underway in the health industry to create and maintain a single electronic patient record. There are two primary concerns: unauthorized access and updates.

The format and delivery of medical services is changing drastically. The distinction between in-patient and out-patient care is very transitory, and as the medical technology changes more services will be offered on an out-patient basis. The trends in ambulatory care centers (ACC) will change the total patient care concept [1]. It is not clear if the innovative approaches will enhance the patient confidence and/or make the services cost effective. Some key considerations are discussed next.

A typical distributed health information system (HIS) includes, long term care, acute care, managed care facilities, and others. The key driver for this distributed information system is also patient and his electronic record. In HIS, potential beneficiaries of the centralized record system are : 1) admissions, discharge and transfer (ADT), 2) medical records reporting, 3) resource scheduling including operating rooms, physicians, nurses, resolution of scheduling conflicts, 4) materials management, 5) payroll, 6) patient assessment, 7) patient plan of care, and, 8) results reporting.

The information common to a patient record gets updated by the attending physician, nursing care staff and administrative staff. The medical history (patient records) get lengthier and complex as the patient continue to age and receives services. A centralized, master patient index accessible from any node of the service provider allow the system to better track patient encounters. Further enhanced nursing care functionally requires implementation of on-line charting. The centralized or *paperless* medical report system require tremendous amounts of data storage and, very high communication speeds to retrieve data. A flexible, integrated clinical?financial data base provides a detailed data on costs and service utilization. This allows tracing of services and a patient's entire episode of illness. The data base would provide a foundation for clinical, operational, and strategic decision making process. It will help setup a nationally accepted ACC standards.

2. Technology Drivers

Patient data consists of text, images and physicians' voice annotations. In the all digital application environment, information generation, processing, distribution are put in the digital format. In the context of this paper, in the digital radiology the X ray images are digitized for transport and get regenerated at the receiving end. As an illustration consider a typical layout of tele-radiology sites: 1) a film digitizer for converting radiographic films into digital representation, 2) interfaces to digital imaging modalities as CT, MR, CR, US, DF, and NM, 3) a gray scale workstation for selecting and monitoring images to be transported and for receiving, reviewing, and annotating displayed images, 4) a hard copy recorder, 5) a computer system for archiving and controlling the system, and, 6) a video-conferencing system. Likewise analog EKG data transferred in digital format to the primary care physician on request. Other forms of digital images can be transported as individual records or as a file. An attending physician may request as much information as possible on a patient and volume of such information depends on the patient age and how readily that information can be accessed. The physician may want to correlate clinical data with patients' age. The driving technology in this case is the ability to archive and retrieve the text, image and voice annotations. The picture archiving and communication systems (PACS) are leading the way in this respect. [2,3]. For now it seems the PACS are expensive form of management of images, however, due to the large volume of medical imaging, it is expected that the costs will drop substantially. According to one report 240 million non-digital X-ray procedures and 80 million digital procedures are performed annually in US [4]. Table 1 provides a list of commonly accepted digital image size estimations, which can be extrapolated to determine the network traffic based on the procedure volume.

As a routine a primary care physician would go to a centralized hospital facility where he could walk through the patient X ray images. Frequent commuting to a centralized location puts additional constraints on his time. The technology needs to make accessible clinically acceptable images at the physicians work place. This requires high speed communication capabilities along with the image digitization, generation and display capabilities. Another critical area of technology is multimedia application data retrieved from different medical services to compile a comprehensive report on a patient. Analogous to the paperless office vision, present drive is to make bedside patient charts free of hand written clumsy sheets. A point-of-care system can be conceptualized as the nerve endings of the hospitals' information body structure.

Table 1: Digital image size estimations

Modality	Resolution	Image Size (MB)
Computer Radiology (CR.)	2048x2056x10	6.25
LASER Film Digitizer (LFD)	1684x2048x10	4.11
Barium Spot Study	2048x2560x10	6.25
Computerized Topography (CT)	512x256x12	0.39[*]
Magnetic Resonance Imaging (MRI)	256x256x12	0.137[**]
Ultrasound	512x512x8	0.25
Fluoroscope	512x512x8	0.25

[*]A CT consists of 20-50 slices; [**]An MRI consists of 80-150 slices

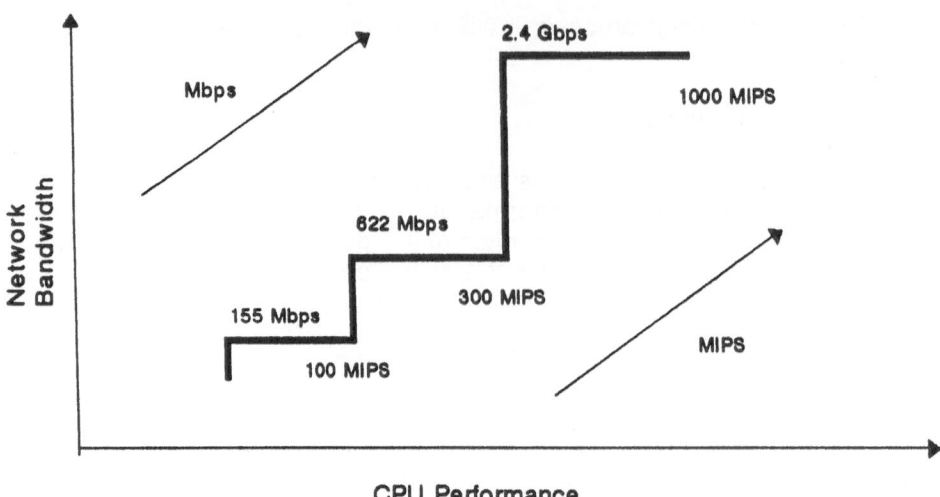

Figure 1: Bandwidth and CPU performance
Source: High-Performance Network Challenge Ethernet,
by Warren Andrews, July 1992, Computer Design

Robust networking products are gaining the confidence of the patient care providers. Robust networking is defined in terms of: 1) transmission, 2) hardware platforms, and, 3) reliable software. The transmission technology is improving rapidly via fiber medium and through the use of diverse routing. The data networks are incorporating many tried and tested technologies evolved in voice networking. Figure 1 illustrates a strong correlation between high-bandwidth communication and instruction processing (MIPS).

One can notice, multi-threaded sessions of high volume protocol packet processing bridges and routers would not have been in the market today without the availability of cost effective high speed processors. For example, RISC chips in routers can process data about twice as fast as comparably priced CISC chips [5]. Modular and object oriented software implementations are making networks easily testable for problem resolution. Large scale network support depends on dynamic reconfiguration capabilities, which to a larger extent depends on software flexibility. Recent trends in product offerings for fault identifications and for network configuration details are example indicators of the trends in software.

3. Networking Requirements

The patient care services have evolved from a simple one on one doctor-patient relationship to a technology driven information sharing industry. There has been an explosive growth in organ donors, transplants, sharing of research activities and expertise. In addition, automation of patient bedside charts, patient accounting, and other patient care related activities around the hospital are evolving rapidly. The automation process leads to increased volume of data that needs to be exchanged between local and remote facilities. This also creates an opportunity for the health care facilitators to offer to the physicians on-line inquiry and access capabilities. For example, a multimedia workstation based bedside system can facilitate the attending physician to review unto the minute patient charts. This would take a cumbersome process of going through the paperwork. As the automation process jump starts, capabilities of bedside workstation platforms will be made available to the physician at the offsite clinics via the public switched broadband ISDN and/or private networks. The business ethics still dictate that the patient history is kept strictly confidential between the doctor and the patient.

As the hospital industry continues to automate for reducing costs and improving the patient care services, two factors become very critical: dependability and accessibility. Unlike other industries, access to critical care information is a dire necessity, else the system will be totally ignored by the physicians and patients alike. The doctor and the patient must be convinced that the data is valid and is readily accessible. If the image data is compressed for storage or transmission reasons, the information needs to be reproduced in its original form with full integrity. This gets further complicated if the image is transmitted to remote locations within the hospital complex or outside the hospital to remote locations. This requires robust networks. The fact that the individual network elements are ultra-reliable, however, does not automatically assure that the total network structure to be robust.

The hospital system works around the clock with workload variations. Needless to say, the information accuracy and delivery are critical at any time of the day. The *clean* backbone network concept arises out of using reliable transmission media such as fiber and putting a concerted effort on supporting only a limited number of application protocols. In order for the backbone to be clean (in the communications sense not corrupted by application protocol problems or by the communication protocols), the protocol specifications and monitoring capabilities must be in place. For example, at the channel multiplexer level, dynamic allocation of channels must be clearly monitored. Other communications protocols in use that need monitoring capability independent of application related issues are TCP/IP, SDLC, DDCMP, SONET, FDDI, Ethernet, Token Ring, SMDS, Frame Relay, and so on. Besides electrical and/or optical characteristics of the transmission media, application protocol issues play an important role in creating a reliable network. The application related requirements, such as file transfer, EDI, electronic mail, Client/Server are some of the common services provided by the network. In addition, hospitals have certain requirements for image transfer, tele-radiology, tele-pathology, etc., which will have considerable impact on the network bandwidth.

Next, we look at the *health* aspects of the backbone network. The healthy networks by definition are robust and reliable. In an ideal situation, application protocols should have a minimum impact on the network ups and down. However, the traffic patterns developed as a result of improper protocol performance could cause networks to go down or force the network to be taken down. Hence, healthy network infrastructure depends on the network configuration (physical layout) and the network reliability. The network layout may consist of redundant paths, alternate routes, back up configurations, and load sharing facilities. The hardware/software failures create new configurations and reliability problems. Dual power supplies or redundant modules are traditional engineering practices used to minimize the failure effects on the network functionality. Given the complexity of these networks, the failure isolation and providing for redundant capabilities is not easy. The degree of reliability and the cost are the major trade-offs in robust networks. This leads to a discussion on network design and architecture issues.

4. Network Justification

The Barnes Hospital/Washington University Medical Center high speed network project evolved through many critical reviews in the process of strategic planning to support increasing demand for network bandwidth. As defined earlier clean

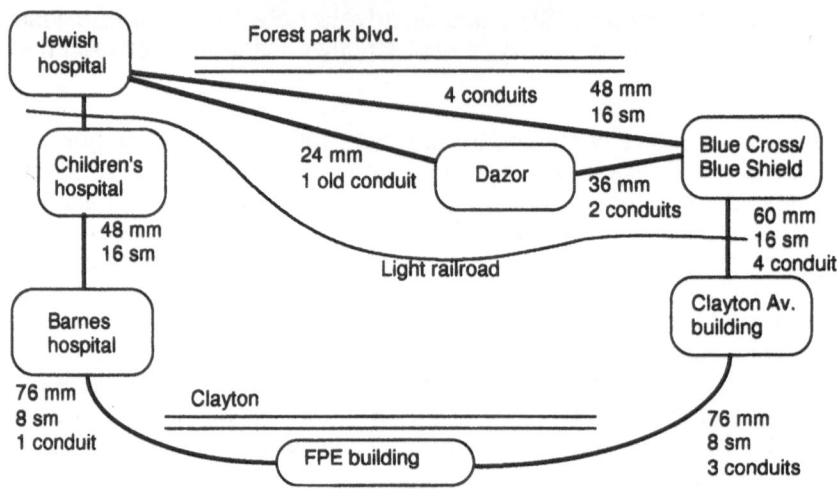

Figure 2: Barnes Hospital/Washington University Medical Center
Planned Fiber Routes

and healthy network is the primary focus. The challenge is to select the technology which meets the criteria set today without locking the current environment into a cost prohibitive evolutionary process. Considering the technology maturity cycles, the recommendation boiled down to a network strategy based on:

a) multimode fiber transmission medium
b) smart hubs
c) collapsed networks

Figure 2 illustrates the layout of the multimode and single mode fiber plant. Additional conduits are provided to pull fiber as needed in the future.

The price differential between single and multimode fiber is minimal. Hence, single mode fiber is installed even though there are no immediate plans to use it today.

A general consensus is evolving in the industry that the interconnected star topologies are more robust than bus based architecture, especially in view of the Asynchronous Transfer Mode (ATM) based technologies. Today, due to a lack of ATM technology the star nodes are based on smart hubs. The intelligent hubs are capable of providing configuration and management of complex backbones [6,7]. A high speed protocol such as FDDI is used to link the star nodes forming the high speed network backbone. A collapsed backbone network strategy is recommended to make the network robust. The sub-networks comprising of Ethernet, Token Ring, and CDDI are collapsed into a star node. The criteria used here is distance and administrative requirements. The network scenario is presented in figure 3. The roles of routers and smart hubs are confusing. As is illustrated in figure 3, configurations I and II, the issue of whether to put router or smart hub functionality on the backbone is discussed next. In the opinion of this author, smart hubs must be placed on the backbone. The hubs, in addition to providing the routing functionality, can also provide fault tolerant features and network management functionality. Relatively speaking, routers are less flexible and less intelligent. It is anticipated that the future high speed networks will evolve into star topologies, and smart hubs of today could be ATM switches of the future.

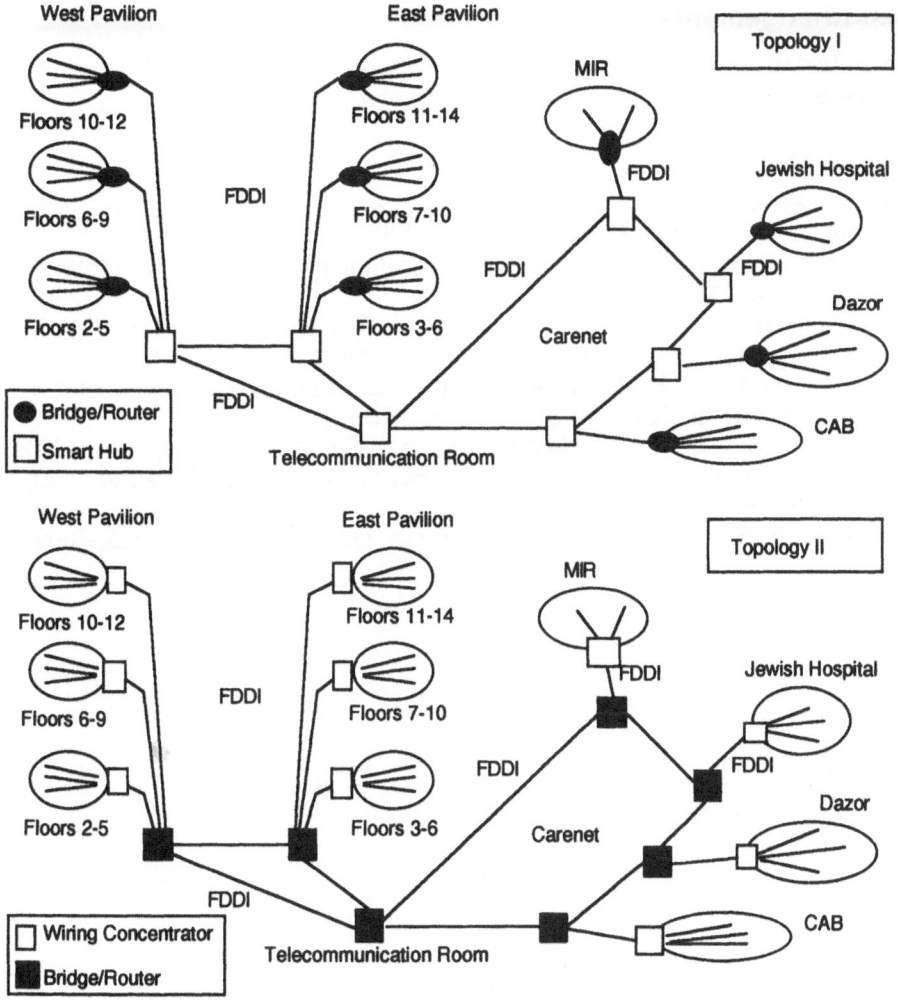

Figure 3: Barnes Hospital/Carenet Topology

The placement of network management stations in figure 3 is more tactical than technical or strategic.

The primary reasons that motivated for choosing the smart hub based star networks warrant further discussion. Various types of smart hubs are available today. The incremental packet throughput capabilities are in the range of 100 Mbps to 1 Gbps. At this stage of planning, it is not clear how to pace the scalability of the node processing capability with the BH/WUMC network traffic growth. The issue that arises is, at what point scaling up a smart hub stops, and replacing it with an ATM switching node proves to be economical.

ATM technology is evolving rapidly. Does it have a place in the hospital environment? Hospitals have a need to support multimedia applications. The basic question that needs to be answered is: does annotated voice need ATM switching capability?

A future paper is planned for discussing the ATM networking issues in the hospital environment.

Acknowledgements

The author would like to express sincere appreciation to David Weiss and Mary Ellen Hulse for letting this paper to be presented.

References

[1] T. A. Matson and M. D. McDougall (Eds.), *Information Systems for Ambulatory Care*, American Hospital Publishing Inc., 1990
[2] S. Fath, *Extending Broadband for Radiological Advances*, Communications News, pp. 8, Oct. 1991
[3] J. Blaine, R. L. Hill, et al., *Image Transmission Studies*, SPIE, v. 918, Medical Imaging II: Image Data Management and Display, 1988
[4] White Paper on High Speed Communications at Barnes Hospital, IBM, Document internal to Barnes Hospital, Oct. 1992
[5] E. M. Hindin, *RISC Comes to Routers*, Data Communications, June 1991
[6] W. Andrews, *High Performance Network Challenge Ethernet*, Computer Design, July 1992
[7] J. Herman, *Smart LAN Hubs Take Control*, Data Communications, June 1991

ON ALLOCATION SCHEMES FOR THE INTERCONNECTION OF LANs AND MULTIMEDIA SOURCES OVER BROADBAND NETWORKS

Mihai Mateescu

GMD FOKUS Berlin
Technical University Berlin
Hardenbergplatz 2
10623 Berlin, Germany

Abstract

One fundamental difference between traditional telecommunications networks and traditional data networks is the level where statistical multiplexing takes place. The LAN world uses broadcast topologies, shared media and statistical multiplexing is performed inside the network. The traditional telecommunications world uses dedicated point-to-point links and the customers compete for resources before entering the network nodes. The current ATM paradigm for B-ISDN inherits a lot from the telecommunications environment but ironically enough its first applications are expected to be the broadband (ATM) LAN and the interconnection of LANs over wide areas. In order to allow for statistical gains, the dedicated links have been transformed into virtual connections which share the physical transmission medium, but without an access coordination mechanism similar to the MAC in LANs. Instead, a host of bandwidth allocation and congestion control schemes are used, which are largely based on a statistical measure of the characteristics of the individual and aggregate traffic streams. The paper presents a critical survey of the tools available to explore the multidimensional allocation space and to characterize the blocking properties of the broadband shared resource environment when carrying traffic generated by heterogeneous sources and in particular by interconnected LANs.

INTRODUCTION

The bandwidth available for digital communications has been growing dramatically over the last twenty years. At the break of data modems day in the beginning of 1970's, communication was possible at 100 bps speeds. In the beginning of 1990's the CRMA LAN[1] reaches 1.13 Gbit/s on fiber connections and it is termed "second generation network" due to the (still) electronic front-ends employed at the network nodes. Third generation "all optical" networks[2] are already emerging, which take on totally new approaches to exploit the properties of fiber, specifically an information carrying capacity of tens of terrabits per second as compared to the few gigabits per second of peak electronic processing speeds. The asynchronous transfer mode (ATM) is seen as the key technology in providing broadband integrated services digital networks (B_ISDN[3]) at 155 Mb/s up to 2.4 Gb/s over local and wide areas and its introduction is expected in the near future. The present report is trying to give an overview of the complicated issue of allocation policies in the broadband environment. Since in the foreseeable future statistical multiplexing is going to play an important role for networks with large populations of users, different levels of traffic control are needed to avoid the degradation of the quality of service (QOS) and to insure network stability. The difficulty arises because of the conflicting worlds which must be brought together. The LAN world uses broadcast topologies, shared media and mechanisms to coordinate the nodes between themselves when accessing the transmission facilities. There is no need to establish in advance a connection and statistical gains are obtained in a controlled manner inside the network. The traditional telecommunications world uses meshed topologies, dedicated point-to-point links established at call set-up time and there is theoretically no need for coordination between contending customers, since the competition for resources takes place before entering the network. The ATM approach for B-ISDN, while keeping with the idea of a point-to-point path, has transformed the dedicated links into virtual connections which effectively share the physical transmission medium. But the end-systems connected to this wide-area shared medium (i.e. the sources multiplexed on a virtual path) don't use a medium access control and therefore a host of bandwidth allocation, bandwidth enforcement and congestion controls are expected to prevent congestion on that path, while allowing for statistical multiplexing. The very different nature of the two worlds becomes most evident in the case of ATM LANs, where the connection oriented nature of the ATM switches strikingly contrasts the connectionless nature of the end-systems connected to them. The remainder of this paper surveys the tools available today to decide how the resources of the broadband network can be shared statistically by heterogeneous traffic sources and in particular by LAN traffic.

A NEW ENVIRONMENT

Numerous   features have been identified which make a distinction between traditional telephony and data networks and the integrated broadband networks of the future. From the traffic management point of view (which is of interest in this paper) a very important distinction between the two environments is the level where statistical multiplexing (and hence congestion) takes place and the extent to which statistical gains are possible. In a circuit-switched network, once a circuit call is accepted, there is a guarantee that the call will receive nothing less and nothing more than the allocated channel (for instance a 64 kb/s DS0 channel). This kind of implicit tight rate control[4] avoids any queueing in the system and therefore allows for congestion-free communication within the network. The price for that is twofold: no statistical gains can be obtained at the link level and the congestion problem is moved at the edge of the network, namely at the call admission control level. In an integrated broadband network, different call sessions between two points in the network can be aggregated and transported over a preassigned path, whose capacity will be better used due to statistical averaging. But this can also cause congestion and therefore connections have to be managed not only at the call level but also at the path level. This can only be achieved by a combination of call admission control and packet scheduling[5,6,7,8,9]. This kind of connection management has been extensively developed for traditional data networks under the form of flow-control, mainly window-based. As the networks evolve towards higher transmission rates, a re-thinking of the traditional paradigms is required and a number of contributions[4,9] have approached these aspects. Even in the wake of unlimited bandwidth, individual services can capture a large fraction of the network resources while starving other services. As shown by Kleinrock[10], there is a critical bandwidth above which the system is latency limited, i.e., more bandwidth will have negligible effect on the response time: Ccrit = b / (1-ρ) *τ, where ρ the system utilization factor, τ the propagation delay, and b is a dimension dependent constant. Supporting real-time traffic (such as voice) in a statistical multiplexing environment requires mechanisms to enforce tight end-to-end delay bounds. Bounding the delay on an end-to-end basis has proved[11] to be a difficult problem. If packets are dropped somewhere on the path, the selective retransmission will require at least the sum of round trip delay and transmission delay and therefore lead to excessive delays.

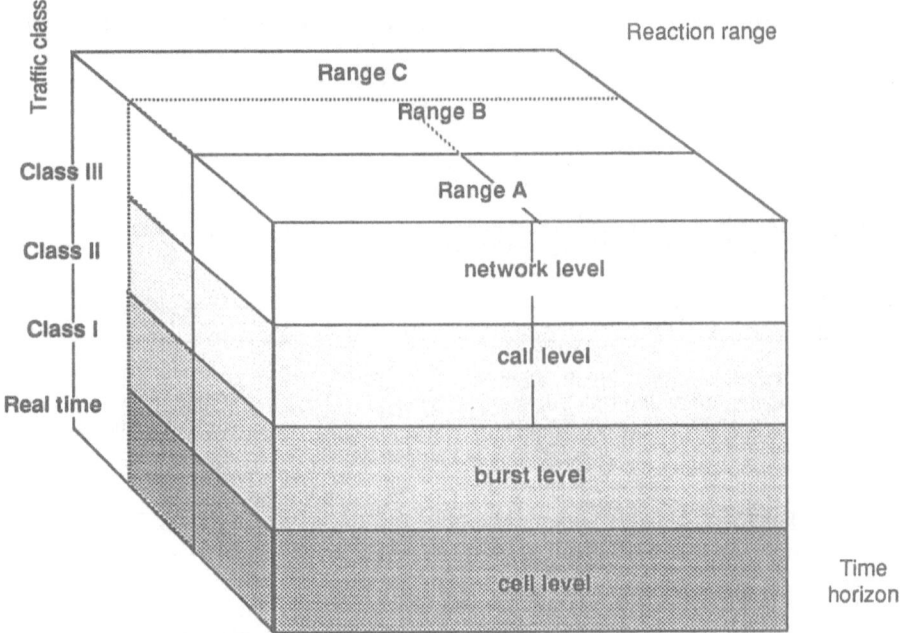

Figure 1. A three dimensional framework for bandwidth management

Due to the latency of the network, congestion avoidance strategies take precedence over reactive control and therefore strict admission policies are enforced at the edges of the network. This will in the first place increase the access delays, but as shown above these are not so important in high speed networks, if retransmissions can be avoided. A second consequence is a less efficient usage of the available bandwidth, but again, enforcing a certain level of QOS takes precedence over maximizing bandwidth utilization in an environment where bandwidth is no longer the bottleneck. This led to the concept of a multilayer bandwidth management[9]. A three-dimensional framework for traffic control, based on traffic type, reaction range and temporal controls has been identified[4] and is shown in Fig.1. The control space is divided into three regions based on the reaction range which is defined as a function of the reaction time inside the network and the so called network time constant[12], an inherent smoothing interval associated with each network. The temporal controls are a set of controls layered in time, which account for the different time horizons associated with the events that must be controlled: network level, call level, burst level and cell level.

NETWORK LAYERS, TIME SCALES, VIRTUAL PATHS AND RESOURCE ABSTRACTION

The role of the time scales at which connections exist at different layers has been recognized as a structuring factor for traffic control[12,13,14]. Distinct switching mechanisms are used in order to accommodate the different life-times of the connections established. At the cell level a "connection" will only last a few microseconds, at the burst or packet level connections last from milliseconds to seconds, while at the call session level, connections last for minutes, use virtual circuits established on a sequence of paths and are much more alike their counterparts from the circuit-switched world. Paths last for hours or days and are configured based on the aggregate traffic. The layering by time scales is a prerequisite for the process of abstraction of network resources[14] and traffic entities which in turn simplifies communication between peer instances by hiding their complexity. In this context, the most important step is to reduce bandwidth management to the management of an abstracted view of the resources allocated at each time scale via a layered notion of equivalent bandwidth.

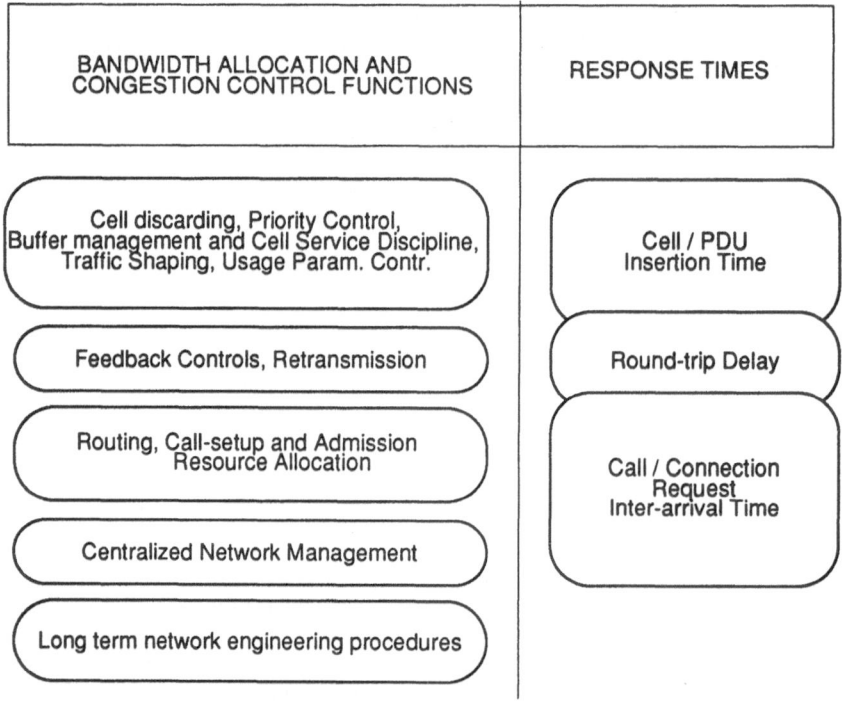

Figure 2. Relationship between events, actions and response times

The ultimate use of this concept is to derive the bandwidth allocation at the path level. The use of virtual paths (a group of connections sharing common paths) reduces more than 10 times the number of software instruction needed to control the set-up of a VCC[15,16]. More generally, switching at one time scale involves the adoption of a subset of alternatives adopted by switching at the next larger time scale. The size of the subset determines the complexity of the functions which are performed repetitively at this smaller time scale (i.e. more often). The chosen subset must also consider the events of blocking at the next smaller time scale and therefore a trade-off between complexity and blocking probability must be reached. This iterative procedure has been termed the principle of layered switching[14] and mathematically formalized.

The different time horizons associated with each traffic entity are reflected by the traditional layering techniques. Different network hierarchies can be iteratively derived[14], for instance within the framework of the seven layer OSI model. There is a relationship between the temporal dimension of the bandwidth management process and each of the iteratively defined layers[12,14,17]. Response times dictate how fast the controls react[18] and therefore structure the controls as shown in Fig. 2. The network level controls are usually processed by a centralized manager[19] using global information with slow time variation (hours or even days).

The call control is of a more distributed nature than the network level control and is based on an agreement between the source, the network access node and the switching nodes on the allocated route, using mainly information which is local to each of the entities involved in the process, and with holding times of the order of minutes. The burst level controls bursts of activity in the range of milliseconds to seconds and relatively long idle periods which lead to under-utilization if resources are reserved for the duration of the session. The cell level control is essentially a local control and enforces policing mechanisms and loss and delay priorities using the parameters passed by the higher control levels. The time constant of these controls is of the order microseconds.

Since each control level is responsible of the next lower level, it should attempt to provide a guarantee for the layer at the smaller time-scale so that the latter can meet its QOS requirements for all services. This is the meaning of the principle of layered switching[14]. A rather far-reaching consequence of this principle is a question which has been only marginally (if at all) addressed until now: if this kind of guarantee can be provided, and no further control actions are needed at the lower level, why do we need this lower level at all? In other words, will the cell format used in the current B-ISDN proposal have any practical significance for the very high speed networks of the future? The recent developments tend to give a negative answer to this question, as it will be shown later.

A striking example is the support of connectionless services in an ATM-based B-ISDN[20]. The typical application of this service is the ATM LAN and the interconnection of LANs in order to perform file-transfer transactions. As is well known it is difficult to predict the size of the files being transferred and the frequency of the transactions. There are at least two ways to deal with such sources. One is a fast reservation protocol[21,22,23] used to reserve bandwidth on the chosen path only for the duration of the bursts (a call within the call), after which the resources are released, without terminating the call session. The second, more radical approach is not to reserve bandwidth at all, the so called "No Bandwidth Reservation" (NBR) service[24]. Instead, loss priorities are defined in order to ensure by selective discarding in the congested nodes that the target loss probability for each traffic class is attained. The first approach is suitable for time-constrained traffic whereas the second approach matches the "store and forward" data traffic which has no stringent end-to-end delay requirements.

THE CELL LEVEL ALLOCATION PROBLEM

The immediate consequence of the principle of layered switching is the need to parametrize the requirement from the network in terms of the characteristics of the source[24]. The resulting measure based on these requirements has been termed equivalent bandwidth (sometimes also capacity) of a connection[4,5,8,14,24,25,26]. The resulting load on the network links is called virtual or equivalent capacity of the multiplexed connections. These two parameters are needed to implement most of the traffic management functionality: call admission control, routing and congestion control heavily rely on some measure of resource utilization in order to provide the target quality of service. In the present paper, the equivalent bandwidth will be the term used for the cell level and the equivalent capacity for the higher levels (burst, call).

Source parametrization

The characterization of traffic sources is an active topic of research[27]. The difficulty is the very different nature of the traffic that must be integrated on a broadband network. The statistical behaviour of an aggregate of multiplexed connections differs significantly from the statistical behaviour of each individual connection. It is therefore of major concern to find ways to represent the metrics of both individual and aggregate statistics. It is generally agreed that the peak rate (R_{peak}) of a source is an indispensable parameter and so also one required by the CCITT standards on B-ISDN[18] to be declared at call set-up time. A number of proposals include the mean burst period b and either the average bit rate m or the utilization factor ρ as further traffic descriptors. The utilization factor is by definition: $\rho = m / R_{peak}$ and together with R_{peak} yields the mean m and the variance σ^2 of the bit rate, and the burstiness R_{peak}/m. The mean of the burst period b is very useful in describing how bursts are generated by the source and helps discriminate between calls which though having the same peak and mean bit rate, display a different behaviour. For the particular case of exponentially distributed burst and idle periods, it has been demonstrated[8] that the vector (R_{peak}, ρ, b) completely identifies the traffic statistics of a bursty connection, as depicted in Fig. 3.

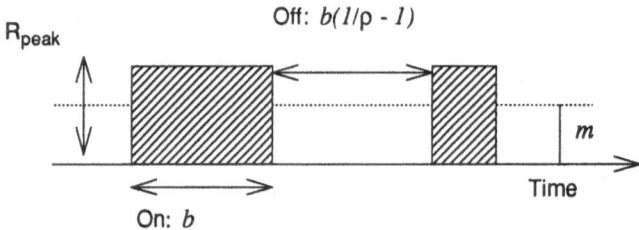

Figure 3. Descriptors for a bursty source

There are basically two ways to derive the equivalent bandwidth of a connection from the traffic descriptors of the source. The first one is to determine by simulation the dependence between the equivalent bandwidth and the parameters which affect bandwidth allocation, such as source characterization, available link rate, queue length and cell loss probability. The simulation typically yields sets of precomputed curves which are to be stored at the bandwidth management centers to determine the equivalent capacity of connections. The advantage of this approach is the absence of any significant real-time computational overhead for determining the bandwidth allocation. The draw-back of this approach is that the precomputed curves may not be able to cope with the actual traffic conditions. The second approach is to find an approximated expression for the equivalent bandwidth, which could be computed in real-time and keep track of the variability of the traffic patterns. The advantage of this approach is that, based on the statistical characteristics of single and multiplexed connections, such an expression of the equivalent bandwidth can provide a dynamic measure of resource utilization which can be used for routing, call control and bandwidth allocation. The disadvantage is that (as is so often the case in the real world) exact solutions are difficult to find in real time- if at all- and the approximations developed to overcome this problem introduce restrictions which might not always be met in a real environment situation. The first approach has produced a quite large number of interesting solutions[25], most of them being limited to a certain range of traffic characteristics. Important contributions have tackled the second approach[8,28]. Between these two basic approaches one can imagine any number of procedures which use combinations of approximated computation and simulation[29], and the literature published is expectably rich.

Traffic models

Whatever the approach, the first step needed to derive a solution for the equivalent bandwidth problem is to model the multiple types of sources which are integrated in the broadband network environment. Several models have been proposed in the literature for the analysis of bursty traffic. The models are supposed to combine reasonable accuracy with tractable computation complexity. Though other, more general source models have been proposed, the above requirements lead to the choice of a two-state continuous time Markov chain as in Figure 4. The choice is quite obvious because of two reasons.

$$\lambda = \rho / b(1-\rho)$$

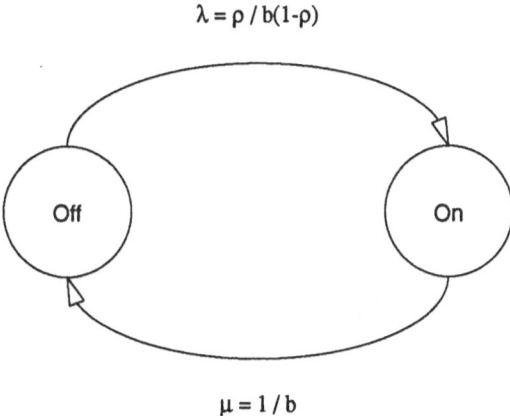

$$\mu = 1 / b$$

Figure 4. A two-state Markov chain source

On one hand, it fits our intuition for a number of widely used traffic sources which have an "on" state (for instance a talk spurt or a burst) and an "off" state (silence or idle). On the other hand, a two-state Markov chain is necessarily a birth-death process, which is a simple and very efficient model for the elementary queuing theory. In terms of the source parameters, the birth and death rates of a source i will be given by

$$\mu_i = \frac{1}{b_i}, \lambda_i = \frac{\rho_i}{b_i(1-\rho_i)} \tag{1}$$

It must be noted here that even with these simplifying assumptions the computational complexity is considerable[30] unless further simplifications are made, especially the assumption of identical bursty sources[8,25]. This seems to be a drastic restriction, but this is not quite so. Namely, the aggregate bit rate for a number of identical video-sources has been modeled[31] as a finite state (M-state) continuous-time Markov process with each state corresponding to a quantization level of the aggregate bitrate. It was found out that this Markov process is equivalent to a process consisting of M independent and identical On-Off sources. This decomposition shows that despite the simplifying assumptions, the On-Off sources still retain a lot of modeling power.

Equivalent bandwidth at the cell level

The equivalent bandwidth[5,8,25] is by definition the bandwidth W assigned to a transmission link with buffer size x in order to satisfy a given grade of service (GOS) requirement for a mix of N distinct sources that share the link. The GOS is defined as the cell loss probability ε due to buffer overflow and has a typical value of 10^{-9}. For $N=1$, W is the equivalent bandwidth w of a single connection. The idea behind this concept is that the equivalent bandwidth of a connection - which might be larger then the mean bit rate m - will hopefully be smaller than the peak bit rate R_{peak}: $m \leq w \leq R_{peak}$. The immediate consequence of this is that for connections with R_{peak} / m close to 1 (low burstiness) w will be close to R_{peak} and no statistical gains can be expected. For N distinct sources:

$$\sum_{i=1}^{N} m_i \leq W \leq \sum_{i=1}^{N} R^i_{peak} \tag{2}$$

The ratio between the equivalent bandwidth and the mean bit rate

$$\zeta = \frac{W}{\sum_{i=1}^{N} m_i} \tag{3}$$

is called expansion factor and measures the effect of traffic burstiness on the bandwidth allocation. For a large number of sources, the statistical averaging of the traffic will lead to a better utilization of the link and for N large enough, the equivalent bandwidth will equal the aggregate mean bit rate, i.e. $\zeta=1$.

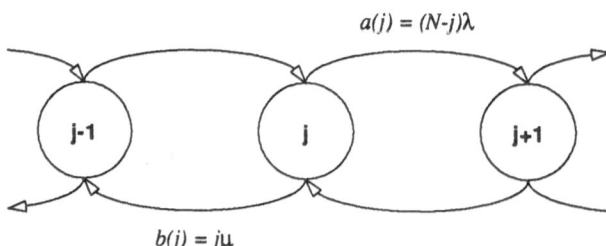

Figure 5. The underlying Markov chain for N identical sources

The equivalent bandwidth is obtained by modeling the statistical multiplexer in relation to the load offered to it. The model is based on a fluid-flow approximation developed in 1982 by Anick, Mitra and Sondhi[32] called the Uniform Arrival and Service (UAS) model and its follow-ups[33,34]. According to the fluid-flow approximation, the bit rate generated by a number of sources multiplexed on a common transmission link is represented as a continuous flow of bits with a rate which depends on the state of the underlying continuous-time Markov chain that models the number of sources in the active state as in Fig. 5. Let $J(t)$ the state of this chain and assume it is a birth - death process. Let $\alpha(j)$ and $\beta(j)$ be the birth and death rates respectively, when $J(t)=j$. As mentioned above, the sources themselves are modeled as On-Off sources with exponentially distributed active and silence periods. In the UAS model a single type of traffic is considered, i.e. the sources are identical with birth rate λ and death rate μ. This gives for the birth-death process $J(t)$

$$\alpha(j) = (N - j) \lambda$$
$$\beta(j) = j \mu \tag{4}$$

Using for λ and μ Eq. (1), the birth and death rates of the process $J(t)$ are obtained as a function of the source parameters. If j sources are active, the intensity of the aggregate bit rate is

$$B = jR_{peak}$$

In a later development, Korsten[33] has proposed analytical and simulation methods which include multiple types of traffic. The multiplexer is modeled as a single queue, single server system with queue length x and fixed service rate c. For an infinite buffer, the queue length distribution is computed using the eigenvalues and eigenvectors of a tridiagonal real matrix, which are obtained in closed form[32]. For finite buffer size, the buffer overflow probability ε is obtained (for given x, c and N) in a similar manner, except that the eigenvalues and eigenvectors have to be computed using numerical methods[34].

The equivalent bandwidth problem depicts the reverse situation: the cell-level equivalent bandwidth W of the multiplexed connections is the smallest value of c for which the buffer overflow probability is smaller than ε. This implies that the expression of the buffer overflow probability must be inverted in order to obtain the minimum service rate c as a function of the other parameters. Unless further approximations are made, this is only possible using numerical methods. An algorithm has been proposed[25] which searches for the appropriate value of ε using a logarithmic interpolation method:

$$\textit{for:} \quad N \cdot m < W < N \cdot R_{peak} \quad \textit{search the interval} \quad 1 > \varepsilon > 0 \tag{5}$$

until the required ε is obtained within a given tolerance. A different approach[5] is the following: c (the fixed link rate) is initially considered big enough to accommodate the N sources with zero cell loss probability, and then a background stream with constant bit rate S is added to the input traffic. S is iteratively increased until the simulator produces a cell loss probability ε with the target value. If S_0 is the rate at which ε reaches its target value, then the equivalent bandwidth of the multiplexed connection will be

$$W = c - S_0 \tag{6}$$

A method has been proposed in order to obtain through inversion from the cell loss probability an approximated analytical expression of the equivalent bandwidth of a multiplexed connection. The method consists of two steps. In the first step, the buffer overflow probability ε produced by a single isolated source when active and transmitting is considered, using queuing theory. Even in this simple case, there is no explicit expression for the equivalent bandwidth w of the single isolated connection. An explicit upper bound can nevertheless be obtained[8] for w:

$$w \cong \frac{\alpha b (1 - \rho) Rpeak - x + \sqrt{(\alpha b (1 - \rho) Rpeak - x)^2 + 4x\alpha b\rho (1 - \rho) Rpeak}}{2\alpha b (1 - \rho)} \tag{7}$$

with $\alpha = ln (1/\varepsilon)$. In the second step, the case of multiple superposed sources is considered, using the asymptotic approximation of infinite link buffer size, which as mentioned above leads to an explicit expression for the queue length distribution. The cell loss probability ε is approximated by the probability $G(x)$ that the queue length exceeds the actual buffer size x, for which in turn only an explicit upper bound can be determined. With this ε and the vector $(R_{peak}, \rho, b)_i$ for each source $i=1,...,N$ the w_i are computed using (7) and then the equivalent bandwidth of the multiplexed connection is obtained as

$$W = \sum_{i=1}^{N} w_i \qquad (8)$$

Due to the approximations undertaken in order to obtain an explicit expression for W, the equation (8) overestimates the equivalent capacity. This problem is tackled by checking Eq. (8) against a stationary estimation[8] derived from a bufferless fluid flow model as shown in Fig. 6.

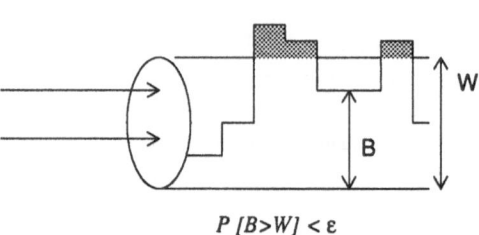

$$P\ [B>W] < \varepsilon$$

Figure 6. The bufferless fluid flow model for the stationary approximation

The equivalent bandwidth W is selected such as to ensure that the aggregate stationary bit rate B exceeds W only with a probability smaller than ε:

$$Pr\ (\ B > W\ ) \le \varepsilon \qquad (9)$$

B is determined from the stationary distribution of the number of active sources (see Eq. (4)) which is modeled by the underlying continuous-time Markov chain. Assuming for the number of active sources j a Gaussian distribution helps estimate the tail of the bit rate distribution and choose W such as that the cumulative tail probability beyond W does not exceed ε:

$$W = m + \alpha'\sigma, \qquad \text{with } \alpha'=(-2\ln(\varepsilon)-\ln(2\pi))^{1/2} \qquad (10)$$

where m is the aggregate mean bit rate ($m = \Sigma_i\ m_i$) and σ the standard deviation of the aggregate bit rate ($\sigma^2 = \Sigma_i\ \sigma_i^2$). It has been shown[8] that this approximation also overestimates the actual equivalent bandwidth. Hence, an explicit estimate for the equivalent bandwidth is chosen as

$$W = min\left\{m + \alpha'\sigma,\ \sum_{i=1}^{N} w_i\right\} \qquad (11)$$

Asymptotic approximations of the cell level equivalent bandwidth

The disadvantage of the approaches presented so far is that they assume as a general rule homogeneous environments, stationary processes, negative exponentials and geometric distributions. These assumptions do not really fit the real world. Even if they would, finding the right model for a particular source is not always easy. A less restrictive approach[35] has been introduced, in which an equivalent bandwidth of a traffic stream without knowledge of an apriori model is derived. Instead of assuming exponentially distributed burst and idle periods, this approach uses as traffic descriptors the mean bit rate and the index of dispersion of the source. These descriptors are estimated for each source type using standard techniques from time-series analysis, i.e. the estimators are obtained by spectral methods[36]. The assumption of stationary sources is maintained, but the Gaussian restriction for the bit rate distribution is dropped. The basic result is that asymptotically in the link buffer size x, the equivalent bandwidth of a source is the sum of its mean bit rate and its index of dispersion. In the following the assumptions of the model[35] are shortly presented. For each class i of traffic ($i=1,...,I$) N_i sources are active. Time is discrete and at each epoch n a source in class i generates R_n^i cells, where $R_1^i, R_2^i,...$ is a stationary process of possibly correlated random variables with mean m_i. The capacity of the link is c and the cell loss probability requirement is ε. The index of dispersion for the source of class i is given by

$$\gamma_i = \lim_{n \to \infty} (\tfrac{1}{n}) E\left[\left(\sum_{k=1}^{n} (R_k^i - m_i)\right)^2\right] = \lim_{n \to \infty} (\tfrac{1}{n}) var\left[\sum_{k=1}^{n} R_k^i\right] \qquad (12)$$

The index of dispersion can be estimated from the data[36] by using its expression in terms of the autocovariance:

$$\gamma_i = \gamma_i(0) + 2 \sum_{k=1}^{\infty} \gamma_i(k) \qquad (13)$$

where $\gamma_i(k)$ is the kth order autocovariance.
  Let

$$w_i = m_i + \frac{\alpha \gamma_i}{2x}, \, \alpha = ln\frac{1}{\varepsilon} \qquad (14)$$

According to the reference, for a large class of stationary processes, the equivalent bandwidth of the mix of f $(N_1,...,N_I)$ sources is approximated by[35]

$$W = \sum_i N_i w_i + o(\tfrac{1}{x}) \qquad (15)$$

where the error $o(1/x)$ tends to zero faster than $1/x$ for x (link buffer size) tending to infinity.

THE BURST LEVEL ALLOCATION PROBLEM

  The approach taken to characterize non-isochronous calls by using an equivalent abstraction of their resource requirement is in a way an attempt to match the unmatchable: the statistical multiplexing gains obtained in shared medium networks (which use a medium access control in order to provide concurrency) with the reservation principle used in the current telecommunication networks, based on dedicated links. After the call has been accepted (on the basis of its equivalent bandwidth) the cells generated on each virtual connection will compete for the link resources in a random way, which hopefully keeps with the call descriptor. Should the latter not be the case, coercive mechanisms (policing, discarding) are enforced at the cell level. But with every type of traffic there is a "natural" traffic entity which is transported, and end-to-end protocols try to recover this entity (a data packet or a graphic object for instance) and not the individual cell. Recently, the need to allocate network resources directly to traffic entities has been recognized as a very important factor in order to efficiently use the available bandwidth, especially when the source peak-rate is a large fraction of the link capacity. This amounts to negotiate the needed resources each time a source becomes active (i.e. a "per-burst call set-up", or "in-call allocation") and release them during each idle period. If the resources are not available the whole "entity" is blocked, if they can be granted, the entity will receive peak allocation. By doing so, the statistical multiplexing is moved to the burst level, which has a coarser granularity. On the other hand, moving controls to a higher (and slower) level reduces the processing load. This approach promises to be of particular interest for managing the interconnection of LANs over B-ISDN.

Fast resource reservation

The basic idea is that stepwise variable bit rate sources and in particular bursty sources can be multiplexed by introducing an access control at the burst level. In one approach[23], the allocation at the burst level is performed end-to-end on the Virtual Path Connection between source and destination. A Reservation Request Cell is sent along the connection path with the required incremental bandwidth value and if all nodes on the path can support it, then the egress node sends a Reservation Accepted Cell back to the source. If for some reason, somewhere on the path the reservation failed or the acknowledgment was lost, a Reservation Denied Cell is sent by a Manager to the requesting entity, which can however continue transmission with the old allocation. Conversely, a Reservation Release is sent each time the last incremental request isn't needed any longer. The draw-back of this approach is that it is very sensitive to the round trip delay in the network, i.e. it is suitable only as long as the round trip delay is negligible compared to the active period of the source. Another proposal[9] uses a pilot reservation cell in front of the burst indicating the required bandwidth. The negotiation is performed in each node on the path and the burst is forwarded to the next node if the allocation at the node succeeds. If it fails, the whole burst is dropped. This procedure is insensitive to the round trip delay of the connection, but by blocking in case of failure the burst with a new requirement, the continuation of transmission with smaller bandwidth (and longer residency time) is not allowed. A variation of this scheme, a fast buffer allocation protocol, has been recently proposed[22]. The burst is segmented in start (**s**), middle (**m**) and end (**e**) cells. The reception of an s-cell activates the allocation state-machine at the switching node and if enough resources are available, the required buffer capacity is allocated and the burst is admitted. On receiving the e-cell or after a certain time-out period, the deallocation is performed (the allocation machine becomes idle). If the allocation can not be performed in the first place, the whole burst is discarded. By buffering the burst, the transmission at lower than requested link rate is enabled (for traffic types which can accept this slowing-down). This approach also has more flexibility, in that it allows for instance the allocation of a constant rate virtual circuit by never sending the e-cell and inhibiting the time-out timer for that circuit. For applications where a certain cell loss is accepted (for instance voice), sending only type **s** cells and the final e-cell will inhibit the "whole burst blocking" feature, since even if the first s-cells are rejected, the next s-cell will restart the allocation machine and the tail of the voice spurt will get through.

Equivalent capacity at the burst level with fast burst reservation

The use of a burst level reservation protocol which allocates the peak rate bandwidth moves the level of statistical multiplexing (and the congestion problem) from the cell layer to the burst layer. As opposed to the approach at the cell level, where the equivalent bandwidth was a measure of the resources needed by a virtual path such that the cell loss probability would be bounded by a target GOS, the equivalent capacity at the burst level is a measure of the carried traffic on the path under a zero cell blocking restriction: a burst is blocked if its arrival causes the total load to exceed the total bandwidth c of the link. Under this assumption, the equivalent capacity $W(t)$ will have the same arrival statistics as the process describing the aggregate bit rate of the bursts offered to the link $B(t)$. In the widely quoted work of Hui[9] an upper bound on the burst blocking probability for a link with capacity c is derived:

$$P(W(t) > c - R_j(t)) \tag{16}$$

(where $R_j(t)$ is the instantaneous bit rate of the burst j (see Fig.6)) using a large deviation theory principle under the assumption of very large number of bursts. The offered aggregate peak rate process $B(t) = \Sigma_k R_k(t)$ ($k=1,...,K$, the number of admitted calls on the link) is modeled by jumps with different amplitudes $a_i > 0$ which arrive with Poisson rate λ_i^{burst} and last b_i (deterministic). For non-Poisson traffic sources, $R_k(t)=a_i$ with steady-state probability p_i. The principal result[9] is to find a sharp bound for the tail distribution of B for a mixture of Poisson and non-Poisson traffic, specifically

$$P(B > y) = \frac{1}{s^* \sqrt{2\pi \mu''_B(s^*)}} e^{-(s^* y - \mu''_B(s^*))} \tag{17}$$

with the moment generating function for B:

$$\mu_B(s) = \sum_k (\mu^k_{POISSON}(s) + \mu^k_{\overline{POISSON}}(s)), and \ \mu'_B(s^*) = y \tag{18}$$

where

$$\mu^k_{POISSON}(s) = \sum_i \lambda_i^{burst} b_i (e^{sa_i} - 1) \tag{19}$$

$$\mu^k_{\overline{POISSON}}(s) = ln \sum_i p_i e^{sa_i} \tag{20}$$

The blocking probability of type i bursts for the lossy system $W(t)$ is related to (17) by

$$P(W(t) > c - a_i) = \frac{1}{1 - P(B(t) > c)} (P(B(t) > c - a_i) - P(B(t) > c)) \tag{21}$$

By putting $y=c$ and solving for s^* in (18), the tail distribution (17) and the blocking probability (21) are obtained. It must be noted that the solution of the implicit equation (18) can only be obtained using numerical methods. Further details can be found in the literature[9]. By imposing that the blocking probability remains under a certain threshold, the carried traffic $W(t)$ can be (at least theoretically) determined. As the reader might have observed, the approach shortly outlined above is somewhat similar to the "bufferless fluid flow model" described before: the required resource is the link capacity and if not available, the arrivals are discarded and not buffered. What happens if the burst is buffered in the multiplexer while being processed by the output port? This will allow for instance to decrease the bandwidth allocated to a connection (an correspondingly increase the residency time in the buffer) without having to block the burst. The burst is blocked when it requires more buffer slots than available. The capacity of the buffered link, seen as the number of active bursts routed through the link and the burst blocking process will depend on both link rate and buffer size. Let us once again remember the notation used earlier: the link rate is c, the (this time discrete) buffer size is x cells, R_i is the peak rate of a given source i and m_i its average rate. Following the reasoning developed by Turner[22], the number of slots in the buffer required by the source i when active will be

$$B_i = \left\lceil \frac{xR_i}{c} \right\rceil \tag{22}$$

and if x_i is the random variable representing the number of slots required at a random instant by the source i, then

$$p_i = P(x_i = B_i) = \frac{m_i}{R_i} \tag{23}$$

and

$$\bar{p}_i = P(x_i = 0) = 1 - \frac{m_i}{R_i} = 1 - p_i \tag{24}$$

When multiplexing n sources on the link, the required buffer capacity will be

$$X = \sum_{i=1}^{n} x_i \tag{25}$$

So this time, the carried traffic is in a certain sense a measure of the available buffer capacity and therefore the equivalent capacity is expressed in terms of buffer size X rather than explicit bandwidth W. In order to determine X, its probability distribution function must be described and a generating function[22] is used for that purpose, assuming the x_i are mutually independent:

$$f_X(z) = \prod_{i=1}^{n} (\bar{p}_i + p_i z^{B_i}) = C_0 + C_1 z + C_2 z^2 + ..., \quad where \quad P(X = j) = C_j \tag{26}$$

In order to decide whether or not to accept a new connection, the probability distribution for X must be maintained and updated in real time. A recursion is found[22] which allows to compute the new coefficients of (26) when a connection with buffer demand x_n is added to the link and the distribution of the new buffer demand $X = X' + x_{n-1}$ is hence determined. How can the burst blocking probability be defined in this buffered model? Two measures are proposed. The first and easy to compute measure is the excess demand probability, i.e. the probability that the instantaneous buffer demand exceeds the available buffer space, as given by Eq. (27). It has been shown that this measure is misleading for the case of connections which become active with a low probability p_i.

$$P(X > x) = C_{x+1} + C_{x+2} + \ldots = 1 - (C_0 + C_1 + \ldots + C_x) \qquad (27)$$

A more computation intensive measure is the contention probability for a given connection i which requires B_i buffer slots with probability p_i, defined by $P(X-x_i > x - B_i)$ and shown[22] to be

$$P(X - x_i > x - B_i) = \frac{1}{p_i} \sum_{k=0}^{\left\lfloor \frac{x}{B_i} - 1 \right\rfloor} \left(\frac{-p_i}{\bar{p}_i}\right)^k P(X > x - (k+1)B_i) + (-1)^{1 + \left\lfloor \frac{x}{B_i} - 1 \right\rfloor} \qquad (28)$$

where Eq. (27) is used to determine $P(X > x - (k+1)B_i)$.

THE CALL LEVEL ALLOCATION PROBLEM

The equivalent bandwidth at the cell level derived with any of the methods (5), (6), (11), (15) - or even (8) and (10) for certain traffic environments - is used for call admission control, i.e. a call is admitted only if the GOS of the call can be provided without compromising the GOS of the calls already in the network. This amounts to make sure that the equivalent bandwidth of the multiplexed connection including the new call doesn't exceed the link capacity. Previous results[5,37] show that the link capacity should actually not be entirely allocated, but rather have an upper bound. A simple solution is to impose a safety coefficient on equivalent capacity usage (for instance 0.9). A pessimistic allocation[25] would be to divide the traffic in K traffic classes and to consider the equivalent bandwidth W_k of the multiplexed connections occupied by sources of type k independently. This has been termed virtual link capacity[26]. Adding up the virtual link capacities (W_k's) will give an upper bound for the link load, since the statistical multiplexing across each traffic class is not accounted for. This pessimistic approach is checked against the equivalent bandwidth obtained by substituting the actual sources with the set of sources with the largest burstiness generating the same average traffic and the minimum is chosen as bound. The procedure is termed class-related rule CRR[37]. It has been shown[25] that CRR displays the best match for equivalent bandwidths computed using a two-state Markov chain for the sources.

Whatever the rule used to compute the maximum channel allocation, the cell-level equivalent bandwidth of the connection must be computed in real-time. This is a real problem, since for instance the solution of the UAS problem requires numerical algorithms with complexity $O(N^3)$ (N is the number of sources) and just as complex will be to determine W. If the characteristics of the sources are known in advance, tables with *a priori* equivalent bandwidths for each class of traffic could be stored in the management entities. This raises the question of traffic classification on one hand and of the size of these tables on the other hand. This is why the explicit expressions of W in (11) and (15) are quite convenient, since they allow the fast computation of the equivalent bandwidth.

Asymptotic equivalent capacity at the call level

By applying a certain call admission policy, a call blocking probability will result as a function of the arrival process at the call layer. The concept of cell-level equivalent bandwidth enables the use of the extensive theoretical work developed for circuit-switched networks[38]. The case of special interest for broadband integrated networks is an environment in which a very large population of heterogeneous users has access to a very large bandwidth network (asymptotic growth assumption). Assuming that for traffic class i the call arrival process is an independent stationary Poisson process with mean λ_i^{call} and that a call of class i requires an amount of bandwidth w_i for a connection time with exponential distribution and mean $1/\mu_i^{call}$, than the asymptotic growth condition is expressed as shown in Eq. (29).

$$
\begin{bmatrix}
c \to \infty \\
\dfrac{\lambda_i^{call}}{\mu_i^{call}} \to \infty \\
\dfrac{\lambda_i^{call}/\mu_i^{call}}{c} \quad const
\end{bmatrix}
\tag{29}
$$

Under this condition, two important theorems are derived[39] for the case where both the call bandwidth and the link capacity are an integer number of bandwidth allocation units (circuits). The *independence theorem* states that the asymptotic behaviour of the blocking probabilities of m different classes of traffic on the same link is as if a call of any given class of traffic, needing k circuits to get established, was set up by grabbing k circuits sequentially and independently. The *separation theorem* states that the asymptotic behaviour of the blocking probabilities of m different classes of traffic on the same link is as if each class of traffic was allocated a part of the link and blocked on this part independently of the other classes of traffic. These two theorems allow to reduce for each traffic class the multitraffic system to an equivalent system with only one traffic class, where the blocking probability is given by the well known Erlang formula[40]. By imposing that this probability is within a certain blocking objective at the call layer, the equivalent capacity C_{eq} (in number of circuits) for that traffic class can be obtained by inverting the expression:

$$
E(\rho, C_{eq}) = \frac{\rho^{C_{eq}}/C_{eq}!}{\sum_{k=0}^{C_{eq}} \rho^k/k!} \leq P_{blocking}^{call}
\tag{30}
$$

where $\rho = \lambda_i^{call}/\mu_i^{call}$

CONCLUSIONS

ATM is praised for providing statistical multiplexing at the cell level for heterogeneous traffic sources and in particular for the interconnection of LANs over wide-area networks. The approach taken in order to achieve this goal is to determine a measure of the requirement from the network in terms of the characteristics of the sources multiplexed on a path. Due to the variability of the traffic types which must be supported and to their different time-scales, there is no unique measure, but rather a particular image the network perceives for each traffic class. The assumptions made to derive computational tractable models for the cell level equivalent bandwidth strongly restricts their applicability. On the other hand, trying to control the cell level might prove extremely expensive in processing power, due to the electronic bottleneck in the switching nodes. Controlling, where appropriate, the higher (and slower) burst level[41] is more natural, easier for the user to parametrize, requires less processing in the switches and flattens the traffic management policies. The price paid for that might be a lesser utilization of the available bandwidth but this must be weighted against the cost of retransmitting whole bursts because the network didn't succeed to provide the desired burst loss rate, which is the parameter of interest for broadband ATM LANs and for the interconnection of LANs over B-ISDN.

Acknowledgements

This paper emerged from work sponsored by the RACE Project R2068 - LACE and from the encouragement and many fruitful discussions with Prof. Radu Popescu-Zeletin, and the helpful suggestions of Thomas Luckenbach, to both of whom I am grateful.

REFERENCES

1. H.R. van As, W. W. Lemppenau, P. Zafiropulo and E. A. Zurfluh, CRMA II: A Gbit/s MAC protocol for ring and bus networks with immediate access capability, *Proc. EFOC/LAN 91* 56:71 London (1991)

2. B. Mukherjee, WDM-based local lightwave networks. Part I: Single-hop systems, *IEEE Network* 12:27 (May 1992)

3. CCITT B-ISDN general network aspects, *Recommendation I.311*, Temporary Document (1992)

4. R. Dighe, C. J. May and G. Ramamurthy, Congestion avoidance strategies in broadband packet networks, *Proc. INFOCOM* 295:303 (1991)

5. M. Decina and T. Toniatti, On bandwidth allocation to bursty virtual connections in ATM networks, *Proc. IEEE-ICC'90*, Atlanta, GA, (Apr. 1991)

6. S. J. Golestani, Congestion-free communication in high-speed packet networks, *IEEE Trans. Commun.*, vol. 39, no. 12 (1991)

7. S. J. Golestani, Duration-limited statistical multiplexing of delay-sensitive traffic in packet networks, *Proc. INFOCOM* 323:332 (1991)

8. R. Guerin, H. Ahmadi, and N. Naghshineh, Equivalent capacity and its application to bandwidth allocation in high-speed networks, in *IEEE JSAC.* 968:981 (1991)

9. J. Y. Hui, Resource allocation for broadband networks, *IEEE JSAC* 1598:1608 (1988)

10. L. Kleinrock, The latency/bandwidth tradeoff in Gigabit networks, *IEEE Commun. Mag.* 36:40 (Apr. 1992)

11. A. Wolisz and R. Popescu-Zeletin, Modelling end-to-end protocols over interconnected heterogenous networks, *Computer Commun.* 11:21 (1992)

12. G. Ramamurthy and R. S. Dighe, Distributed Source Control: A network access control for integrated broadband packet networks, *IEEE JSAC* 990:1002 (1991)

13. J. Filipiak, Analysis of automatic network management controls, *IEEE Trans. Commun.* 1776:1786 (1991)

14. J. Y. Hui, M. B. Gursoy and N. Moayeri, R. Yates, A layered broadband switching architecture with physical or virtual path configurations, *IEEE JSAC* 1416:1426 (1991)

15. J. Burgin and D. Dorman, Broadband ISDN resource management: The role of virtual paths, *IEEE Commun. Mag.* 44:48 (Sept. 1991)

16. E. Tirtaatmadja and R. A. Palmer, The application of virtual paths to the interconnection of IEEE 802.6 Metropolitan Area Networks, *Proc. XII ISS Stockholm* 133:137 (1990)

17. I. Chlamtac, A. Ganz and G. Karmi, Transport optimization in broadband networks, *Proc. INFOCOM*, 49:58 (1991)

18. CCITT Traffic control and congestion control in B-ISDN, *Recommendation I. 371*, Temporary Document (1992)

19. M. Mateescu, Management of DQDB based MANs: State of the art, presented at the *5th IEEE Workshop on Metropolitan Area Networks*, Taormina, Italy, May 1992.

20. CCITT Support of broadband connectionless data service on B-ISDN, *Recommendation I.364*, Temporary Document, (1992)

21. P. Boyer, A congestion control for ATM, presented at the *7th ITC Specialist Senminar*, session 4, Morriston, NJ, (Oct. 1990)

22. J. S. Turner, Managing Bandwidh in ATM Networks with Bursty Traffic, *IEEE Network* 50:58 (Sept. 1992)

23. P. E. Boyer and D. P. Tranchier, A reservation principle with applications to the ATM traffic control, *Computer Networks and ISDN Systems* 321:334 (1992)

24. I. Cidon, I. Gopal and R. Guerin, Bandwidth management and congestion control in plaNET, *IEEE Commun. Mag.* 54:64 (Oct. 1991)

25. J. A. Monteiro, Bandwidth allocation in broadband integrated services digital networks, *Technical Report CSD-900018*, UCLA Comp. Sc. Dpt. (1990)

26. T. Murase, H. Suzuki, S. Sato and T. Takeuki, A call admission control scheme for ATM networks using a simple quality estimate, *IEEE JSAC* 1461:1470 (1991)

27. K. Kawashima and H. Saito, Teletraffic issues in ATM networks, *Computer Networks and ISDN Systems* 369:375 (1990)

28. F.P. Kelly, Effective bandwidths at multi-class queues, *Queueing Systems* 5:16, vol. 9 (1991)

29. COST 224, "Performance Evaluation and Design of Multiservice Networks", J.W. Roberts, ed., Comm. of European Communities, Brussels (1992)

30. H. Kobayashi and Q. Ren, A mathematical theory for transient analysis of communication networks, to appear in *IEICE Trans. Commun.* Japan, Dec. 1992

31. B. Maglaris, D. Anastassioou , P. Sen, G. Karlson, and J. Robbins, Performance models of statistical multiplexing in packet video communications, *IEEE Trans. Commun.* 834:844 (1988)

32. D. Anick, D. Mitra and M. M. Sondhi, Stochastic theory of data handling systems with multiple sources, *The Bell System Technical Journal* 1871:1894 (1982)

33. L. Kosten, Liquid models for a type of information storage problems, *Delft. Prog. Rep.: Math. Eng., Math. and Inform. Eng.* 71:86 (1986)

34. R. C. F. Tucker, Accurate method for analysis of a packet-speech multiplexer with limited delay, *IEEE Trans. Commun.* 479:483 (1988)

35. C. Courcoubetis, G. Fouskas, Using Asymptotic Techniques in Call Acceptance Management for ATM Networks, *in* "ERCIM Workshop on Network Management", Heraklion, Crete (1992)

36. D. Cox and P.Lewis, "The Statistical Analisys of Series of Events", Chapman and Hall (1966)

37. M. Decina, T. Toniatti, P. Vaccari, and L. Veri, Bandwidth assignement and virtual call blocking in ATM networks, *Proc. INFOCOM*, 881:888 (1990)

38. J. Kaufman, Blocking in a shared resource environment, *IEEE Trans. Commun.* 1474:1481 (1981)

39. J.P. Labourdette, G.W. Hart, Blocking probabilities in multitraffic loss systems: insensivity, asymptotic behavior, and approximations, *IEEE Trans. Commun.* 1355:1366 (1992)

40. L. Kleinrock, "Queueing Systems", Vol.I: Theory, John Wiley & Sons, New York (1975)

41. B. Doshi, Performance of in-call buffer-window allocation schemes for short intermitent file transfers over broadband packet networks, *Proc.INFOCOM* 2463:2472 (1992)

A SUPERPOSITION OF BURSTY SOURCES IN A LAN INTERCONNECTION ENVIRONMENT

Johan M Karlsson

Department of Communication Systems
Lund Institute of Technology
P.O. Box 118
S-221 00 LUND
Sweden

ABSTRACT

Designed as a solution for future public networks, ATM is now quickly being adapted to low cost LANs and private networks. As a true international standard, it is already widely supported by the communications and computer industries as a way of ensuring interoperability and protecting existing and future investments. The rapid changes in technology and service requirements make it increasingly important to find an efficient way to evaluate different network configurations. The problem with a different evolution of raw computing power, memory size and communication rates is pointed out. A LAN interconnection scenario is discussed, and a model (IBP) for bursty arrivals is presented. Finally, a queuing model with superposition of these bursty sources as arrival process is developed in order to capture the behaviour of a LAN interconnection environment. The resulting arrival process has two important characteristics: firstly, it captures the periodical behaviour due to the deterministic character of a single active source and secondly, the new process is modulated by the number of sources connected.

INTRODUCTION

The evolution of telecommunications is towards a multi-service network fulfilling all user needs for voice, data and video communications in an integrated way. Up to now, new networks were developed whenever a new service became relevant. This hardly seems an efficient and cost effective way to meet emerging communication

Local Area Network Interconnection, Edited by R.O. Onvural
and A.A. Nilsson, Plenum Press, New York, 1993

needs. Some of these applications are computer related (e.g. communication among remote supercomputers performing jointly a task, others involve the transmission of images and video signals). While fiber optic technology provides the necessary bandwidth for transmission purposes, the creation of a network that can provide high bandwidth services to the users remains a challenge. The problems that evolve are mainly in switching and resource sharing. Since such a network will carry all applications in an integrated fashion it is an important task to be able to switch according to the bandwidth allocation algorithm. This is to be done in a packet switching mode which offers greater flexibility than circuit switching in handling the wide diversity of data rates and latency requirements resulting from the integration of services. Today, the world is thinking about this network as an ATM (Asynchronous Transfer Mode) network. The advancement of ATM techniques saw the first set of B-ISDN Recommendations in 1990 [1]. Succeeding efforts toward the completion of B-ISDN techniques have raised them to a level sufficient for network providers to introduce ATM techniques into their telecommunication networks for the first time. This introduction is essential because demand for high speed multimedia communication has increased significantly. In this context, we are considering first applying ATM techniques in the provision of cost effective and flexible multimedia ATM leased line services to interconnect private networks, LANs and MANs.

EVOLUTION OF SYSTEM COMPONENTS

The capacity of the equipment connected to various networks has shown a rather high growth in terms of processing speed and memory size. However, neither of these could alone make up a system, essential requisites are a high processing speed, massive storage and fast communications. With the raw computing power on the rise, system performance is bumping into network and storage limits. These problems indicate that in the near future (if no main contributions to network speed are done) the old concept of back-end and front-end network would arise. Capabilities in data communications have simply not kept up with those of the computer manufacturing industry [2]. The next bottlenecks will be in the communications area. This evolution is cyclic as we could see from the past. We have to remember that for a computing system to work optimally, the capabilities of all its major components must be in balance. In Figure 1 the growth of the peak potential processing power, memory size of supercomputers and local area network communication rates have been compared. The parameters shown are normalized to 1980 figures[1]. The peak potential processing power as well as the memory size have increased by three orders of magnitude. During that same time, the communication rate of local area networks has increased by less than two orders of magnitude.

TRAFFIC CHARACTERISTICS

To describe the traffic characteristics for not existing services, as well as existing but not yet commonly used, or even worse not known, is one of the major problems that we have to deal with. It could be discussed which new services that are going to

---

[1]100 million floating point operation/s, 8 Mb and 50 Mb/s

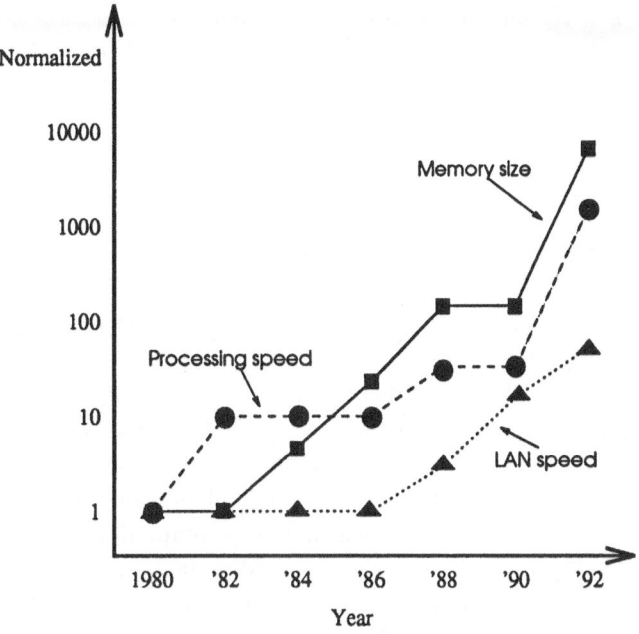

Figure 1: *Growth in local area network communication rates has fallen behind advances in processing power and memory capacity.*

evolve, and to what extent they and already existing services will be used. In spite of the fact that it theoretical is possible, to allocate any bandwidth to a user, for ATM applications any multiple of 53 octets, it is undoubtedly some 'standardized' sizes that would be predominant. If we try to focus on some of these sizes, they would correspond to services like HDTV (50-100 Mbps), picture telephony (64-128 kbps (CCITT H.261)) , Hi-Fi sound and group 4 telefax (64 kbps), lower speed data (-64 kbps) and some other services related to office based communication devices. Coding technic is going to be essential for all graphic, image and video information services. Technics standardized by JPEG (Joint Photographic Experts Group) and MPEG (Motion Picture Experts Group), sponsored by ISO and CCITT, provides for compression ratios up to 1:200. For further information about line transmission of non-telephone signals, the reader is referred to [3].

Many of the applications or services are going to be of a more bursty nature than we are used to from POTs (Plain Old Telephone). Therefore we need source generating models that take this into account, one suggestion is to be found in the modelling part of this paper.

The superposition of on/off sources has been studied, notably, in the context of packetized voice. It has then been established that the suitable and handy approximation that a superposition of independent sources yields a Poisson stream can lead to quite inaccurate results [4].

Underlying the field of broadband telecommunications are the concepts of universal interface and bandwidth-upon-demand: all types of traffic are presented in a common packet format and are distinguished only on the basis of the frequency with which the packets are generated. Therefore it is possible to model the arrivals as a superposition of all the above mentioned traffic types. Due to the different frequencies that they are sent by the user, a bursty nature of this arrival process is, in most cases, a worst case scenario compared to a smooth arrival process. In our case, looking at a LAN interconnection model, that seems to be a natural way of modelling the problem. However, in a node in the longhaul network (see Figure 2) the burstiness has been leveled out due to the number of interacting traffic flows. However, it is, as discussed above, still not to be characterized as a Poissonian traffic stream.

LAN INTERCONNECTION

B-ISDN have to utilize the existing infrastructure of the digital network. This means that, due to already invested time and money, the physical configuration has to follow the current fiber cabling and offices. Therefore, it is going to consist of regional networks and longhaul networks, as could be seen in Figure 2. The regional network include the access network and the transfer network. The latter is the connection between the users 'home-network' and the longhaul network, used to reach other access networks in the system. The local nodes accommodate customers, and the gate node has a hub function in one region and a gate function in the longhaul network. One of the main purposes of computer networks is to allow hosts and other equipment connected to the different networks to intercommunicate. Mostly, the hosts are connected to LANs, and will continue to be if this internetwork model applies. This, due to several reasons; a connection to a public ATM network will be expensive, most traffic is local, existing LANs should still be able to operate and so on. Vendors are already producing small ATM switches to be used in interconnection with LANs. Such a scenario is shown in Figure 3 where different kinds of equipment connected to the LAN also serve as gateways to the local ATM node. The ATM node supplies in this way the whole LAN with circuit mode, packet mode and pure ATM mode services. All these different switching modes are then combined in the switch to one common outlet, on which an ATM cell stream is transferred. As could be seen in the picture the stream consists of a mixture of the different switching modes, i.e. voice, data and image cells are transmitted sequentially on the same link. One of the benefits with ATM is a high capacity with possibilities to be multiplexed to several Gigabits (Terabits) per second. Another advantage is the prospect to create virtual networks, with the servers geographically distant but at the same time

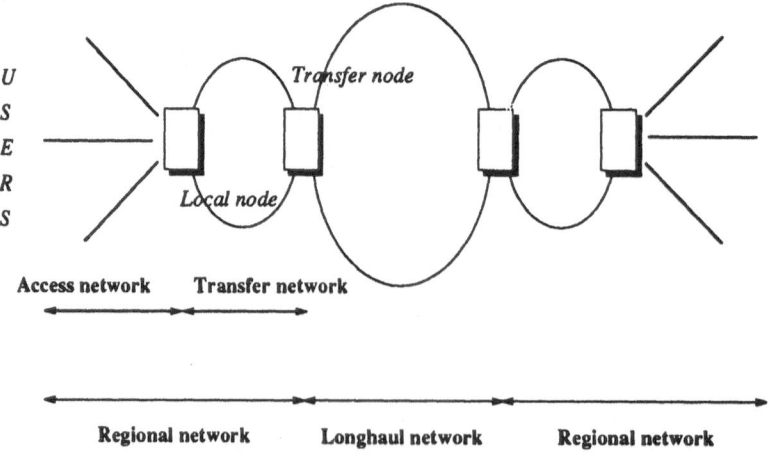

Figure 2: *A scenario of a general system structure.*

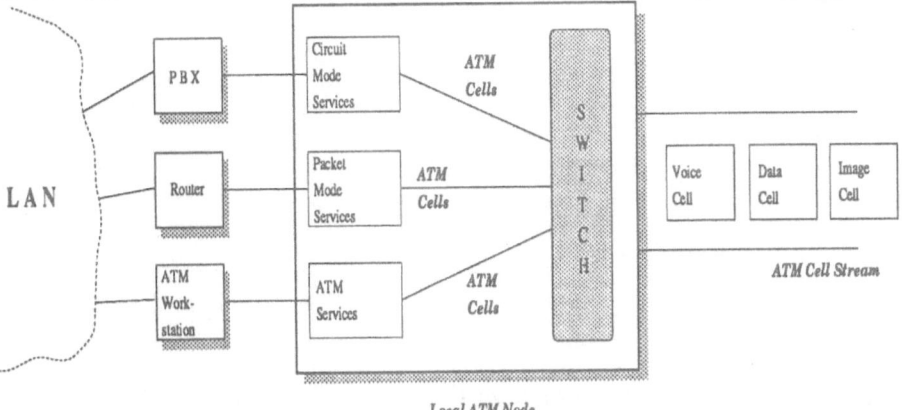

Figure 3: *The LAN is connected to the local ATM node through already existing equipment connected to the LAN.*

'directly' connected to the same network. With Management programs, changes in the configuration could be performed by rather easy manipulations. In the beginning, ATM will be used as a bearer service by the local area networks between segments and hubs, further on it could be used as a switch between different local area networks. The final context is to use ATM all the way to the users.

However, a lot of unsolved questions remain unanswered. So far, there are yet no standard on ATM LAN addressing. The ATM addresses could be administrated by public operators who then have to assign addresses to LANs to be interconnected. The LANs could also be connected by fully ATM integrated devices, in which case we have to use some kind of addressing on higher layers. This solution is presumably going to be used in a lot of applications, in particular for older LANs connected to a public ATM network.

MODELLING LAN INTERCONNECTION

As discussed in previous sections, we have to take some special considerations into account dealing with this model. They are:

- the sources are of bursty nature,

- the bearer service is cell oriented, i.e. we model a slotted system with the time slots equal to a cell size,

- there will be a variety of different applications, mixed and transferred on the same lines,

- the Poisson distribution is not valid, not even after superposition of several of these traffic streams,

- in a lot of LAN applications, there would be a common device for internetwork traffic to and from the LAN.

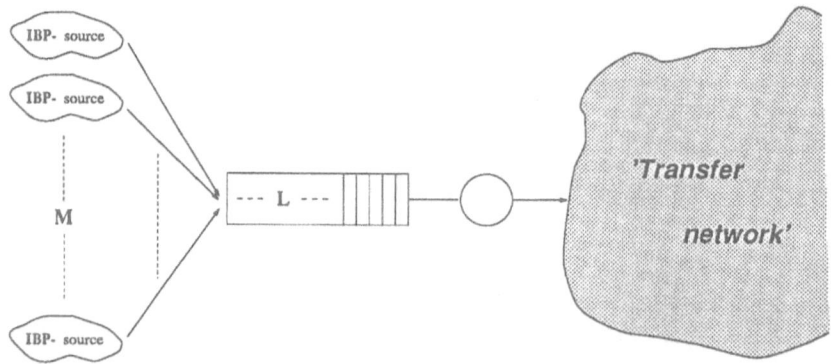

Figure 4: *A brief outline of the LAN interconnection model used.*

Due to these circumstances, we are going to adopt a bursty arrival rate model called Interrupted Bernoulli Process (see the section about the arrival process below). This model tries to capture the bursty behaviour of the sources. Further, since there is one common device for incoming and outgoing traffic between the LAN and the transfer network (or directly to a public ATM network), we model this as one single server queue with finite space. Several users could simultaneously make use of this local node device. Hence, a number (M) of IBP sources is connected in the queueing model, see Figure 4.

The Arrival Process

Since it is concluded that the arrival streams are going to be bursty (discussed above), consisting of different information services, the Poissonian assumption is no longer valid. To be able to make a performance analysis we have to use some other model that reflects these bursty arrivals.

One way of achieving that end is to use a model with arrival periods (active state) and silence periods (idle state). The duration of these periods has a geometrical distribution and the two periods alternate for ever. This type of arrival process can be seen as being the discrete equivalent to the Interrupted Poisson Process (IPP). This process, called a discrete time Interrupted Bernoulli Process (IBP), is governed by a two-state Markov chain in which the time unit is equal to a slot of the incoming link to the local node. This implies that given the process is in active state, it will remain in this state with a certain probability p or will change to idle state with probability $1 - p$. If the process is in idle state, it will remain in that state with probability q, or will change to active state with probability $1 - q$. To summarize, we could describe the arrival process by the following Markov chain, which could be seen in Figure 5, where the time unit is equal to a slot of the incoming link to the local node. To generalize the process, we say that when in active state a slot will contain a cell with probability α. We define two time parameters to calculate the mean, variance and squared coefficient of variation of the interarrival time between two cells. Let $\tilde{\xi}_A$ be the time interval from a slot in the active state to the time of the next arrival of a cell, and $\tilde{\xi}_I$ be the time interval from a slot in the idle state to the time of the next arrival. Then, if the process is in active state and stays in active

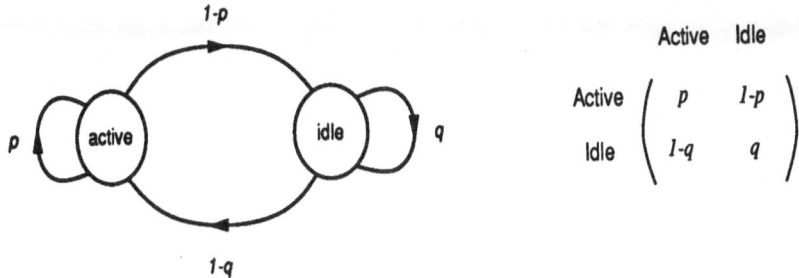

Figure 5: *The Markov chain for the arrival IBP to the buffer.*

for the next slot, and there is an arrival in that slot, $\tilde{\xi}_A$ is equal to 1 with probability $p\alpha$. From the memoryless property we get that if the process stays in active state but that slot is empty $\tilde{\xi}_A$ is equal to $1 + \tilde{\xi}_A$ with probability $p(1-\alpha)$. Finally, if the process changes to idle state in the next slot, $\tilde{\xi}_A$ is equal to $1 + \tilde{\xi}_I$ with probability $1 - p$. Applying similar arguments we also get the equations for $\tilde{\xi}_I$

$$
\tilde{\xi}_A = \begin{cases} 1 & p\alpha \\ 1 + \tilde{\xi}_A & p(1-\alpha) \\ 1 + \tilde{\xi}_I & 1 - p \end{cases} \tag{1}
$$

$$
\tilde{\xi}_I = \begin{cases} 1 + \tilde{\xi}_I & q \\ 1 & (1-q)\alpha \\ 1 + \tilde{\xi}_A & (1-q)(1-\alpha). \end{cases} \tag{2}
$$

Using the above equations we get

$$
E(z^{\tilde{\xi}_A}) = zp\alpha + zp(1-\alpha)E(z^{\tilde{\xi}_A}) + z(1-p)E(z^{\tilde{\xi}_I}) \tag{3}
$$

$$
E(z^{\tilde{\xi}_I}) = zqE(z^{\tilde{\xi}_I}) + z(1-q)\alpha + z(1-q)(1-\alpha)E(z^{\tilde{\xi}_A}). \tag{4}
$$

We define $G(z)$ as the generating function of the probability distribution of the interarrival time, $\tilde{\xi}_A$:

$$
\begin{aligned} G(z) &= E(z^{\tilde{\xi}_A}) \\ &= \frac{z\alpha(p + z(1-p-q))}{(1-\alpha)(p+q-1)z^2 - (q + p(1-\alpha))z + 1}. \end{aligned} \tag{5}
$$

From this generating function we can obtain the mean interarrival time $E(\tilde{\xi}_A)$ and the second moment of the interarrival time $E^2(\tilde{\xi}_A)$,

$$
E(\tilde{\xi}_A) = G'(1) = \frac{(2-p-q)}{\alpha(1-q)} \tag{6}
$$

$$
\begin{aligned} E^2(\tilde{\xi}_A) &= G''(1) + G'(1) = (8 - 4\alpha + (5\alpha - 8)(p+q) + \\ &\quad (2-\alpha)q^2 + 2(1-\alpha)p^2 + (4-3\alpha)pq)/(\alpha^2(1-q)^2). \end{aligned} \tag{7}
$$

From the above equations, we also obtain the squared coefficient of variation of the interarrival time, C^2,

$$
\begin{aligned} C^2 &= \frac{Var(\tilde{\xi}_A)}{[E(\tilde{\xi}_A)]^2} \\ &= 1 + \alpha\left(\frac{(1-p)(p+q)}{(2-p-q)^2} - 1\right). \end{aligned} \tag{8}
$$

71

The probability ρ, that any slot is busy (i.e. it carries a cell) is equal to the mean number of cells transmitted during the active period (which is equal to the mean length of the active period) over the mean length of the silence and active periods. This is equal to $1/G'(1)$, or

$$\rho = \frac{\alpha(1-q)}{2-p-q}. \tag{9}$$

To simplify the calculations an α equals to 1 could be used, i.e. we have an arrival each time slot the arrival process is presence in the active state. For reasons of not encumber all readers with this calculations they are to be found in an Appendix.

Queuing Model

We model the system as an imbedded Markov chain, with the status after each slot as the state. Just before time slot k's status, at t_k, two possible actions could take place. Departure at $t_k - \varepsilon_1$ and arrivals at $t_k - \varepsilon_2$, where $\varepsilon_1 > \varepsilon_2$. In our case we get a two dimensional chain (x, y). With the two parameters representing the number in the buffer ($0 \le x \le L$ (buffer length)) and the number of active sources ($0 \le y \le M$ (number of sources)). All the sources generating traffic in this model are set to be equal, in the sense that they all have the same parameters. The transition between two states ((x_1, y_1) and (x_2, y_2)) could then be calculated as (Figure 6):

$$\gamma(x_1, y_1, x_2, y_2) = r_{y_1 y_2} \cdot s_{x_2 - \max(x_1, 1) + 1, y_1} \tag{10}$$

where

$$r_{ij} = Pr(\text{transition from } i \text{ active sources to } j \text{ active sources}) \tag{11}$$

$$s_{ij} = Pr(\text{having exactly } i \text{ arrivals in a slot} \mid j \text{ active sources}). \tag{12}$$

The two parts of the transition probability, could be more thorough investigated. We then find the following expressions to be calculated:

$$r_{ij} = \sum_{k=\max(0, i-j)}^{\min(i, M-j)} \binom{i}{k} \beta^k (1-\beta)^{i-k} \cdot$$
$$\cdot \binom{M-i}{k-i+j} \eta^{k-i+j} (1-\eta)^{M-k-j} \tag{13}$$

where (compare Figure 5)

$\beta = Pr(\text{transition from state } active \text{ to } idle \text{ for one source}) = 1 - p$

$\eta = Pr(\text{transition from state } idle \text{ to } active \text{ for one source}) = 1 - q$

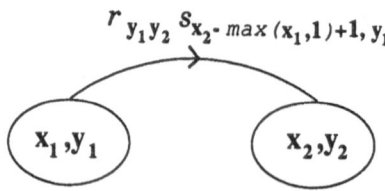

Figure 6: *The transition between state (x_1, y_1) and (x_2, y_2), in the imbedded chain.*

and

$$s_{ij} = \begin{cases} \binom{j}{i}\alpha^i(1-\alpha)^{j-i} & j \geq i \geq 0 \\ 0 & otherwise \end{cases} \tag{14}$$

where

$$\alpha = Pr(\text{arrival} \mid \text{source in active state}).$$

Total Model

Finally, we numerically solve the total system by using the set of equations obtained in the 'Queueing Model section' above. From these we get the state probabilities of the system being in a specific state (x, y). Knowing the state probabilities, we could obtain some system performance measures, like time blocking and call blocking. Using the same notation as before, i.e. x meaning the number of cells in the system and y the number of active sources, the following expressions are derived (note that L is the maximum queue size and M the number of superposed IBP sources):

$$\text{Time Blocking} = T_B = \sum_{y=0}^{M} Pr(L, y) \tag{15}$$

$$\begin{aligned}
\text{Call Blocking} = C_B = \\
= \frac{\sum_{x=0}^{L} \sum_{y=L-x}^{M} \sum_{k=0}^{y} \max(0, k - (L - x) - \min(1, x)) s_{ky} Pr(x, y)}{\sum_{x=0}^{L} \sum_{y=1}^{M} \sum_{k=0}^{y} k s_{ky} Pr(x, y)}.
\end{aligned} \tag{16}$$

Another, performance measure to be obtained is the total time in the system for a cell, which further could be divided into waiting time and service time. In [6] the following well known relation, called *Little's result*, could be found and proved:

$$\overline{N} = \lambda_A \overline{T}. \tag{17}$$

where $\overline{N}$ is the mean number in the system, i.e.,

$$\overline{N} = \sum_{x=0}^{L} \sum_{y=0}^{M} x Pr(x, y) \tag{18}$$

and λ_A is the actual input rate, defined as

$$\lambda_A = M\rho\,(\overline{x})^{-1}\,(1 - C_B) \tag{19}$$

where ρ for an individual source is obtained from Equation 9. Further, the mean service rate for a customer, defined below, is one time slot. Moreover, $\overline{T}$, the total time spent in the system by a customer, could now be calculated combining the equations above.

Finally, we could find the mean waiting time $(\overline{W})$ as

$$\overline{W} = \overline{T} - \overline{x}$$

and since the service time $(\overline{x})$ is deterministic, equal to one slot, we have

$$\overline{x} = 1 \text{ slot}.$$

RESULTS

The model shown in Figure 4 was evaluated for two different traffic cases, the first case representing smooth traffic arrivals with a rather high mean arrival rate, while the other case shows arrivals of a more bursty nature but at a considerable lower mean arrival rate (see Table 1). The same behaviour could be seen for both cases, shown in Figures 7 and 8, i.e. the traffic loss decreases rather rapidly while initially increasing the number of buffer spaces and then levels out as the number of buffer spaces is further increased.

The curves in Figures 7 and 8 show a characteristic bend, especially for increasing burst length. This effect indicates that the behaviour is governed by two different effects, which could be analyzed in two different time scales. If we use a short buffer, short term variations (the time scale is in the order of the minimum cell interarrival time) in the cell arrival process are dominant. Within this period a negative correlation of the cell arrival process can be observed. This means that if many cells arrive within a given time slot, fewer cells will arrive during the following time slot, and vice versa, due to the minimum spacing between consecutive cells from a given connection.

When the buffer becomes larger, the long time variations in the cell arrival process becomes dominant. In this time scale the cell arrival stream is generated by an alternation between active and idle state for the different sources. The duration of these phases is longer than the minimum cell interarrival time. In contrast to the short time variations, the correlation in this time scale is positive, i.e. if we have a high intensity of cell arrivals in a time cycle of a rather large number of consecutive cells the next time cycle will also have a high intensity of cell arrivals, due to the fact that many sources are in an active state.

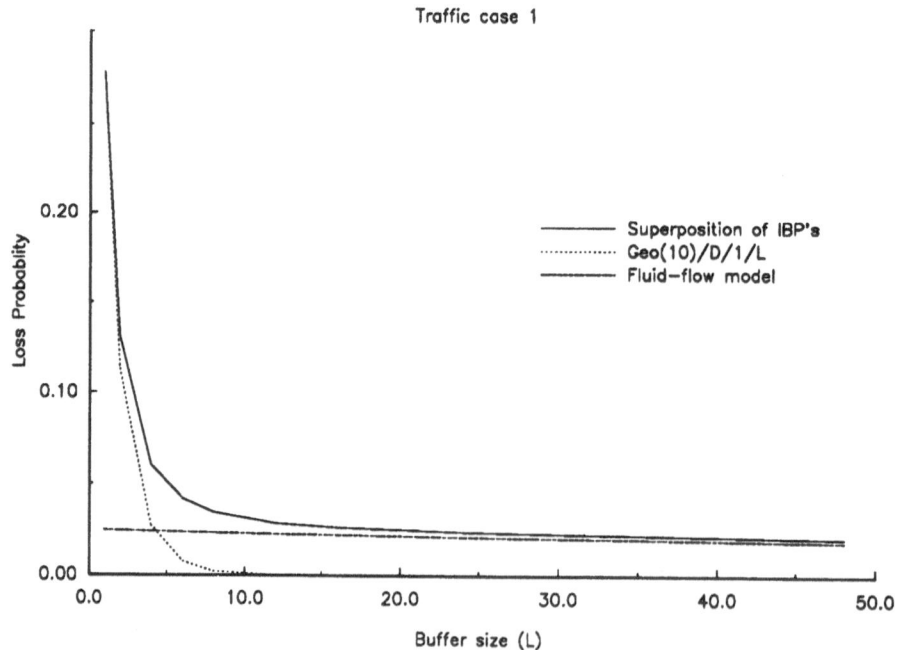

Figure 7: *The superposition of IBP sources compared to two other models, representing the cell scale and burst scale blocking, for traffic case 1.*

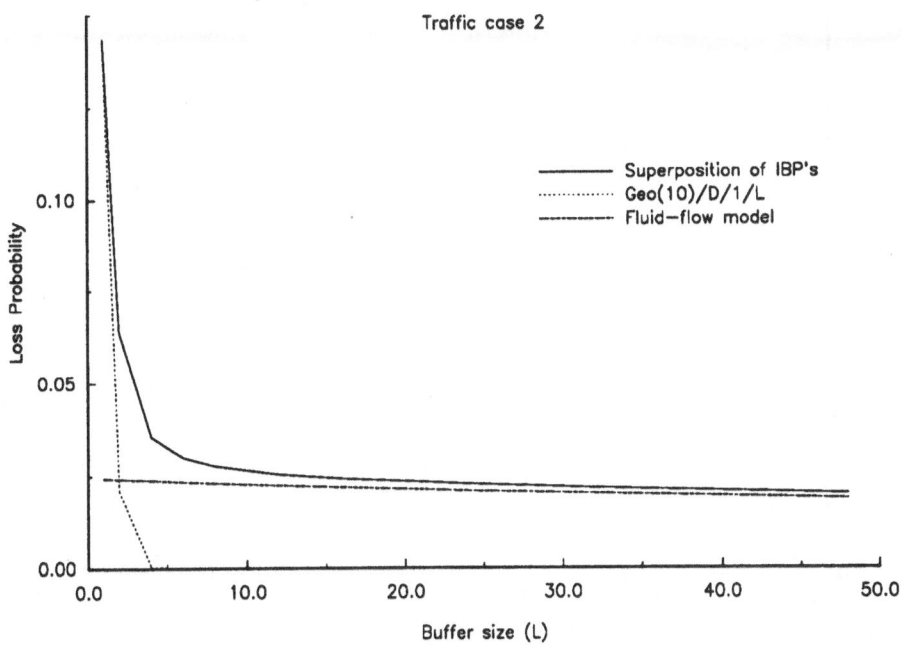

Figure 8: *The superposition of IBP sources compared to two other models, representing the cell scale and burst scale blocking, for traffic case 2.*

The model presented in this paper manages to capture the sharp bend at the transition from cell level to burst level statistics in the loss curve, a tractable issue for these kinds of models. The queueing behaviour with repect to the discussed time scales can be described by different queueing models. The results from the calculated model are compared with two other curves representing the cell scale and burst scale of the incoming traffic. The cell scale is represented by the *Geo(10)/D/1/L*-model which follows the curve for the new model for small buffer sizes. When the buffer size increases in the model, the burst scale is the dominant source for cell loss, an the approximative model used is the Fluid-flow model [7], which as can be seen, captures this behaviour rather well. The *Geo(10)/D/1/L*-model is analyzed with the same mean arrival rate as for the new model. The departure rate is also the same, that is one cell departs per slot if the system is non empty, otherwise the departure process is null. The other model (Fluid-flow model) is analyzed with the same number of users, same mean arrival rate and the same burstiness of the arrival stream. In Figure 10 the mean number of customers in the system, for traffic case 1, is shown. As could be expected, the mean number of customers increases with an increasing number of buffer spaces in the model, as the buffer is able to capture more of the fluctuations of

Table 1: *The parameters for the different traffic cases analyzed, traffic case 1 is less bursty than traffic case 2, but with considerable higher mean arrival rate.*

Traffic case	Mean arrival rate/source	Number of sources	Source parameters			System
			p	q	α	ρ
1	0.075	10	1/5000	1/5000	0.15	0.75
2	0.035	10	1/2500	1/22500	0.35	0.35

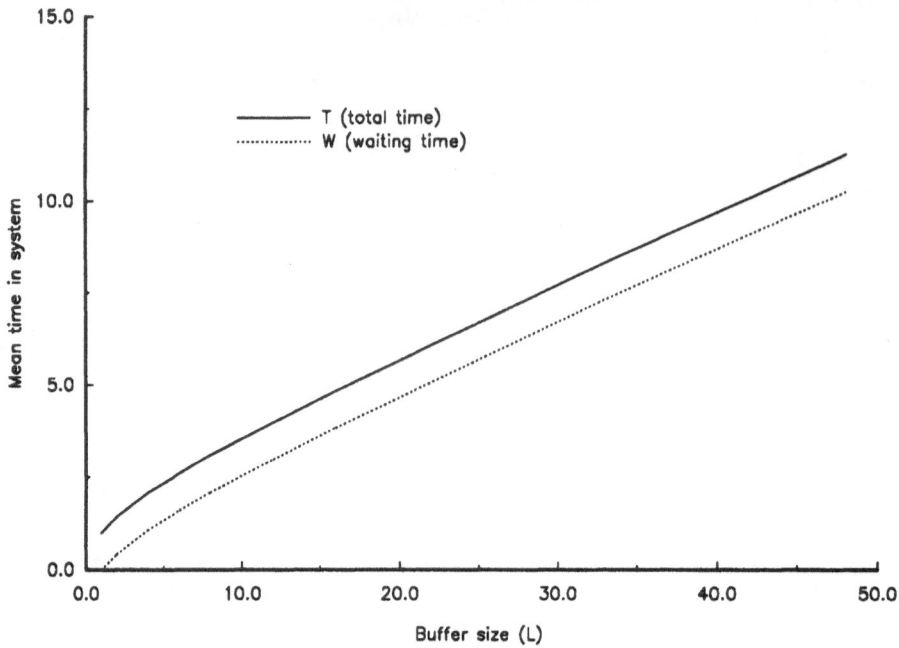

Figure 9: *Mean total time and mean waiting time in the system vs. buffer size, for traffic case 1.*

the arrival process. The same arguments could be applied to the mean system time, shown in Figure 9, also shown for traffic case 1 for reason of comparison. As discussed in Equation 17, the two measures are highly correlated. The quotient $\frac{N}{T}$ is λ_A, which is an increasing quantity as the number of buffer spaces increases. This could be seen by combining Equation 19 and Figure 7, and also by a direct comparison between

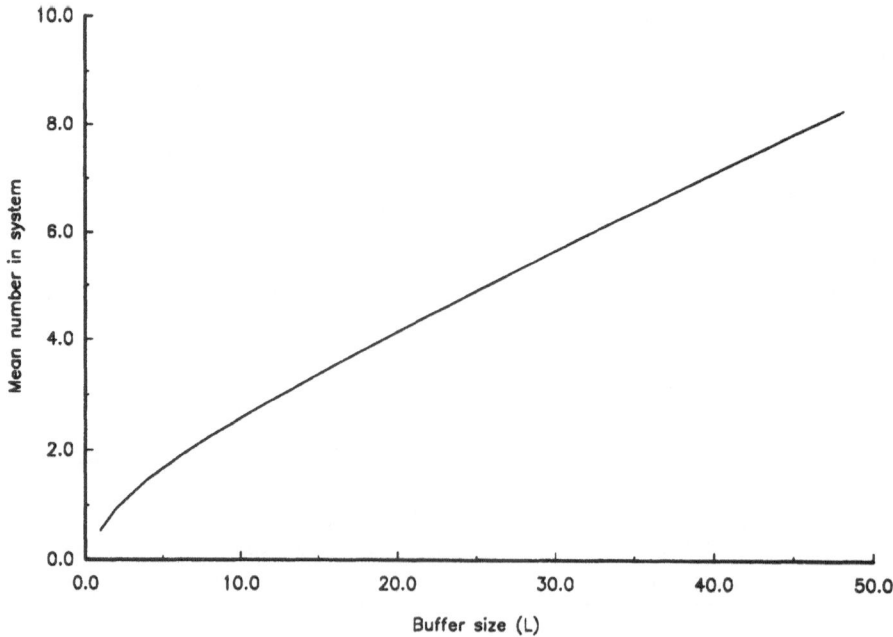

Figure 10: *Mean number in the system vs. buffer size, for traffic case 1.*

76

the actual value of the mean number in the system (Figure 10) and the total system time (Figure 9) for an increasing value of the number of buffer spaces in the model.

CONCLUSIONS

B-ISDN have to utilize the existing infrastructure of the digital network. This means that, due to already invested time and money, the physical configuration has to follow the current fiber cabling and offices. Therefore, it is going to consist of regional networks and longhaul networks. Designed as a solution for future public networks, ATM is now quickly being adapted to low cost LANs and private networks. As a true international standard, it is already widely supported by the communications and computer industries as a way of ensuring interoperability and protecting existing and future investments. The rapid changes in technology and service requirements make it increasingly important to find an efficient way to evaluate different network configurations. A lot of things change over time, some of these important issues for communication systems are discussed in the paper. The different evolution of the system components, the changing traffic characteristics and the connectivity of different communication networks are among things that are brought up to discussion. In this paper a model is evaluated that takes these things into account, i.e. a bursty traffic pattern, interconnectivity and the possibility to use already installed equipment. The arrival model is a superposition of several independent bursty sources, each modelled by an Interrupted Bernoulli Process. The model manages to capture the sharp bend at the transition from cell level to burst level statistics in the curve for call blocking probabilities, a tractable issue for these kinds of models. It also corresponds very well with other models that are able to capture either the cell level or the burst level statistics. Finally, the number of customers, as well as the total time and waiting time in the system are shown. They all, as expected, increase with larger buffer sizes, due to the fact that the buffer is more likely to capture the bursty arrival stream.

APPENDIX

The calculations could be simplified by letting the α value be equal to 1 (one). This implies that an arrival occurs each time slot that the arrival process is in the active state. Which, probably, could be a good approximation for some applications. If we model the system under this assumption ($\alpha = 1$) the calculations carried out above will instead give us the following relations (to be compared with Equations 8 and 9):

$$\rho = \frac{1-q}{2-p-q} \qquad \text{and} \qquad C^2 = \frac{(p+q)(1-p)}{(2-(p+q))^2}.$$

Now, we can alter the characteristics of the traffic by simply playing with the values of p and q. By varying these two parameters we could obtain a set of different traffic cases. In particular we have the relations shown in Table 2. We see, therefore, that by varying p and q we can increase the traffic and at the same time we can change the burstiness (C^2) of the arrival process. We note that the above process becomes geometric if we set $q=1-p$ ($p+q=1$).

Table 2: *The impact of different p and q on the probability that a slot contains a cell (ρ) and the burstiness (C^2), under the assumption that $\alpha = 1$.*

p and q	ρ	C^2
p $\to$ 0, q $\to$ 0	$\rho \to 0.5$	$C^2 \to 0$
p $\to$ 0, q $\to$ 1	$\rho \to 0$	$C^2 \to 1$
p $\to$ 1, q $\to$ 0	$\rho \to 1$	$C^2 \to 0$
p $\to$ 1, q $\to$ 1	$\rho \to 0.5$	$C^2 \to \infty$

If we on the other hand want to analyze a system with the ρ and C^2 values for the arrival processes given, we could easily get the p and q values as,

$$p = \frac{C^2 - 1 + 3\rho - 2\rho^2}{1 - \rho + C^2}$$
$$q = \frac{1 - 2\rho + p\rho}{1 - \rho}.$$

References

[1] CCITT Recommendation, I Series (B-ISDN), Nov, 1990.

[2] C. E. Catlett, "Balancing resources", *IEEE Spectrum, Sept, 1992.*

[3] CCITT Recommendation, H and J series, Nov, 1988.

[4] K. Sriram and W. Whitt, "Characterizing superposition arrival processes in packet multiplexers for voice and data", *IEEE Journal on Selected Areas in Communications, vol. 4, no. 6, Sept, 1986.*

[5] J. S. Kaufman, "Blocking in a Shared Resource Environment", *IEEE Transactions on Communications, Vol 29, No. 10, October 1981*

[6] J. D. C. Little, "A Proof of the Queueing Formula $L = \lambda W$", *Operations Research, No. 9, 1961.*

[7] D. Anick, D. Mitra and M. M. Sondhi, "Stochastic theory of a data handling system with multiple sources", *Bell Syst. Technical Journal, vol. 61, 1982.*

INTERCONNECTING LANs FOR REAL TIME TRAFFIC APPLICATION *

Imrich Chlamtac
Electrical & Computer Engineering Department
University of Massachusetts
Amherst, MA 01003

ABSTRACT

This work deals with the flow control of real time traffic in communication systems interconnecting multiple LAN-s through a gateway or over a high speed backbone. Multiple LAN-s connected to a WAN, or over a MAN are typical examples of systems belonging to these categories. The systems under consideration, LAN-s or gateway, have a finite set of buffers, and packets destined to the gateway or a destination LAN and that cannot be stored are considered lost. We propose and investigate flow control which can minimize probability of loss, without packet retransmission. Traffic control is studied for geographically adjacent systems, as well as for interconnected systems in which the packet transfer time is nonnegligible. An approximate Analytical model is introduced, which is used to set optimal control parameters, and derive the interconnected system performance in terms of packet loss probability, throughput and the average packet delay. It is shown that the use of the proposed control mechanism can significantly reduce the packet loss probability.

1 INTRODUCTION

Real time traffic imposes strict Quality of Service (QOS) requirements on the transport mechanism in terms of delay, and loss probability. When real time data has to traverse an interconnected network environment, consisting of multiple interconnected (LAN) networks, maintaining the QOS becomes a non trivial task. With real time traffic, as in multimedia applications for instance, the flow control should keep packet loss below a required level, even when, from data timeliness considerations, no retransmission of packets is possible.

In this paper we study the control of real time traffic in a gateway connecting multiple LAN-s to a wide area or MAN system which carry packets to a remote desti-

*The work was supported in part by the National Science Foundation under NCR-0104317, and by the Army Research Office under contract DAAL03-91-LG-0070.

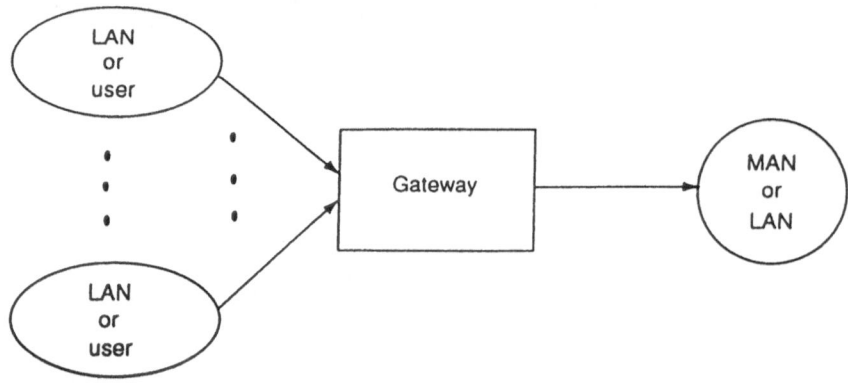

Figure 1: Gateway interconnected system

nation. A set of LAN-s interconnected to an ATM network through a gateway is one such example. A generic interconnected system is shown in Figure 1. In this case the system can be considered hierarchical, with control executed at the gateway. A flat system, representing a "direct" interconnection of multiple LAN-s over a backbone has been studied in [6].

In an interconnected LAN system, even when admission control is practiced at each LAN, packets arriving from multiple streams (LAN-s) to the gateway may be rejected due to the existence of a finite number of buffers in the system. Since distance between each LAN and the gateway may be significant, rejected packets are either lost, or if retransmitted they may fail to meet the deadline of the real time application. For effective interconnection through a gateway, we shall therefore consider control which minimizes the probability of losing real time packets arriving at the gateway, when no packet retransmission is used as a method for preventing packet loss. The proposed control mechanism can also be used to balance the flow of traffic from different LAN-s, and maximize the global system throughput.

Several earlier works on LAN interconnection dealt with related issues by simulation, or by assuming that the packet transfer time between the LAN-s and the gateway are negligible [1-3,6-8]. The work presented here extends the earlier results by accounting for arbitrary transfer delays between users and the gateway. A family of policies for managing the gateway is derived and analytical models are introduced to study the performance of the interconnected system under these policies for different network configurations, as well as to numerically determine the best policy setting. It is shown that the use of the proposed control mechanism can significantly reduce the packet loss probability.

2   NETWORK MODEL AND OPERATION

We consider a gateway mechanism connecting an arbitrary number of LAN-s or individual users (hosts). Packets from the users (input LAN) buffer queues are being transmitted to the (output) buffer queue of the gateway. For simplicity, we shall refer to the input and the output buffer queues simply as input and output queues, respectively. Each queue, input or output, has a finite number of buffers. Figure 2 shows the representations of the input/output buffer configuration and the system model adopted. To formally describe the gateway operation, we shall define:

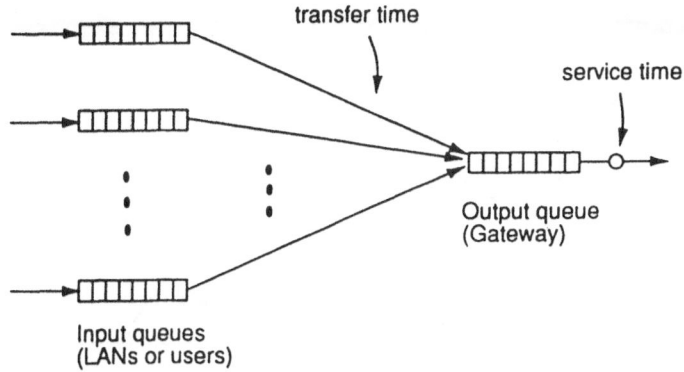

Figure 2: Model used to determine the optimal control

Gateway Control: determines how many packets will be allowed to be transferred from which queue and at what time. It is a combination of a control strategy and a control policy:

Control Strategy: determines the input queue from which a packet will be transferred to the output queue given a transfer permission was granted. The input rate to each input queue, the input queue length and the distribution of packets in the input queues are the main factors determining the control strategy.

We consider (a)*Random Strategy* (RN), in which a packet will be chosen randomly from one of the input queues, and (b) the *Longest Queue First Strategy* (LQ), in which a packet will be chosen randomly from one of the input queues among those having the least available space.

Control Policy: determines when a permission is granted to transfer a packet from an input queue, and how many packets overall are to be transmitted, to maintain a desired number of packets in the output queue. We define a policy P_m (with policy parameter m, $m > 0$) as one which maintains m packets in the gateway output queue.

We shall consider (a) the *Earliest Transfer Policy* (EX), in which packets from the input queues will be transferred into the output queue as early as possible. Observe that in this policy m equal to the size of the output queue. We shall also observe (b) the *Delayed Transfer Policy* (DX), in which packets from the input queues will be transferred into the output queue as late as possible, i.e. policy with $m = 1$.

For simplicity of notation when referring to a given Gateway Control we shall use the combined notation of policy and strategy to represent the associated control. For example, LQDX will thus stand for a control consisting of the LQ strategy and DX policy.

2.1 Gateway Flow Control

Although the control policy defined below can be applied to a continuous time system in which packets and information between the input and the output queues can be exchanged at arbitrary times, for convenience of representation and modeling in this paper we consider the following discrete system operation: Let T be the maximum transfer time, consisting of packet transmission time and link propagation delay, between the input queues and the output queue. We assume that the output queue (gateway) knows the size of the buffer at each input queue. We consider the

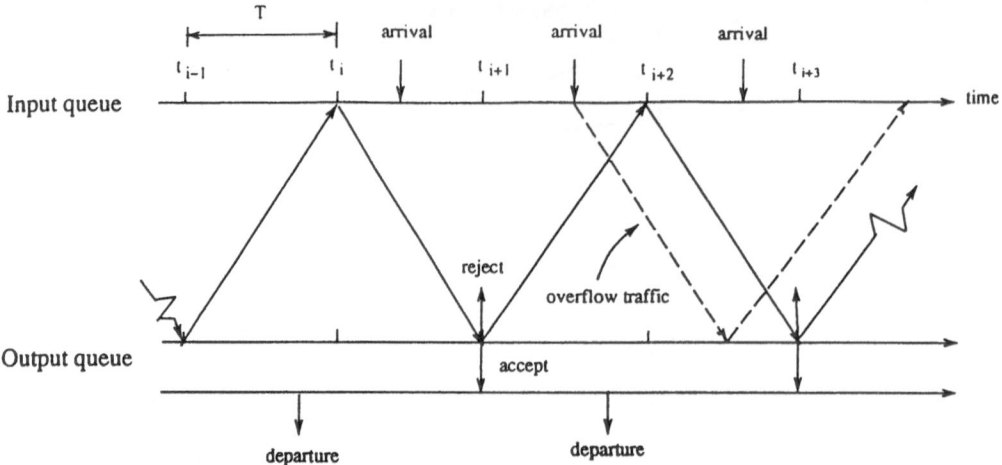

Figure 3: Traffic flow mechanism

time sequence, denoted by $t_{i+1} - t_i = T$, $t_0, t_1, ..., t_i, ...$ At instances $t = t_i$ all input queues transmit the information on the number of queued packets to the output queue. The output queue collects the queue length information, and at $t = t_{i+1}$ sends out a permit to each input queue j to transmit n_j packets. This number is determined by the policy and strategy in use. At times $t = t_{i+2}$ each user (input queue) can transmit to the output queue the requested $n > 0$ packets. Whenever a new arrival occurs to a full input queue, the head of the line packet will be transmitted immediately. In this system a packet will be considered lost if at arrives at a full output queue. Figure 3 demonstrates the actual operation of the system, where an overflow in an input queue results in an immediate transfer of the packet to the output queue. Figure 4 represents the "discrete" system traffic flow model mechanism in which packet transfers occur only at times t_i.

2.2 Optimal Gateway Control

There are two factors affecting the decision to transfer a packet so that packet loss is minimized. One, the policy should keep as many buffers available in the output

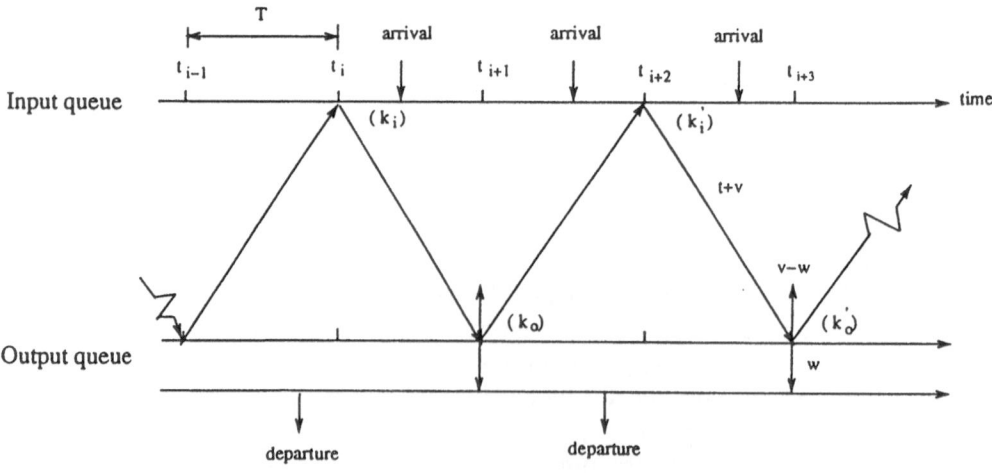

Figure 4: Approximate model mechanism

queue as possible. For this purpose, packet transfer from input queues should be done as late as possible. Intuitively, this leads to a smaller probability of overflow in the output queue, since more room is reserved for packets arriving to full input queues. The second counterbalancing factor is that in order to minimize the packet loss (and maximize system throughput as well) we wish to keep the server in the output queue as busy as possible. For systems with nonnegligible packet transfer times, and in systems in which service completion times are not well known, e.g. exponentially distributed service time, the gateway cannot predict when the transmission of its last packet will be completed, and it cannot access packets in the input queues immediately after the completion. Therefore, contrary to the preceding consideration, in order to reduce the probability that the output queue's server remains idle while there are packets in the input queues, the output queue should be kept as full as possible. For systems where the input and output processes are stochastic and the transfer delays are non negligible an optimal policy thus needs to balance the above two factors deciding on the optimal number of packets to be kept in the output queue.

From the preceding discussion it is easy to see why in systems where the packet transfer delays are zero the optimal policy is DX as shown in [1,2]. In other words, for negligible transfer times the optimum control policy delays transfers from an input LAN to the gateway until the last moment. It can also be shown that, in general, LQ is the optimum strategy for systems with identical Poisson arrivals in the input queues when the transfer delays are equal between all input queues and the output queue. In other words, the optimal control should transfer packets from the queue with the largest number of packets, for arbitrary data transfer times. However, it is also easy to see from the preceding discussion, that for non negligible packet transfer times DX is no longer the optimal policy. In subsequent sections we develop an analytical model which allow us to numerically determine the best policy as a function of system configuration and packet transfer time.

3 MODELING INTERCONNECTED SYSTEM PERFORMANCE

In studying the performance of the interconnected system four different cases can be determined, given by the combination of system configuration and packet transfer times [6,7]. In terms of system configuration, we distinguish between a hierarchical, gateway controlled, system, and a multiple output LAN, flat configuration, interconnected system. In terms of data transfer times, a negligible and non-negligible models exist. In this paper we focus on the analysis of a gateway controlled system with arbitrary packet transfer times, a model representation of which is shown in Figure 2. We shall consider systems using policy P_m. We shall assume the mean service time of the output queue, $1/\beta$, to be exponentially distributed. The arrival process for input queue i, with rate λ_i, is assumed to be Poisson. The total arrival rate λ is therefore equal to $\sum_{i=1}^{K} \lambda_i$. $B_i, i = 1, ..., K$ and B_0 denote the numbers of buffers in input queue i and in the output queue, respectively.

For the negligible transfer time case an exact model can be developed with the state defined by the number of data packets in each of the input and output queues. We have

$$\mathbf{k} = (k_0, k_1, ..., k_i, ..., k_K) \tag{1}$$

where $0 \leq k_i \leq B_i$, for $i = 0, 1, .., K$. k_0 is the number of data packets in the output queue and $k_i, i = 1, ..., K$, is the number of data packets in the ith input queue. Note that we indicate the state $(k_0, ..., k_i \pm 1, ..., k_K)$ later simply by $\mathbf{k}_i \pm 1, \forall i$. We can then build a Markov chain and compute the transitions probabilities as given in [7].

For larger systems an exact model becomes nontractable even for negligible packet transfer times. In this case a closed-form approximate solution can be developed for symmetric systems. When the size of the queues and the arrival rates are equal for all queues it is possible to group all *individual* input queues into an *aggregate* input queue, leading to a system state represented by the total number of packets. By using Urn models a simplified tractable model can be obtained[4].

For non-negligible packet transfer times the state description used in the negligible transfer time models above, does not summarize the system in terms of packets in transit between the input and output queue and their arrival time. Including this information in the state description makes the system analysis intractable. Therefore, we provide a new approximate analysis using a discrete-time dicrete-state Markovisn model, in which we again can use Urn models [4] to reduce the size of the state space. We shall deal with a discrete time, synchronous and symmetric system. We shall assume the packet transfer times to be equal between all input queues and the output queue. When a packet overflow occurs in the input queue we hall assume that the overflowed packet waits for transfer until the up-coming discrete packet transfer time. The traffic flow mechanism for the approximate model is shown in figure 4. We shall consider the random strategy. We define the state to be the total number of packets in the system. Let k_i and k_i' be the total numbers of packets in all input queues (after packet transfer) at t_i and t_{i+2}, respectively. Let k_o and k_o' be the number of packets in the output queue (after receiving packets from input queues) at t_{i+1} and t_{i+3}, respectively. Let $P_A(a)$ be the probability that there are a packets arriving to the input queues from t_i to t_{i+2} and $P_D(d)$ be the probability that there are d departures from the output queue from time instance t_{i+1} to t_{i+3}. Hence, we have

$$P_A(a) = \frac{(2\lambda T)^a}{a!} e^{-2\lambda T} \quad a = 1, 2, ... \tag{2}$$

and

$$P_D(d) = \begin{cases} \frac{(2\beta T)^d}{d!} e^{-2\beta T} & 0 \le d \le k_o - 1 \\ 1 - \sum_{i=0}^{k_o-1} \frac{(2\beta T)^i}{i!} e^{-2\beta T} & d = k_0. \end{cases} \tag{3}$$

For the sake of analysis we shall name packets transferred due to overflow in input queues, overflowed packets and the remaining packets, normal packets. Let t be the number of normal packets transferred at time t_{i+2}, given by

$$t = \begin{cases} 0 & k_o \ge m \\ m - k_o & k_o < m \ and \ k_i > m - k_o \\ k_i & otherwise. \end{cases} \tag{4}$$

Let v be the number of the overflowed packets transferred at t_{i+2} and assume that w of them are accepted to enter the output queue at t_{i+3} (the remaining packets being lost) where w is given by

$$w = \begin{cases} v & v < B_o - k_o + min[\lfloor \frac{d}{2} \rfloor, -t + d] \\ B_o - k_o + min[\lfloor \frac{d}{2} \rfloor, -t + d] & otherwise. \end{cases} \tag{5}$$

where $\lfloor \frac{d}{2} \rfloor$, the largest integer smaller or equal to $d/2$, is a correcting term since, in the real system, overflowed packets do not wait until the normal packet transfer time.

We next consider the time interval from t_i to t_{i+2}. Let n_p be the number of packets in the input queues at the pth arrival packet (beginning from t_i) comes to the input queues and b_p be the variable with the value 0 or 1. $b_p = 1$ signifies that an overflow occurred due the pth packet arrival, while $b_p = 0$ signifies the contrary. Assuming that there are n_p packets in the K input queues, the conditional joint probability of the individual occupancies for $b_j = 0 \ \forall j < p$ (i.e. there is no overflow prior to the pth packet arrival) can be approximated as

$$\Pr(\vec{N}| \sum_{l=1}^{K} n_l = n_p) = 1/C_B(n_p, K), \ \vec{N} \in S(n_p, K), \ 1 \le n_p \le KB. \qquad (6)$$

Let the probability that a particular input queue is full due to the pth packet arrival be $p_1(n_p)$ and let the probability that queue is not full be $p_0(n_p)$ $(=1 - p_1(n_p))$. Hence, for $b_j = 0 \ \forall j < p$, we have

$$p_1(n_p) = \begin{cases} 0 & 0 \le n_p \le B - 1, \\ C_B(n_p - B, K - 1)/C_B(n_p, K) & B \le n_p \le KB. \end{cases} \qquad (7)$$

On the other hand, for $\exists b_j = 1, \ j < p$ (i.e. there is at least one overflow prior to the pth packet arrival), since there is at least one full input queue. The conditional joint probability of the individual occupancies becomes

$$\Pr(\vec{N}| \sum_{l=1}^{K} n_l = n_p) = 1/[C_B(n_p, K) - C_{B-1}(n_p, K)], \ \vec{N} \in S(n_p, K),$$
$$1 \le n_p \le KB. \qquad (8)$$

Hence, for $\exists b_j = 1, \ j < p$, we have

$$p_1(n_p) = \begin{cases} 0 & 0 \le n_p \le B - 1, \\ \frac{C_B(n_p - B, K - 1)}{C_B(n_p, K) - C_{B-1}(n_p, K)} & B \le n_p \le KB. \end{cases} \qquad (9)$$

Therefore, we have the conditional probability

$$P_V(v|a, k_i) = \sum_{\forall b_p : \sum_{p=1}^{a} b_p = v} [\prod_{p=1}^{a} p_{b_p}(n_p)] \qquad (10)$$

where $n_{p+1} = n_p + 1 - b_p$ and $n_1 = k_i$. Note that $\forall b_p : \sum_{p=1}^{a} b_p = v$ means all possible combination of b_p's such that the total number of overflowed packets in the input queues due to a packet arrivals from t_i to t_{i+2} is equal to v.

Moreover, we have the relationships between k_i and k_i' and between k_o and k_o', respectively, as

$$\begin{aligned} k_i' &= k_i - t + a - v \\ k_o' &= k_o + t + w - d. \end{aligned} \qquad (11)$$

The probability that from k_i and k_o to k'_i and k'_o is given by

$$P_{(k_i,k_o),(k'_i,k'_o)} = \sum_{a,d \text{ and } v} P_A(a)P_D(d)P_V(v|a,k_i) \tag{12}$$

for all possible values of a, d, and v such that (37) holds.

We model the system as a discrete Markovian chain with duration $2T$ and state (k_i, k_o). Assume that the system is stationary. Then, we can relate the state probabilities, $P(k_i, k_o)$'s, by the state transition probability $P_{(k_i,k_o),(k'_i,k'_o)}$. Therefore, we set up $(KB+1)(B_o+1)-1$ linear independent equations. Together with the normalization equation, $P(k_i, k_o)$ can be solved.

The loss probability is then given by

$$P_{loss} = \sum_{k_i=0}^{KB} \sum_{k_o=0}^{B_o} \sum_{a=1}^{\infty} \sum_{d=0}^{k_o} \sum_{v=0}^{a} \frac{v-w}{a} \frac{P_A(a)}{1-P_A(0)} P_D(d)P_V(v|a,k_i,k_o)P(k_i,k_o) \tag{13}$$

where w is given by (31) and $(v-w)$ is the number of packet loss in the output queue.

The throughput is given by

$$S = \lambda(1 - P_{loss}) \tag{14}$$

The average delay is given by

$$D = \frac{\sum_{k_i=0}^{KB} \sum_{k_o=0}^{B_o} (k_i + k_o)P(k_i,k_o)}{\lambda(1 - P_{loss})} + T \tag{15}$$

where T is the packet transfer times.

4   RESULTS

In this section, we present the performance of the interconnected system using different combinations of policies and strategies for negligible and non-negligible packet transfer times, deriving the best control in each case. In addition, we consider the benefits and relative efficiency of the use of the proposed gateway control by comparing the performance in terms of packet loss with that of an uncontrolled system, in which packets arriving at the input queues will be transferred to the output queue immediately and packets will be lost if the output queue is full. The uncontrolled system can be modeled by an M/M/1/K queue with queue length equal to the queue length of the output queue.

Since in most cases an approximate model is needed for tractability, we first evaluating the quality of the approximation for the zero and non-zero transfer time cases, considering loss probability and the average packet delay. The quality of throughput approximation follows from the validation of packet loss probability. For the negligible transfer times we refer to the model given in [7], for the non-negligible case the analytical model introduced in the preceding section is used. We choose $B_0 = 4$, $B = 3$, $K = 3$, and $\lambda = 1$. For the zero transfer time case, tables 1 and 2 show that for all policies P_m, the error between the exact and the approximate models is below 2.5 %. For the non-zero case, tables 3 and 4 show that the error of the approximate solution stays below 10 %.

Table 1: Comparison of loss probabilities for zero case

m	Exact $(\times 10^{-2})$	Approx. $(\times 10^{-2})$	Error (%)
1	7.28	7.25	0.41
2	7.46	7.44	0.27
3	7.92	7.97	0.63
4	9.09	9.28	2.09

Table 2: Comparison of average packet delay for zero case

m	Exact	Approx.	Error (%)
1	6.89	6.91	0.29
2	6.75	6.77	0.30
3	6.40	6.38	0.31
4	5.63	5.53	1.80

Table 3: Comparison of loss probabilities for non-zero case

m	Simul. $(\times 10^{-1})$	Approx. $(\times 10^{-1})$	Error (%)
1	1.83	1.67	9.6
2	1.49	1.39	7.2
3	1.16	1.13	2.7
4	1.10	1.21	10.0

Table 4: Comparison of average packet delay for non-zero case

m	Simul.	Approx.	Error (%)
1	13.26	13.45	1.4
2	11.39	11.92	4.7
3	8.98	9.73	8.4
4	7.23	7.83	8.3

Figure 5 shows the loss probability against offered load with zero transfer time for four control cases and one uncontrolled case. In addition the figure shows the "ideal" system behavior, in the sense that incoming packets can "share" all input queue buffers. We consider a system with 3 LAN-s, each having 3 (input) buffers, and a three unit buffer at the gateway. As expected, the performance of the controlled cases is significantly better. Among the different policy and strategy combinations, the LQDX control has the best performance, being also close to the performance bounds given by the "shared buffer" system model. In this bound we consider all LAN queues as being shared between all LAN-s [7], which indicating that the LQDX control is highly efficient. As seen from this figure, if e.g. for a given real time application, the allowable loss probability is less than 1 to 0.75, while without control the throughput would have to be limited to a mere 0.25. Lastly, this figure also shows that the difference between the performance of RNEX (or RNDX) and LQEX (or LQDX) is not very significant in the symmetric system configurations. Fig.6 investigates a larger interconnected system. It can be seen that with an increased number of the LAN-s (input queues), the benefits of gateway control become more pronounced. In fact, additional performance results (not shown here for lack of space) suggest that, more generally, the

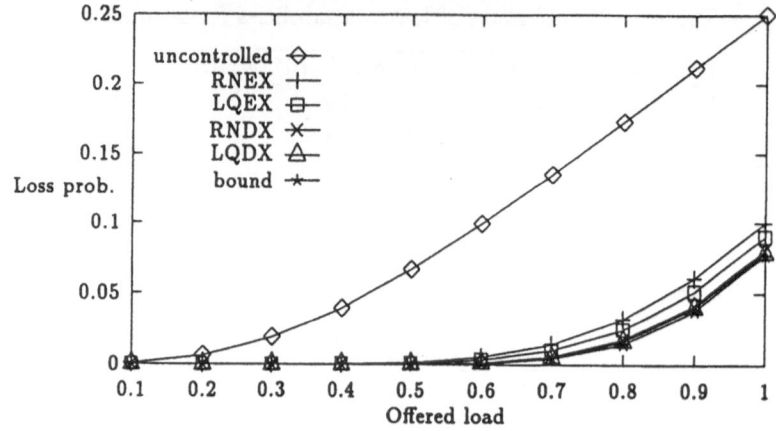

Figure 5: Loss prob. vs. offered load, $K = B = B_o = RTTT = 0$.

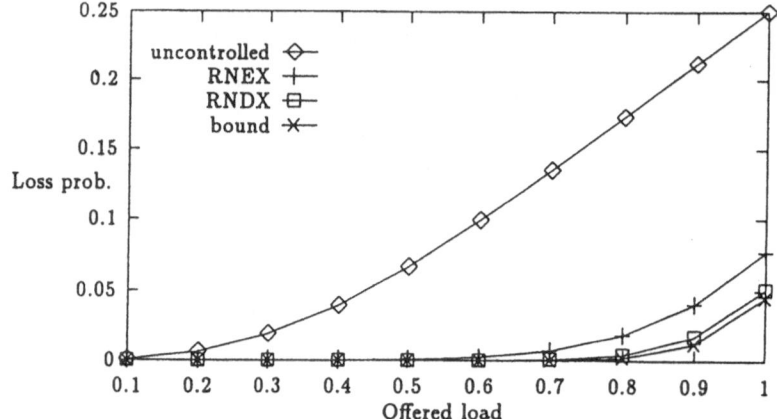

Figure 6: Loss prob. vs. offered load, $K = 6$, $B = B_o = 3$, $RTTT = 0$

larger the system in terms of the total number of input buffers, or the higher the ratio between the combined number of all input buffers to the number of output buffers, the more significant the improvement obtained by the proposed control. Figure 7 shows packet loss probability as a function of the policy parameter, m for different values of non-negligible round trip transfer time (RTTT). This figure shows that, consistent with the control motivation given in section 3, the optimal value of m increases along with the transfer time. In other words, the larger the RTTT value, the larger the number of packets that need to be kept in the output queue to ensure that the output queue server (gateway to WAN) will remain occupied when the input queues are not empty. Secondly, Figure 7 shows that the smaller the transfer time the more significant the reduction in packet loss probability compared to the uncontrolled case. Figure 8 considers the effects of system size on packet loss probability. It can be seen that the reduction in packet loss is more significant, similarly to the zero transfer time case, when the system becomes larger. Taken together Figures 7 and 8 show that the importance of control policy grows with system size, but diminishes with increasing transfer time values. Intuitively, with larger RTTT values the information used at the gateway about the state of the input queues becomes outdated, so that the control becomes gradually less effective. On the other hand, when system size grows (in terms of total number of buffers) the benefits of using control will remain significant even for large transfer times. This suggests that for all but very small systems with very large transfer times the use of control based on time sensitive feedback is well justified. Figure 9 considers the throughput-delay behavior of the system. Here, RN policy is used with policy parameter $m = 5$ for RTTT=1 and $m = 7$ for RTTT=3. For RTTT=0, we see that the delay for the controlled case is larger compared to the uncontrolled case and the difference becomes more pronounced as throughput increases. In other words, we can observe a packet loss and delay trade off between the controlled and uncontrolled cases. However, we see that as RTTT changes, the delay of the controlled case is offset from the uncontrolled case by a "constant" value equal approximately to one and a half times the round trip transfer time for most of the arrival rates. This delay is due to the synchronization delay introduced by the need to exchange information between input and output queues using the discrete control operation. Lastly, Figure 10 presents the performance of an asymmetric system

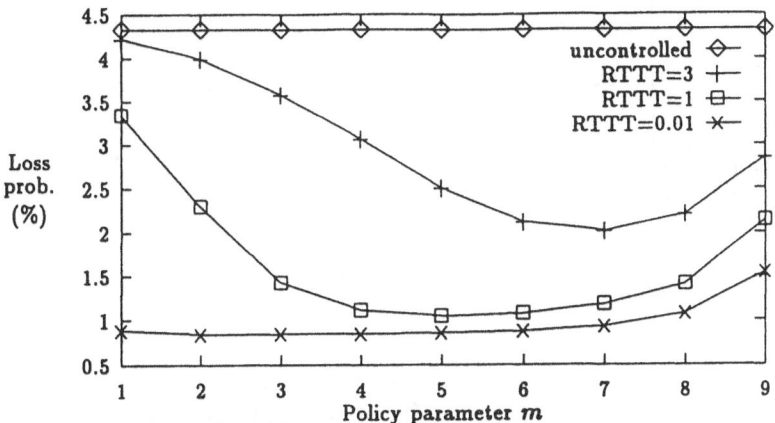

Figure 7: Loss prob. vs. parameter,
$\lambda = 0.85$, $K = B = 3$, $B_o = 9$, $RTTT > 0$.

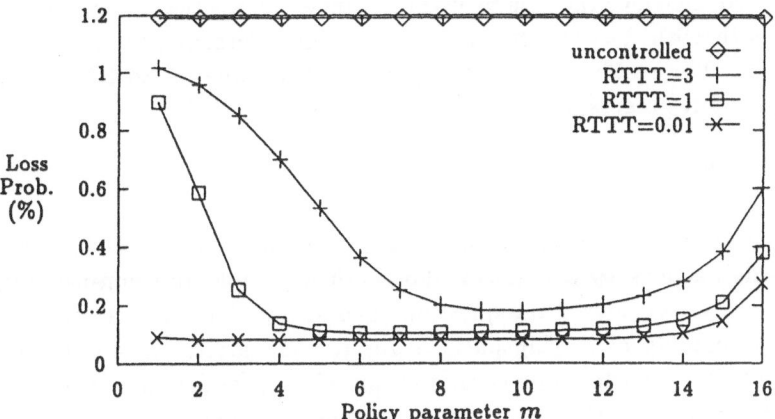

Figure 8: Loss prob. vs. policy parameter,
$\lambda = 0.85$, $K = B = 4m$ $B_o = 16$, $RTTT > 0$.

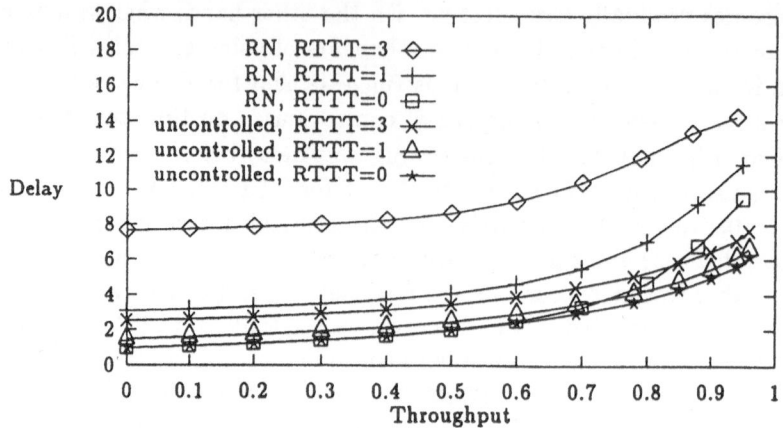

Figure 9: Delay vs. Throughput, $K = B = 3, B_o = 9$.

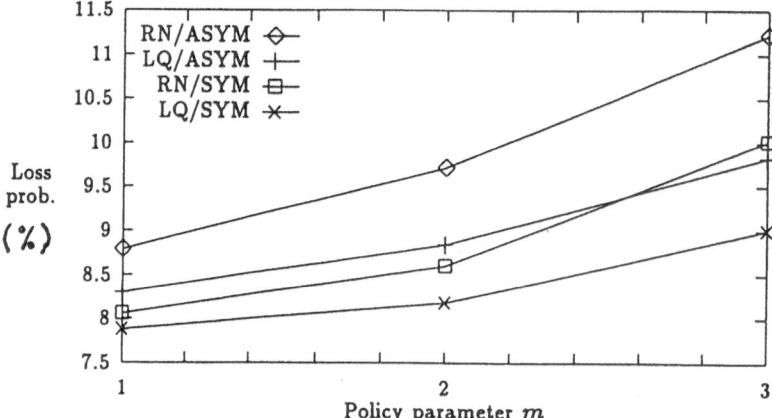

Figure 10: Loss prob. under different imput processes,
$K = B = B_o = 3, \ RTTT = 0.$

obtained by using the analytical models for asymmetric loads under zero propagation delay case. We consider the loss probability under symmetric and asymmetric input processes both with $\lambda = 1$. For asymmetric input process, we chose $\lambda_1 = \lambda_2 = 0.2, \lambda_3 = 0.6$. The results show that LQ strategy is superior to RN strategy under asymmetric input process, as expected intuitively.

5 SUMMARY

This study focused on the problem of controlling a system of LAN-s connected to an outside network by a gateway. For both negligible and non-negligible packet transfer times we considered the question of control policies and strategies that minimize the probability of loosing packets due to their arrival to a full gateway. For the non-negligible transfer time case, an approximate models was provided to study the control parameter setting to optimize system performance. It was shown that for the non-zero case, while longest queue strategy remains the best strategy, the delayed transfer policy is no longer optimal. For arbitrary transfer times, the optimal policy parameter m, one which minimizes packet loss and maximizes throughput, was shown to provide significant performance gains. Furthermore it was shown, that the relative benefits of the proposed policies, become more pronounced as the number of LAN-s and their buffering capacities grow.

REFERENCES

1. L.N. Wong and M. Schwartz, "Flow Control in Metropolitan Area Networks", Proc. INFOCOM'89, 1989.
2. D. Towsley, S. Fdida and H. Santoso, "Design and Evaluation of Flow Control Protocols for Metropolitan Area Networks", Proceedings of NATO Workshop on High Speed Network, Sophia, Antipolis, France, June 1990.
3. W. Bux and D. Grillo, "Flow Control in Local-Area Networks of Interconnected Token Rings", IEEE Transaction on Communications, Vol COM 33, October 1985.
4. A. Ganz and I. Chlamtac, "Finite Buffer Queueing Model for P-persistent CSMA/CD Protocol", Performance '87, Brussels, Belgium, December 1987.
5. A.W. Marshall, I. Olkin, Inequalities: Theory of Majorization and Its Applications, Academic Press, 1979.
6. I. Chlamtac, E.W.M. Wong , " Interconnecting Multiple Input and Output LAN-s over a High Speed Gateway", IEEE International Conference on Communications, ICC, Geneva, Switzerland, May 1993.
7. I. Chlamtac and E. Wong, "Configuration and Control of Gateways", SUPER-COMM/ICC, Chicago, IL, June 1992.
8 . J.J. Barrett and E.F. Wunderlich, "LAN Interconnect Using X.25 Network Services", IEEE Network Magazine, Sept. 1991.

AN ANALYTICAL MODEL FOR ATM BASED NETWORKS WHICH UTILIZE LOOK-AHEAD CONTENTION RESOLUTION SWITCHING

James V. Luciani and C.Y. Roger Chen

Department of Electrical and Computer Engineering
121 Link Hall
Syracuse University
Syracuse, NY 13244-1240

Abstract

Interconnection of multiple media sources will be handled by ATM based networks in B-ISDN systems. These ATM networks act as a backbone for subnetworks (such as LANs, MANs, etc). These subnetworks hold individual media or wholly integrated subsets of the necessary data. This paper will present a new analytical model describing the behavior of ATM based interconnection networks which utilize a scheme called *Look-Ahead Contention Resolution (LCR)* to reduce packet collision and thereby increase throughput. To the knowledge of the authors, this is the only analytical model in the current literature that addresses the *LCR* problem. The switches modeled in this work are output queued and are non-lossy. The analytical model gives a set of closed-form equations which lead to an iterative solution for normalized bandwidth and normalized delay for networks in which the queues are searched to a depth of k ($DOS = k$) where $1 \leq k \leq sizeof(queue)$ in order to find a successful candidate for packet transmission during a network cycle. The model provides results which are very accurate compared to simulation results even under very high loads.

Index Terms— Asynchronous Transfer Mode, Bypass Queue, Communication Networks, Crossbar, Interconnection Network, Look-Ahead Contention Resolution, Packet Switching, Performance Evaluation.

1   Introduction

The demand for multimedia systems and services is becoming an increasingly important driving force in the communication market. Media such as full motion video demand that the underlying communication pathways cause minimal delays to data packets (especially under high load conditions) while maintaining a high overall bandwidth. Local and metropolitan area networks are likely to continue to hold the raw individual media source but this leaves open the question of integration and transport of the media. Emerging standards such as Broadband Integrated Services Digital Network (B-ISDN) are attempting to address these questions.

Asynchronous Transfer Mode (ATM) will be the mechanism used in the implementation of B-ISDN because it is able to efficiently deal with multimedia applications via statistical multiplexing of fixed length packets (also known as cells) [SIGE91]. Nodes on ATM networks communicate via a fabric of multistage interconnection networks (MIN) which permit one or more paths between input nodes and output nodes of the fabric. Communication networks of the kind discussed in this work will be of the form shown in Figure 1. Such networks will be assumed to have a single unique path between an input node and an output node. For a MIN operating as a backbone, these input and output nodes represent connections to local and metropolitan area networks which act as routers for traffic leaving the network in order to be integrated into a multimedia stream. In this work we will not address the integration issues, but rather, we will discuss an interesting model which uses *Look-ahead Contention Resolution* to maximize the transport of data between local area networks that are connected via a MIN operating as a backbone.

When the underlying communication pathways are packet switched (as is the case in ATM switches), a solution to the competing requirements of low delay and high bandwidth can be approached through a technique known as *Look-ahead Contention Resolution (LCR)*. When packet switching is employed, usually FIFO queues exist at either the input or the output of the individual data switches which make up the network. However instead of using a strictly FIFO discipline, when *LCR* is employed, first the head element of a queue is offered for transmission and if that packet is unable to proceed (due to collision with competing packets or due to a full receiving queue), the element of the queue that is directly behind the head element is then offered. This procedure of offering subsequent packets continues until either the entire queue is searched or until some architecturally designated maximum *Depth Of Search (DOS)* is reached. If no packet is found to be acceptable then that queue does not transmit during that cycle.

An example of *LCR* is shown in Figure 2. The larger boxes represent crossbars while the smaller boxes within the larger boxes represent slots in a queue. The figure shows a *DOS=2* search for 2×2 crossbars where the number within the smaller boxes represents a destination address within the receiving crossbar. In Figure 2a the head of the queues of the sending crossbars are both attempting to access the same queue of the receiving crossbar. As a result, only one is chosen (at random) and the other is rejected. In Figure 2b, the rejected queue then offers the next packet within it (i.e., a packet a *DOS=2*) which is subsequently accepted since there are no conflicts.

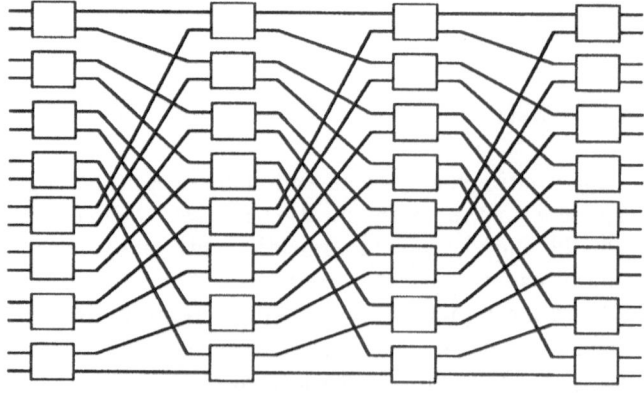

Figure 1: Unique Path Multistage Interconnection Network

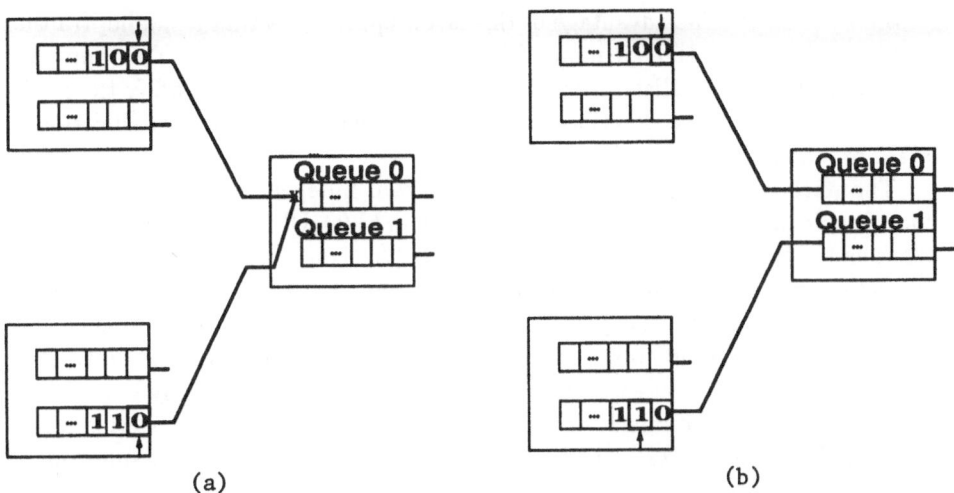

Figure 2: Example of *DOS=2* Search with Collision at *DOS=1*

To the knowledge of the authors, the analytical model presented in the next section is the only existing analytical model that addresses the *LCR* problem in the current literature. The solutions which the analytical model yields are very accurate yet the equations which make up the model are reasonably simple. The equations presented will lead to an iterative solution for *DOS=k* where $1 \leq k \leq$ *sizeof(queue)*. The systems that will be investigated can be used in ATM networks which are finite-output-buffered, packet-switched, multistage interconnection networks utilizing cross-bar switches.

Most work done in the area of output-queued, packet switched networks assumes the use of nonblocking switches [JBM91, KSW84, KSW86, KSW88, KHM87, HK88]. This paper will not make that assumption since such a switch requires that costly, extremely high speed internal buses are used within the individual switches.

The layout of this paper is as follows: Section 1 provides an introduction to the work along with some motivation; Section 2 provides the definition of the analytical model along with description of the model's components; Section 3 provides a comparison between two simulation models and the analytical model in order to verify the accuracy of the analytical model; and Section 4 provides a brief summary of the paper.

2   The Analytical Model

The analytical model presented here contains elements which have their origins in a model devised by Jenq [Jen83]. In this model a packet *can* move forward if it wins contention among packets attempting to go to the same output. Further, the winning packet *will* move forward if one of the following occurs: 1) the receiving queue is not full; 2) the receiving queue is full and a packet from the receiving queue moves forward during the same transfer period (thus leaving an open buffer in which the packet of interest will be received.) This approach was taken by [Jen83, YLL90] in their ground breaking work on the investigation of single and finite buffered MINs respectively. An alternative approach is to neglect the second case mentioned above so that a winning packet *will* move forward only if the receiving queue is not full regardless of whether or not a buffer slot will be vacated at the receiver during the same cycle. This approach was taken by [DB91]. The advantage of this latter approach is that a smaller transfer cycle is necessary even though more "blocking events" may occur. This approach would seem to be more realistic from a design point of view and to modify the model to neglect the second case described above merely requires a small modification to the state probabilities.

The work presented will make the well known Independence Assumption and Uniform Traffic Model Assumption as was done in [YLL90, DB91, DJ81, Jen83, Pat81, KS83]. The Uniform Traffic Model says that packet/cell output node destinations will be generated with a uniform distribution at an input node of the system. The independence assumption says that packet destinations are uniformly distributed at each stage (not just at the input to the interconnection network) and that the destination address is generated anew at each cycle. This will lead, as the previously noted works have clearly stated, to optimistic results. However, it will be shown in Section 3 that when using LCR with $DOS > 1$ the error due to the independence assumption is small.

The following definitions will be used from this point forward:

$p_j(k,t) \equiv$ *the probability that a queue of a crossbar at stage k at time t will have j buffers full*

$S_C(k,t) \equiv \sum_{j=0}^{C} p_j(k,t)$

$r(k,t,d) \equiv$ *the probability that a packet at stage k at time t at depth d of a crossbar* **can** *move forward given that the queue has a packet at depth d*

$rt(k,t,d) \equiv$ *the sum of $r(k,t,l)$ as l goes from 1 to d*

$R(k,t,d) \equiv rt(k,t,d)$ *times the probability that the receiving queue at stage $k+1$* **will** *accept the packet from stage k of the sending queue at a depth less than or equal to d*

$q(k,t,d) \equiv$ *the probability that at time t a packet is ready to come to a buffer of a queue at stage k from a queue at stage k-1 t and that that packet is at depth d of the queue at stage k-1*

$qt(k,t,d) \equiv$ *the sum of $q(k,t,l)$ as l goes from 1 to d*

$\overline{z} \equiv 1.0 - z$ *for some arbitrary probability z*

$a \equiv$ *number of inputs*

$b \equiv$ *number of outputs*

$n \equiv$ *number of stages*

$m \equiv$ *offered load at each input node*

The definition of rt gives the probability that a packet, at a depth d or less, *can* move forward since it wins contention over rival packets destined for the same queue given that such a packet exists at that depth of the sending queue. Thus R is the probability of winning contention and of the receiving queue being willing to take the packet given that there is a packet to send at depth d or less. That is, R is the probability that a packet at depth d or less *will* move forward given that such a packet exists.

The definition of qt describes the probability that a particular queue will be receiving a packet which is at a depth d or less in a sending queue at the previous stage.

A symmetric crossbar structure will be studied (i.e., $a = b$); however, the extension to a non-symmetric structure will usually be somewhat straightforward. The equation for each probability will be presented and the roots of its components will then be clarified. While the work presented here looks at significantly different systems than those studied by [Jen83, YLL90], their work does correspond roughly to the case of $DOS=1$ since their investigation looked at modeling of MINs with a strictly FIFO queueing discipline.

$$q(k,t,1) = \underbrace{\sum_{j=0}^{a} \binom{a}{j} \overline{p_0}(k-1,t)^j p_0(k-1,t)^{a-j}}_{(A)} \underbrace{\left(1 - (1 - \frac{1}{a})^j\right)}_{(B)}$$

$$= 1 - \sum_{j=0}^{a} \binom{a}{j} \left(\overline{p_0}(k-1,t)(1 - \frac{1}{a})\right)^j p_0(k-1,t)^{a-j}$$

$$= 1 - (1 - \frac{\overline{p_0}(k-1,t)}{a})^a \qquad (1)$$

The equation for q can be broken into two parts. Part (A) shows the probability of there being j inputs from the sending queues each of which has at least one packet in it. Part (B) shows the probability that the particular output queue of the receiving crossbar will receive a request from one of those j inputs. Part (B) was derived in a slightly different context in [Pat81] and would contain $\frac{1}{b}$ rather than $\frac{1}{a}$ in the case of an $a \times b$ crossbar.

$$qt(k,t,1) = q(k,t,1) \qquad (2)$$

This is as expected as a result of the definition of qt.

$$r(k,t,1) = \underbrace{\sum_{j=0}^{a-1} \binom{a-1}{j} \overline{p_0}(k,t)^j p_0(k,t)^{a-1-j}}_{(A)} \underbrace{\sum_{l=0}^{j} \binom{j}{l} (\frac{1}{a})^l (1 - \frac{1}{a})^{j-l}}_{(B)} \underbrace{\frac{1}{l+1}}_{(C)}$$

$$= \sum_{j=0}^{a-1} \binom{a-1}{j} \overline{p_0}(k,t)^j p_0(k,t)^{a-1-j} \frac{a}{j+1}(1 - (1 - \frac{1}{a})^{j+1})$$

$$= \frac{1}{\overline{p_0}(k,t)} \underbrace{\sum_{j=0}^{a} \binom{a}{j} \overline{p_0}(k,t)^j p_0(k,t)^{a-j} \left(1 - (1 - \frac{1}{a})^j\right)}_{(D)}$$

$$= \frac{1}{\overline{p_0}(k,t)}(1 - (1 - \frac{\overline{p_0}(k,t)}{a})^a)$$

$$= \frac{q(k+1,t,1)}{\overline{p_0}(k,t)} \qquad (3)$$

For a packet to be able to move forward, it must win contention over all other packets at its depth in competing queues. So for the head of the queue of interest to move forward, given that the queue being looked at is not empty, it must win the contention over up to $a - 1$ other potential candidates. Part (A) describes the probability that there are j additional candidates. Part (B) describes the probability that of those j candidates, l of them are bound for the same receiving queue for which the packet under scrutiny is bound (note that the instances of $\frac{1}{a}$ would be $\frac{1}{b}$ in the non-symmetric case). Part (C) gives the probability that the packet of interest wins contention among the $l+1$ competing packets. Part (D) of this equation is exactly the same result as for q except that this result would be for stage $k+1$ as is shown in the last equality. As it turns out, this relationship between q and r holds any time both have the same parameter values.

$$rt(k,t,1) \;=\; r(k,t,1) \tag{4}$$

This is a direct result of the definition of rt.

$$R(k,t,1) \;=\; rt(k,t,1)(\overline{p_{max}}(k+1,t) + p_{max}(k+1,t)R(k+1,t,DOS)) \tag{5}$$

In the calculation for R at depth 1, R equals the probability of winning contention times the probabilities that either the receiving queue is not full or that if the receiving queue is full then a packet will leave from it at time t as well (this would leave an open slot in the receiving queue for the sender's packet). The subscript max represents the maximum number of buffers in which a packet can reside for the size of the queue being examined.

$$
\begin{aligned}
r(k,t,2) \approx\; & \underbrace{(1-p_0(k,t))(1-R(k,t,1))}_{(A)} \times \\[6pt]
& \underbrace{\sum_{j=0}^{a-1} \binom{a-1}{j} (1-p_0(k,t))^j p_0(k,t)^{a-1-j}}_{(B)} \times \\[6pt]
& \underbrace{\sum_{l=0}^{j} \binom{j}{l} (1-R(k,t,1))^l R(k,t,1)^{j-l}}_{(C)} \underbrace{\left(\frac{a-(j-l)}{a}\right)}_{(D)} \times \\[6pt]
& \underbrace{\sum_{h=0}^{l} \binom{l}{h} (1-p_0(k,t)-p_1(k,t))^h (p_0(k,t)+p_1(k,t))^{l-h}}_{(E)} \times \\[6pt]
& \underbrace{\sum_{i=0}^{h} \binom{h}{i} (\tfrac{1}{a})^i (1-\tfrac{1}{a})^{h-i}}_{(F)} \underbrace{\frac{1}{i+1}}_{(G)} \\[10pt]
=\; & \frac{1}{(1-p_0(k,t)-p_1(k,t))} \times \\[4pt]
& \{1 - [1 - \frac{(1-p_0(k,t))(1-R(k,t,1))(1-p_0(k,t)-p_1(k,t))}{a}]^a - \\[4pt]
& R(k,t,1)(1-p_0(k,t)) \times \\[4pt]
& (1 - [1 - \frac{(1-p_0(k,t))(1-R(k,t,1))(1-p_0(k,t)-p_1(k,t))}{a}]^{a-1})\} \\[10pt]
=\; & \frac{1}{(1-S_1(k,t))} \times \\[4pt]
& \{1 - [1 - \frac{\overline{S_0}(k,t)(\overline{R}(k,t,1))\overline{S_1}(k,t)}{a}]^a - \\[4pt]
& R(k,t,1)(\overline{S_0}(k,t)) \times \\[4pt]
& (1 - [1 - \frac{\overline{S_0}(k,t)(\overline{R}(k,t,1))\overline{S_1}(k,t)}{a}]^{a-1})\}
\end{aligned}
\tag{6}
$$

The equation for r at depth 2 has essentially seven parts to it. We initially choose a sending queue and will compute the contribution to the probability that as result of a packet at $DOS=2$ that a packet can move forward from the queue we chose. Parts (E), (F), and (G) should look very similar to the exposition of r at depth 1 except that in this case we are talking about packets at $DOS=2$. Parts (B) and (C) modify this probability based on how many packets were blocked and how many packets will move from $DOS=1$. Part (D) gives the probability that the packet under scrutiny at $DOS = 2$ is destined for an output that is not currently active as a result of a packet moving forward from $DOS=1$. Since (B) through (G) do not prevent the case where a packet moves from both $DOS=1$ and $DOS=2$, we must take this into account by multiplying by 1-$R(k,t,1)$. Also, since we do not condition part (E) on the existence of packets at $DOS=1$, we must prevent the case where a packet moves from $DOS=2$ when there is no packet at $DOS=1$. These last two points make up part (A) of the equation.

$$
\begin{aligned}
q(k+1,t,2) &= r(k,t,2)(1 - p_0(k,t) - p_1(k,t)) \\
&= r(k,t,2)(S_1(k,t))
\end{aligned} \tag{7}
$$

After a great deal of manipulation, it can be shown that q at a depth of 2 for stage $k+1$ is equal to r at a depth of 2 at stage k times the probability that the sending queue (i.e., the queue at stage k) has an element at the second buffer slot. Thus Equation 7 is the result.

$$
qt(k+1,t,2) = \begin{cases} q(k+1,t,2) + qt(k+1,t,1) & \text{if maximum } DOS \geq 2 \text{ else} \\ qt(k+1,t,1) & \text{otherwise} \end{cases}
$$

$$
rt(k,t,2) = \begin{cases} r(k,t,2) + rt(k,t,1) & \text{if maximum } DOS \geq 2 \text{ else} \\ rt(k,t,1) & \text{otherwise} \end{cases}
$$

The above equations are valid for any depth. The meaning of the maximum DOS in this context is the maximum value to which the architecture being studied will search in order to find a candidate packet for transmission. Note that qt and rt are being defined recursively with respect to the current depth parameter (i.e., the third parameter) so that, for example, rt at depth 2 is defined in terms of rt at depth 1 and r at depth 2.

$$
\begin{aligned}
r(k,t,3) \approx\ & (1 - p_0(k,t) - p_1(k,t))(1 - R(k,t,2)) \times \\
& \sum_{j=0}^{a-1} \binom{a-1}{j} (1 - p_0(k,t) - p_1(k,t))^j (p_0(k,t) + p_1(k,t))^{a-1-j} \times \\
& \sum_{l=0}^{j} \binom{j}{l} (1 - R(k,t,2))^l R(k,t,2)^{j-l} (\frac{a-(j-l)}{a}) \times \\
& \sum_{h=0}^{l} \binom{l}{h} (1 - \sum_{g=0}^{2} p_g(k,t))^h (\sum_{g=0}^{2} p_g(k,t))^{l-h} \times \\
& \sum_{i=0}^{h} \binom{h}{i} (\frac{1}{a})^i (1 - \frac{1}{a})^{h-i} \frac{1}{i+1} \\
=\ & \frac{1}{(1 - \sum_{g=0}^{2} p_g(k,t))} \times \\
& \{1 - [1 - \frac{(1 - \sum_{g=0}^{1} p_g(k,t))(1 - R(k,t,2))(1 - \sum_{g=0}^{2} p_g(k,t))}{a}]^a -
\end{aligned}
$$

$$R(k,t,2)(1 - \sum_{g=0}^{1} P_g(k,t)) \times$$

$$(1 - [1 - \frac{(1 - \sum_{g=0}^{1} P_g(k,t))(1 - R(k,t,2))(1 - \sum_{g=0}^{2} P_g(k,t))}{a}]^{a-1})\}$$

$$= \frac{1}{(1 - S_2(k,t))} \times$$

$$\{1 - [1 - \frac{\overline{S_1}(k,t)(\overline{R}(k,t,2))\overline{S_2}(k,t)}{a}]^a -$$

$$R(k,t,2)(\overline{S_1}(k,t)) \times$$

$$(1 - [1 - \frac{\overline{S_1}(k,t)(\overline{R}(k,t,2))\overline{S_2}(k,t)}{a}]^{a-1})\} \tag{8}$$

The above equation for r at $DOS=3$ should follow reasonably easily from r at $DOS=2$. Further, r can be readily generalized in its open form for an arbitrary value of DOS through the same reasoning given for r at $DOS=2$. However a generalized closed form exists and is given as follows:

$$r(k,t,d) \approx (\frac{1}{1 - S_{d-1}(k,t)}) \times$$

$$\{1 - [1 - \frac{(\overline{S_{d-2}}(k,t))\overline{R}(k,t,d-1)(\overline{S_{d-1}}(k,t))}{a}]^a -$$

$$R(k,t,d-1)(\overline{S_{d-2}}(k,t)) \times$$

$$(1 - [1 - \frac{(\overline{S_{d-2}}(k,t))(\overline{R}(k,t,d-1))(\overline{S_{d-1}}(k,t))}{a}]^{a-1})\} \tag{9}$$

The previous equations which were not shown in their generalized form will now be generalized. Here d represents a current depth in the preparation of the probabilities.

$$q(k+1,t,d) = r(k,t,d)(1 - \sum_{i=0}^{d-1} p_i(k,t)) \tag{10}$$

$$qt(k+1,t,d) = \begin{cases} q(k+1,t,d) + qt(k+1,t,d-1) & \text{if maximum } DOS \geq d \text{ else} \\ qt(k+1,t,d-1) & \text{otherwise} \end{cases}$$

$$rt(k,t,d) = \begin{cases} r(k,t,d) + rt(k,t,d-1) & \text{if maximum } DOS \geq d \text{ else} \\ rt(k,t,d-1) & \text{otherwise} \end{cases}$$

$$R(k,t,d) = rt(k,t,d)(\overline{p_{max}}(k+1,t) + p_{max}(k+1,t)R(k+1,t,DOS)) \tag{11}$$

The following are the state probability equations. They represent the probability that a queue at stage k has a particular number of packets in it. These equations are valid for values of $1 \leq k \leq n$.

$$p_0(k,t+1) = \overline{qt(k,t,DOS)}(p_0(k,t) + p_1(k,t)R(k,t,1)) \tag{12}$$

$$p_1(k,t+1) = qt(k,t,DOS)(p_0(k,t) + p_1(k,t)R(k,t,1)) +$$
$$\overline{qt(k,t,DOS)}(p_1(k,t)\overline{R(k,t,1)} + p_2(k,t)R(k,t,2)) \tag{13}$$

$$p_j(k,t+1) = qt(k,t,DOS)(p_{j-1}(k,t)\overline{R(k,t,j-1)} + p_j(k,t)R(k,t,j)) +$$
$$\overline{qt(k,t,DOS)}(p_j(k,t)\overline{R(k,t,j)} + p_{j+1}(k,t)R(k,t,j+1))$$
$$2 \leq j \leq max - 1 \tag{14}$$

$$p_{max}(k,t+1) = qt(k,t,DOS)(p_{max-1}(k,t)\overline{R(k,t,max-1)} +$$
$$p_{max}(k,t)R(k,t,max)) +$$
$$p_{max}(k,t)\overline{R(k,t,max)} \qquad (15)$$

The subscript max represents the maximum number of buffers in which a packet can reside for the size of the queue being examined. Also, DOS in this context refers to the maximum depth to which the architecture under investigation will search in order to find a candidate for transmission. It becomes clear that q and qt need only be a single value defined from rt at a depth of DOS. They are included primarily to facilitate understanding of the model. Additionally, rt and r have some redundancy associated with them since R is calculated for a given depth from an accumulated value (i.e., rt) of previous depth values of r. This too is a mechanism to facilitate understanding of the model and for ease of computation.

It is also necessary to discuss certain boundary conditions for calculation of R, r, rt, q, and qt. Most analytical work assumes that the system being investigated is input queued. Here, however, the system is output queued which brings about some interesting boundary conditions for the transition probabilities. These boundary conditions are presented now.

$$R(n,t,d) = rt(n,t,d) = r(n,t,d) = 1, \forall d \qquad (16)$$

$$q(1,t,d) = qt(1,t,d) = 1 - (1 - \frac{m}{a})^a, \forall d \qquad (17)$$

$$r(n-1,t,d) \approx (1-p_0(n-1,t)) \times$$
$$\sum_{j=0}^{a-2} \binom{a-2}{j} (\overline{S_{d-2}(n-1,t)})^{j+1}(S_{d-2}(n-1,t))^{a-2-j} \times$$
$$\sum_{l=0}^{j} \binom{j}{l} (1-R(n-1,t,d-1))^{l+1}R(n-1,t,d-1)^{j-l} \times$$
$$(\frac{(a-1)-(j-l)}{a}) \times$$
$$\sum_{h=0}^{l} \binom{l}{h} (\overline{S_{d-1}(n-1,t)})^h(S_{d-1}(n-1,t))^{l-h} \times$$
$$\sum_{i=0}^{h} \binom{h}{i} (\frac{1}{a})^i(1-\frac{1}{a})^{h-i}\frac{1}{i+1}$$
$$= (\frac{1-p_0(n-1,t)}{1-S_{d-1}(n-1,t)}) \times$$
$$\{1-[1-\frac{\overline{S_{d-2}(n-1,t)}(\overline{R(n-1,t,d-1)})\overline{S_{d-1}(n-1,t)}}{a}]^{a-1} -$$
$$R(n-1,t,d-1)(\overline{S_{d-2}(n-1,t)}) \times$$
$$(1-[1-\frac{\overline{S_{d-2}(n-1,t)}(\overline{R(n-1,t,d-1)})\overline{S_{d-1}(n-1,t)}}{a}]^{a-2})\}$$
$$= R(n-1,t,d)$$
$$\qquad (18)$$

The first boundary equation expresses the fact that at the last stage (i.e., stage n), if the queue is not empty then the output node receiving the packet will always accept a packet. The second boundary equation says that q for the first stage is calculated based purely on the probability of request at the input (i.e., m) node.

The last boundary equation takes a bit of explaining. As a result of the first boundary equation there will never be more than a single packet in a buffer of a crossbar at stage n. Thus given the definition of R we see that $R(n-1, t, d) \equiv r(n-1, t, d)$ since the buffer at stage n will never be full. Also from this, we see that it will *never* be the case that *all* packets attempting to access a crossbar at stage n will be blocked since at least one will always move forward; i.e., at least one will win contention and since it cannot be blocked due to an already full receiving queue it will move forward. This then means that the equation for r must be modified to take this into account. This is done by noting that there are now at most $a-2$ potential rivals for the packet under scrutiny at $DOS \geq 2$ and there are at most *a-1* potential outputs to receive it. With these last two points in mind the generalized equation at stage *n-1* follows directly. There is, however, a premultiplier which stipulates that there was a packet at *DOS=1* in the first place for all cases where $DOS \geq 2$. This is necessary in order to take into account what happens if no packet at all is attempting to get to the receiving crossbar.

With the previous equations, the normalized delay and normalized bandwidth can be easily computed.

$$NormalizedBandwidth = \overline{p_0}(n) \tag{19}$$

$$NormalizedDelay = \frac{1}{n} \sum_{j=1}^{n} \frac{1}{R_{ave}(j)} \tag{20}$$

$$R_{ave}(j) = \sum_{l=1}^{max} \frac{p_l(j) R(j, l)}{(l) \overline{p_0}(j)} \tag{21}$$

The equations for normalized bandwidth and normalized delay make use of the steady state values of the system and thus no time reference is necessary. These equations are slight modifications of those used by [Jen83, YLL90, DB91]. Normalized bandwidth is just the probability that the last stage's queue is not empty since if a packet is in the queue, the receiving node will accept it. Note since no more than one packet can arrive in a cycle, there will always be either exactly 1 or exactly 0 packets in the last stage's output queue. The effects of this were discussed earlier.

3 Analytical Model Validation

For this paper, two simulators were created for the purpose of verifying the analytical model. An in-depth simulation study was performed by [Sar91]. Both of the simulators used in this paper make use of the Uniform Traffic model as in [YLL90, DB91, DJ81, Jen83, Pat81, KS83] and one simulator makes use of the Independence assumption. The model that makes the independence assumption will be referred to as the *Independent Model Simulator (IMS)* whereas the other will be referred to as the *Realistic Model Simulator (RMS)*. The Uniform Traffic Model and Independence assumption were defined in Section 2.

Once again, the system being modeled is an ATM network which is packet-switched, output queued, and maintains a single unique path between an input and an output node of the network. The switch mechanisms are crossbars which incorporate *LCR* as described in Section 1.

The system under scrutiny does not have buffers at the input nodes and thus the *LCR* has only a single packet to choose from at the input node of the network (i.e., there is effectively no *LCR* at the first stage).

It will become instructive to define a quantity called the *Effective System Load*. The effective system load is defined to be the probability that a packet will win contention in an attempt to enter the output queue of a crossbar at the first stage (this makes no statement as to whether the receiving buffer is full or not). This quantity can be seen to be described by Patel's equation [Pat81]

$$1 - (1 - \frac{m}{a})^a$$

for the normalized bandwidth for an $a \times a$ crossbar switch. The effective system load is .75 when the applied load m is 1.0. Note that LCR is not involved in this case.

From the previous assertion, it is clear that in a system such as that described here, the maximum normalized throughput is .75 even when LCR is employed (i.e., one cannot get more out of the system, in the long run, than is put in to the system). With this bit of data in mind the discussion will now turn to the results of the comparison.

The number of buffer slots at each queue will be designated by the term *buffers* for the remainder of this paper. So, for example, *buffers=4* means there are 4 buffers in each queue of every crossbar in the MIN.

3.1 Effects on Normalized Bandwidth of LCR with $DOS = 2$

In this subsection, results from the two simulators (i.e., the IMS and RMS models described above) and the analytical model will be compared. The comparisons will be made for the $DOS=2$ case. The simulation has shown the performance of $DOS > 2$ to be both similar in nature and nearly equivalent in actual value to the performance of $DOS=2$ in many cases and as a result there is no loss of generality.

Figure 3 shows a comparison between the three models when the system being modeled has a small number of stages and a small number of buffers. This test is a good indicator of the fidelity of the analytical model because of the small number of buffers. As can be seen, all models examined give essentially the same results.

Figure 4 shows a comparison of the three models with a large number of stages and a small number of buffers. This particular test shows the performance difference due to making the independence assumption. This particular case is quite extreme because it exacerbates the negative effects on performance of the correlation between packets inherent in the RMS. The reason for this is as follows: if there is a blocking event at time *t-1* in the RMS, the probability that there will be a packet bound for

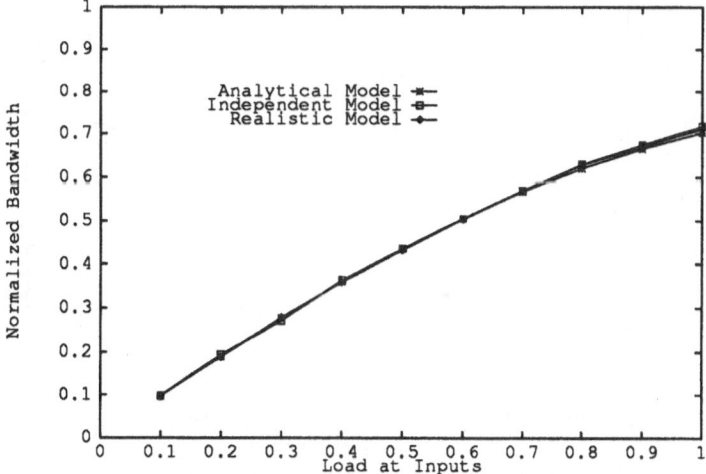

Figure 3: Comparison of Models for $n=2$, $DOS=2$, $buffers=2$ for 2×2 Crossbars

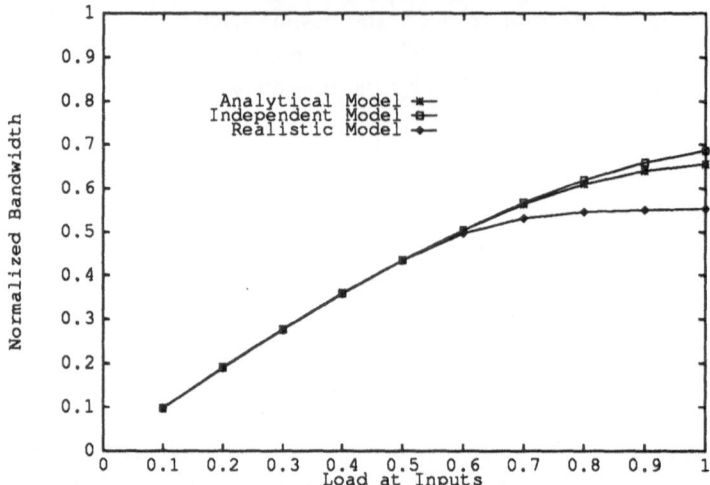

Figure 4: Comparison of Models for *n=8, DOS=2, buffers=2* for 2 × 2 Crossbars

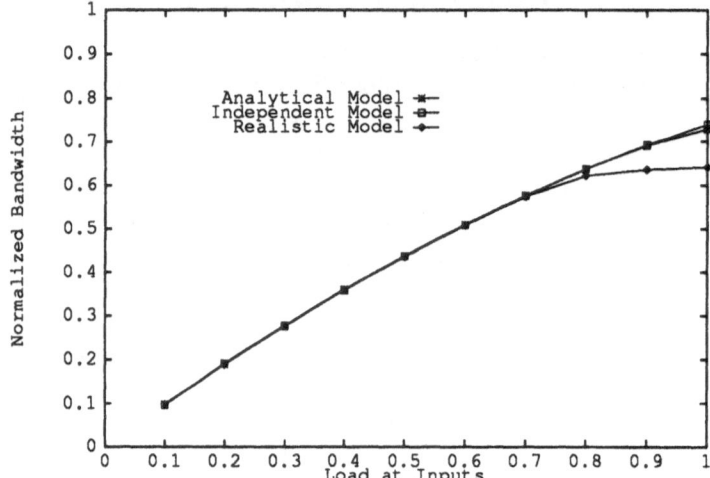

Figure 5: Comparison of Models for *n=8, DOS=2, buffers=4* for 2 × 2 Crossbars

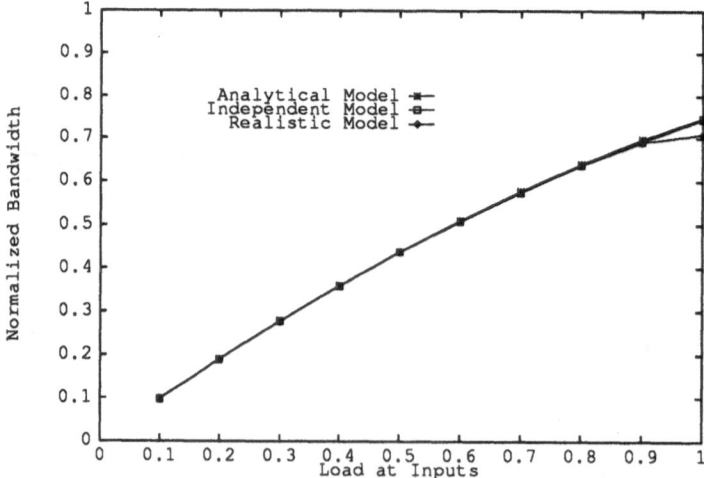

Figure 6: Comparison of Models for *n=8, DOS=2, buffers=8* for 2 × 2 Crossbars

that same destination at time t is 1.0 regardless of the destination of the other sending queues. Thus the probability of blocking is higher in the RMS model. As can be seen from Figure 5, the effects of the correlation are significantly ameliorated for as few as 4 buffers and are almost non-existent for 8 buffers as can be seen from Figure 6. In fact, at 8 buffers, the RMS performed similarly as in Figure 6 for as many stages as was possible to simulate given the output node constraints of the system on which the simulation was run (this was about 14 stages).

Finally, Figure 7 shows that with a small number of stages and middling number of buffers the performance of the three models is indistinguishable. The general rule here seems to be that the performance of the three models is indistinguishable when $buffers > n$.

3.2 Effects on Normalized Delay of LCR width $DOS = 2$

From Figures 8 and 9 we see that the three models are almost indistinguishable when $buffers > n$. From Figure 11 it can be seen that when both n and $buffers$

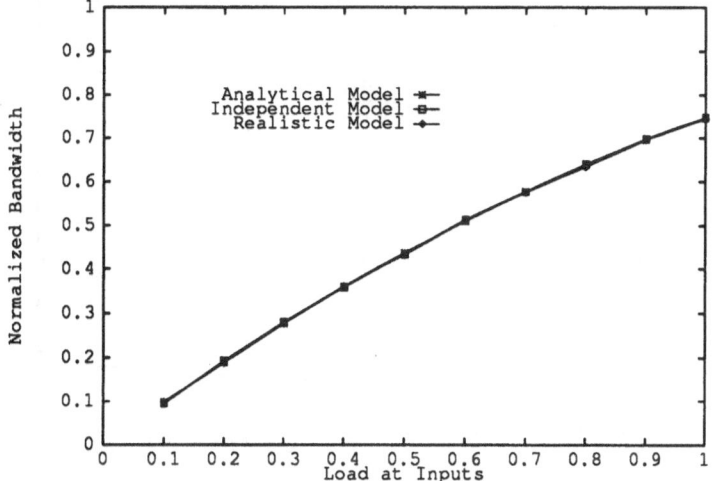

Figure 7: Comparison of Models for $n=2$, $DOS=2$, $buffers=8$ for 2×2 Crossbars

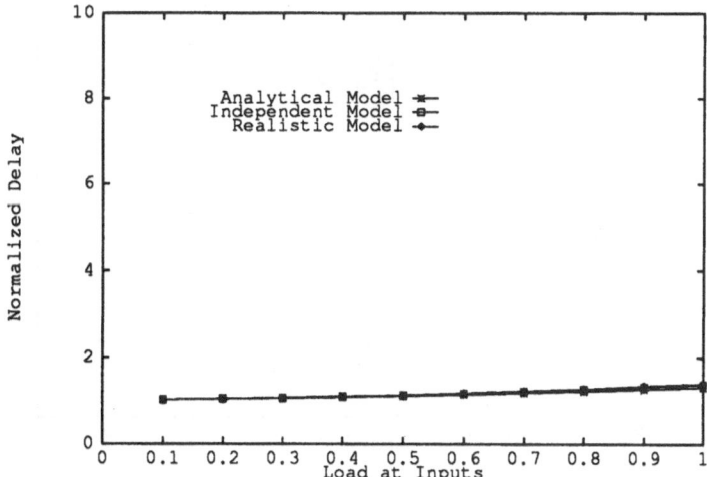

Figure 8: Comparison of Models for $n=2$, $DOS=2$, $buffers=2$ for 2×2 Crossbars

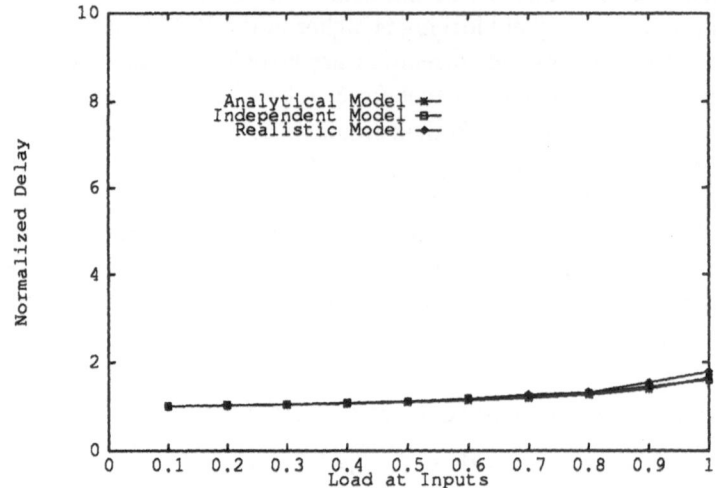

Figure 9: Comparison of Models for *n=2, DOS=2, buffers=8* for 2 × 2 Crossbars

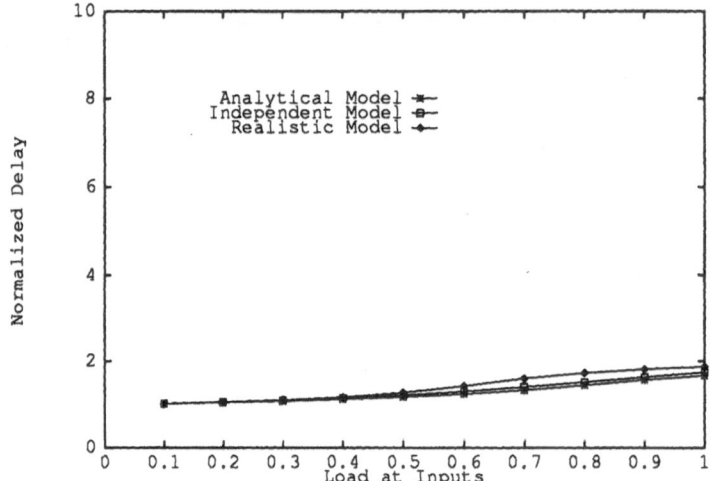

Figure 10: Comparison of Models for *n=8, DOS=2, buffers=2* for 2 × 2 Crossbars

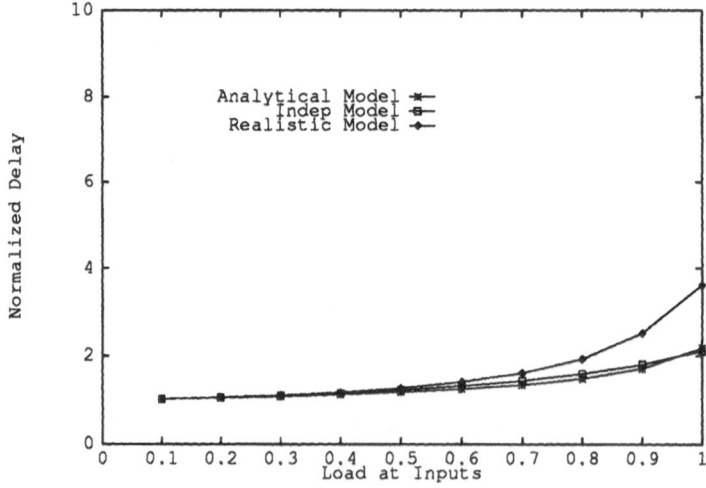

Figure 11: Comparison of Models for *n=8, DOS=2, buffers=8* for 2 × 2 Crossbars

get larger, the RMS tends to separate from the other two models. This is due to the negative effects on performance due to packet correlation within the RMS model (this point is discussed somewhat in the subsection on normalized bandwidth).

The simulation has shown that when *DOS* is large and *buffers* is large, there tends to be clustering of packets bound for the same destination within a single queue which has a significantly negative effect on both normalized delay and normalized bandwidth in the RMS system. Moreover, even if normalized performance metrics (i.e., averaged values) are not severely affected, an individual packet may suffer extreme delays due to this clustering effect in the RMS model. However, the clustering plays a smaller role when *DOS* is small.

4 Summary

In summary, an analytical model for a backbone, crossbar-based MIN utilizing *LCR* has been developed for an arbitrary level of *DOS*. To the best of the knowledge of the authors, this is the only analytical model dealing with the *LCR* problem in the current literature. The model were shown to be quite accurate even when it comes to the performance of a system which does not ignore packet destination correlative effects (i.e., when the independence assumption is used). It is also noted that the error due to the use of the independence assumption becomes much less significant for *DOS > 1* by comparison to the error for *DOS=1*.

References

[DB91]   Jianxun Ding and Laxmi N. Bhuyan, "Performance Evaluation of Multistage Interconnection Networks with Finite Buffers", In *Proceedings of the 1991 International Conference on Parallel Processing*, pages I-592 – I-599, St. Charles, IL, August 1991, CRC Press, Inc.

[DJ81]   Daniel M. Dias and J. Robert Jump, "Analysis and Simulation of Buffered Delta Networks", *IEEE Transactions on Computers*, C-30(4):273–282, April 1981.

[HK88]   Michael G. Hluchyj and Mark J. Karol, "Queueing in High-Performance Packet Switching", *IEEE Journal on Selected Areas in Communications*, 6(9):1587–1597, DEcember 1988.

[JBM91] Hong Jiang, Laxmi N. Bhuyan, and Jogesh K. Muppala, "MVAMIN: Mean Value Analysis Algorithms for Multistage Interconnection Networks", *Journal of Parallel and Distributed Computing*, (12):189–201, December 1991.

[Jen83] Yih-Chyun Jenq, "Performance Analysis of a Packet Switch Based on Single-Buffered Banyan Network", *IEEE Journal on Selected Areas in Communications*, SAC-1(6):1014–1021, December 1983.

[KHM87] Mark J. Karol, Michael G. Hluchyj, and Samuel P. Morgan, "Input Versus Output Queueing on a Space-Division Packet Switch", *IEEE Transactions on Communications*, COM-35(12):1347–1356, December 1987.

[KS83]   Clyde P. Kruskal and Marc Snir, "The Performance of Multistage Interconnection Networks for Multiprocessors", *IEEE Transactions on Computers*, C-32(12):1091–1098, December 1983.

[KSW84] Clyde P. Kruskal, Marc Snir, and Alan Weiss, "On the Distribution of Delays in Buffered Multistage Interconnection Networks for Uniform and Nonuniform Traffic", In *Proceedings of the 1984 International Conference on Parallel Processing*, pages 215–219, Bellaire, MI, August 1984.

[KSW86] Clyde P. Kruskal, Marc Snir, and Alan Weiss, "The Distribution of Waiting Times in Clocked Multistage Interconnection Networks", In *Proceedings of the 1986 International Conference on Parallel Processing*, pages 12–19, University Park, PA, August 1986, IEEE Computer Society Press.

[KSW88] Clyde P. Kruskal, Marc Snir, and Alan Weiss, "The Distribution of Waiting Times in Clocked Multistage Interconnection Networks", *IEEE Transactions on Computers*, C-37(11):1337–1352, November 1988.

[Pat81] Janak H. Patel, "Performance of Processor-Memory Interconnections for Multiprocessors", *IEEE Transactions on Computers*, C-30(10):771–780, October 1981.

[Sar91] Ken W. Sarkies, "The Bypass Queue in Fast Packet Switching", *IEEE Transactions on Communications*, 39(5):766–774, May 1991.

[SIGE91] Yoshito Sakurai, Nobuhiko Ido, Shinobu Gohara, and Noboru Endo, "Large-Scale ATM Multistage Switching Network with Shared Buffer Memory Switches", *IEEE Communications Magazine*, (1):90–96, January 1991.

[YLL90] Hyunsoo Yoon, Kyungsook Y. Lee, and Ming T. Liu, "Performance Analysis of Multibuffered Packet-Switching Networks in Multiprocessor Systems", *IEEE Transactions on Computers*, 39(3):319–327, March 1990.

TRANSIENT ANALYSIS OF NONHOMOGENEOUS CONTINUOUS TIME MARKOV CHAINS DESCRIBING REALISTIC LAN SYSTEMS

A. Rindos, S. Woolet and I. Viniotis*

IBM Corp.
P.O. Box 12195
Research Triangle Park, NC 27709

Dept. of Electrical and Computer Engineering (*)
North Carolina State University
Raleigh, NC 27695-7911

Abstract

Most actual local area network (LAN) systems experience surges in the number of users that vary in magnitude over time. In addition, these surges may often be approximated by a periodic process (with a period of one day or one week). Therefore, a nonhomogeneous continuous time Markov chain (CTMC) may be the best model for such a system, at least for those periods of the day when these surges vary rapidly in magnitude. The transient analysis of time-varying linear systems is highly advanced in the field of systems and control theory. We present a review of some useful results, and then apply them to the analysis of nonhomogeneous CTMCs (especially periodic ones). One of the results of this analysis is a very simple method for transforming any nonhomogeneous CTMC (not just a periodic one) to an equivalent homogeneous CTMC that is then amenable to such homogeneous methods as uniformization. We also present what we feel is a more realistic transient model for the discrete changes in user population that are observed.

Introduction

The transient behavior of a continuous time Markov chain (CTMC) is defined by the well known linear system of first order differential equations:

$$\dot{\pi}(t) = \pi(t)Q(t) \tag{1a}$$

$$\sum_{i=1}^{m} \pi_i(t) = 1 \tag{1b}$$

where $\pi^T(t)$ is a vector, the ith element of which, $\pi_i(t)$ ($i = 1, , \dots, m$), represents the probability of finding the system in state i at time t, while $Q(t)$ ($\in \mathbb{R}^{m \times m}$) represents the infinitesimal generator matrix that is constructed from the rates into and out of each state. We assume throughout that $\pi(t)$ is of finite dimension m. Because the infinitesimal generator is a function of time, eqs.(1a) and (1b) represent the more general case of a nonhomogeneous or time-varying CTMC.

The special case in which $Q(t) = Q$, i.e., in which the infinitesimal generator matrix is time-invariant, has been the focus of most studies on the transient behavior of CTMCs (see, e.g., [1] for a comparison of numerical methods in obtaining a solution; it should be noted that the numerical integration methods examined are just as applicable for time-varying systems). However, several studies of nonhomogeneous systems exist. One of the most significant (and earliest) is that by Kendall [2] in which he derives explicit expressions for the state probabilities (as well as the population mean and variance) as functions of time of a birth-death process in which the birth and death rates are arbitrary functions of time multiplied by the population size (assuming an initial population size of 1). These results are extended in [3] and [4]. An excellent review of these and other results (especially a closed form solution for the transient behavior of a special time-varying process, the Polya process) is presented in [5]. A more comprehensive review and comparison of methods for obtaining the transient behavior of time-varying systems is given in [6]. A different approach, using fluid flow models to obtain transient behavior of mean statistical quantities directly, is given in [7].

Time-varying linear systems have been extensively studied and characterized within the systems/control theory literature (see, e.g., [8], [9], and the many references therein). Such systems often arise during the construction of actual control systems when a nonlinear system is linearized along a desired trajectory. The solution to systems whose dynamics are defined by eq.(1a) is characterized by the *state transition matrix*, a generalization of the well known matrix exponential solution for time-invariant linear systems. Although obtaining an explicit value for a given system using numerical methods is as difficult as that of computing the matrix exponential (see [1], [10]), the general properties of this matrix are well known and can provide enormous insight into the behavior of nonhomogeneous CTMCs.

As argued by Tipper in [7], observations of the behavior of real systems, such as local and wide area networks, lead one to realize that most actual systems exhibit time-varying arrival rates. For example, data for an actual LAN over the course of several days [11] indicated that a surge in the number of job arrivals always occurred in the morning, probably due to a large number of people first arriving at work and turning on their system. Such surges of varying magnitudes occurred at other very characteristic and repeatable times throughout the day, with the level of activity sometimes changing rapidly in a short period of time. It is reasonable to assume that the service time distributions might also change as the nature of jobs presented to the system varies with changes in the user population. Such systems therefore may frequently exhibit nonstationary behavior that should be considered during their design. Standard equilibrium analysis methods may be useful for "worst case" design analyses of such systems. However, such analyses may lead to costly overdesigned systems, and may be of limited use in developing effective dynamic control systems, such as flow control mechanisms. Furthermore, because these fluctuations in the arrival statistics often followed a cyclic pattern that is repeated from one day to the next (or from one week to the next, with weekends possibly breaking the workday pattern), the behavior of many real systems might be reasonably approximated by that of a periodic system. Many results for periodic time-varying systems are available from systems/control theory [8], [9].

In this paper, we will first review some basic properties of the state transition matrix for time-varying linear systems. We will then examine the behavior of periodic nonhomogeneous CTMCs, and describe a well known technique from control theory for periodic systems, the method of Floquet [8], [9], that can simplify the computation of transient probabilities. We will extend this technique to aperiodic systems, providing an exact and computationally simple method for converting any nonhomogeneous Markov chain to an equivalent homogeneous one that can then be solved by standard methods that assume a time-invariant system, such as uniformization. It is computationally much simpler than the approach developed by Van Dijk [12]; this complexity forced the author to suggest a finite-grid approximation in order to solve for the transient behavior. Finally, we will present a simple model that we feel captures the discrete arrival process changes in real systems and that can be readily solved with standard techniques for homogeneous systems.

The State Transition Matrix

As stated previously, the behavior of a nonhomogeneous CTMC can be completely characterized by eqs.(1a) and (1b). If $Q(t)$ is piecewise continuous (or more weakly, integrable), then there will exist a unique solution to eq.(1a) of the form

$$\pi(t) = \pi(t_0)\Phi(t, t_0) \tag{2}$$

$\Phi(t, t_0)$ is the state transition matrix ($\in \mathbb{R}^{m \times m}$) mapping $\pi(t_0)$ (the initial condition) into $\pi(t)$. A general explicit form for $\Phi(t, t_0)$ exists for any integrable $Q(t)$, and is given by the Peano-Baker series

$$\Phi(t, t_0) = I + \int_{t_0}^{t} Q(\tau_1)\, d\tau_1 + \int_{t_0}^{t} Q(\tau_1) \int_{t_0}^{\tau_1} Q(\tau_2)\, d\tau_1\, d\tau_2 + \dots \tag{3}$$

Eq.(3) does not, in general, yield a useable expression for $\Phi(t, t_0)$, but it can be used to determine a number of very useful properties that can simplify the analysis of time-varying systems. For example, the state transition matrix exhibits the *semigroup property*

$$\Phi(t_0, t_1) = \Phi(t_0, t_2)\Phi(t_2, t_1) \tag{4}$$

This, together with the requirement that $\Phi(t, t) = I$, immediately yields the existence of an inverse given by

$$\Phi^{-1}(t_0, t_1) = \Phi(t_1, t_0) \tag{5}$$

Under certain conditions, a simpler expression for $\Phi(t, t_0)$ may be obtained. If $Q(t)$ and $\int_{t_0}^{t} Q(\tau)\, d\tau$ commute (always true for a scalar $Q(t)$, but not true in general), then

$$\Phi(t, t_0) = e^{\int_{t_0}^{t} Q(\tau)\, d\tau} \tag{6}$$

Because a nonhomogeneous CTMC must satisfy both the forward and backward Chapman-Kolmogorov equations, its state transition matrix will always be given by eq.(6) [13]. From eq.(6), we may further observe that if $Q(t) = Q$, a constant, then

$$\Phi(t, t_0) = e^{Q(t - t_0)} \tag{7}$$

The matrix exponential, $e^{Q(t - t_0)}$, is given by the Taylor series expansion of the exponential, i.e.

$$e^{Q(t - t_0)} = \sum_{i=0}^{\infty} \frac{Q^i(t - t_0)^i}{i!} \tag{8}$$

Several methods exist for obtaining the transient behavior for homogeneous CTMCs. For an M/M/1/K system, the ij th element of the matrix exponential is given by [14], [15]

$$\phi_{ij}(t, \lambda, \mu) = \frac{(1 - \rho)\rho^j}{1 - \rho^{K+1}} + \frac{2\rho^{(j-i)/2}}{K+1} \sum_{l=1}^{K} \frac{\alpha_{il}\,\alpha_{jl}\,e^{\beta_l}}{1 - 2\sqrt{\rho}\,\cos\left(\dfrac{l\pi}{K+1}\right) + \rho} \tag{9}$$

where $\rho = \lambda/\mu$ $(\lambda \neq \mu)$ and

$$\alpha_{il} = \sin\left(\frac{il\pi}{K+1}\right) - \rho^{1/2} \sin\left(\frac{(i+1)l\pi}{K+1}\right)$$

$$\beta_l = -(\lambda + \mu)t + 2t\sqrt{\lambda\mu}\,\cos\left(\frac{l\pi}{K+1}\right)$$

λ is the arrival rate and μ is the service rate. A similar complex expression exists for $\lambda = \mu$. Such an expression can be obtained because of the special tridiagonal structure of Q for the M/M/1/K system. The matrix eigenvalues and eigenvectors may be explicitly defined in terms of λ and μ, so that the matrix can be diagonalized and e^{Qt} easily obtained.

Many techniques exist for obtaining the matrix exponential for more general homogeneous systems (see, e.g., [10]). One method that is often used to obtain $\pi(t)$ is uniformization [1]. The original CTMC is reduced to a discrete time Markov chain (DTMC) subordinated to a Poisson process, and solved by reexpressing $\pi(t)$ as an infinite series that is readily truncated (depending upon the desired accuracy), i.e.

$$\pi(t) = \sum_{i=0}^{\infty} \Omega(i) e^{-qt} \frac{(qt)^i}{i!} \tag{10a}$$

where $q \geq \max_j |-q_{ii}|$ is the largest magnitude diagonal element of the matrix Q and

$$Q^* = \frac{1}{q} Q + I \tag{10b}$$

$$\Omega(i) = \Omega(i - 1)Q^* \tag{10c}$$

$$\Omega(0) = \pi(0) \tag{10d}$$

Numerical techniques for integration of the original first order differential equation can also be used [1], [6], [7]. These methods have the advantage of being equally applicable to nonhomogeneous systems.

Periodic Systems

As can be seen above, obtaining $\pi(t)$ for a homogeneous system can be extremely computationally intensive. The transient behavior of a nonhomogeneous CTMC is typically even more difficult to obtain. However, some simplification is possible for a periodic nonhomogeneous system.

Given that $Q(t) = Q(t + T)$, $\forall t$, where T is the period, we may immediately show that

$$\int_{t_0}^{t} Q(\tau_1)\, d\tau_1 = \int_{t_0 + T}^{t + T} Q(\tau_1)\, d\tau_1$$

which from eq.(6) yields

$$\Phi(t, t_0) = \Phi(t + T, t_0 + T) \qquad \forall t, t_0 \tag{11}$$

(Note that eq.(11) is in fact valid for any periodic time-varying system, not just for one satisfying eq.(6).) The following useful relationship may be obtained from eq.(11) and the semigroup property:

$$\Phi(kT, 0) = \Phi(T, 0)\Phi(2T, T) \dots \Phi(kT, (k-1)T) = \{\Phi(T, 0)\}^k \tag{12}$$

However, note that none of the results above implies that $\pi(t) = \pi(t + T)$, $\forall t$, i.e., that the state probabilities are periodic with period T. This would require that $\Phi(t + T, t_0) = \Phi(t, t_0)$, which does not, in general, follow from the periodic nature of $Q(t)$.

We may, however, present a necessary and sufficient condition for $\pi(t)$ to be periodic with period T, given that $Q(t)$ is periodic with period T. We will first prove an intermediate result that we will require in determining these conditions.

Lemma 1: $\pi(t) = \pi(t + T)$, $\forall t$ if and only if there exists a t^* such that $\pi(t^*) = \pi(t^* + T)$

Trivially, if $\pi(t) = \pi(t + T)$, $\forall t$, then we may immediately substitute a specific t^* for t above, thereby demonstrating the existence of such a time. To prove the converse, we may immediately obtain from eq.(2) for an arbitrary t

$$\pi(t) = \pi(t^*)\Phi(t, t^*) \tag{13a}$$

$$\pi(t + T) = \pi(t^* + T)\Phi(t + T, t^* + T) \tag{13b}$$

Given $\pi(t^*) = \pi(t^* + T)$, together with eq.(11), we may immediately show that eqs.(13a) and (13b) are equal for an arbitrary t, i.e., $\pi(t) = \pi(t + T)$, $\forall t$.

We may now state the necessary and sufficient conditions for $\pi(t)$ to be periodic with period T (given a periodic $Q(t)$).

Theorem 1: $\pi(t) = \pi(t + T)$, $\forall t$, if and only if $\Phi(T, 0)$ has an eigenvalue of 1 and $\pi(0)$, the initial state probability vector, is the corresponding (normalized) left eigenvector.

First, if $\Phi(T,0)$ has an eigenvalue of 1 and $\pi(0)$ is the corresponding left eigenvalue, then $\pi(T) = \pi(0)\Phi(T,0) = \pi(0)$. This result, combined with Lemma 1, immediately implies that $\pi(t) = \pi(t + T)$, $\forall t$. To prove the converse, we are given $\pi(t) = \pi(t + T)$, $\forall t$, which immediately implies $\pi(T) = \pi(0)$. If this latter relationship is true, then there must exist a unique state transition matrix $\Phi(T,0)$ satisfying the relationship $\pi(0) = \pi(0)\Phi(T,0)$. Trivially, this will be true if $\pi(0) = 0$, which is impossible for a system satisfying eq.(1b). More generally, $\pi(0) = \pi(0)\Phi(T,0)$ if and only if $\Phi(T,0)$ has an eigenvalue of 1 and $\pi(0)$ is the corresponding left eigenvector (which includes the case where $\Phi(T, 0) = I$).

Because the columns of $Q(t)$ must always sum to zero, $Q(t)$ has an eigenvalue of 0 $\forall t$, $0 \leq t \leq T$. Therefore, $\Phi(T, 0)$ has an eigenvalue of 1 $\forall t$, $0 \leq t \leq T$, and it will always be possible for our state probabilities to be periodic with period T, given the appropriate initial condition stated above. However, Lemma 1 indicates that this is the only way possible for the state probabilities to be periodic. The system cannot move from being aperiodic to periodic with period T if $Q(t)$ follows its original periodic trajectory.

However, the state probabilities can converge in the limit as $t \to \infty$ to a periodic cycle with period T, as, for example, recently demonstrated in [16]. From the state transition matrix over a single period, $\Phi(T,0)$, we can determine whether or not the system will converge for an arbitrary initial condition, and if so, at what rate. First, if $\|\pi((k + 1)T) - \pi(kT)\|$ approaches zero as k approaches infinity, then $\|\pi((k + 1)T + \delta) - \pi(kT + \delta)\|$ also approaches zero at a proportional rate since

$$\|\pi((k + 1)T + \delta) - \pi(kT + \delta)\| \leq \|\pi((k + 1)T) - \pi(kT)\| \|\Phi(\delta, 0)\|$$
$$\leq \|\pi((k + 1)T) - \pi(kT)\| e^{\|Q^*\|T} \quad (14)$$

where Q^* is the value $Q(t)$ having the maximum norm over the interval $(T,0)$. (We assume $Q(t)$ is bounded over a period.) Therefore, the state probabilities will converge to a periodic limit cycle if $\pi(kT)$ approaches a limit as k approaches infinity. The dynamics of $\pi(kT)$ are given by the discrete time system

$$\pi(kT) = \pi(0)\Phi(kT,0) = \pi(0)\{\Phi(T,0)\}^k \quad (15a)$$

A well known result for discrete time systems from systems/control theory is that the above system will converge to an equilibrium value if the eigenvalues of $\Phi(T,0)$ are on or within the unit circle in the complex plane (with at most one eigenvalue on the unit circle). Divergence will occur if this condition is not met.

Any irreducible and aperiodic homogeneous discrete-time Markov chain (DTMC; please note that we are not considering our original CTMC here) will converge to an equilibrium distribution [13]. Therefore, for most realistic systems, even unstable ones, the state probabilities of our original CTMC with periodic $Q(t)$ will converge to a periodic limit cycle with period T, because $\pi(kT)$ will approach the left eigenvector corresponding to the unity eigenvalue of $\Phi(T,0)$. The rate of convergence is easily obtained. Rewriting $\pi(0)$ using the left normalized eigenvectors of $\Phi(T,0)$, $\{q_i\}$, as a basis, and then substituting this into eq.(15a), yields

$$\pi(kT) = \pi(0)\Phi(kT,0) = \left\{\sum_{i=1}^{m} \alpha_i q_i\right\} \{\Phi(T,0)\}^k = \sum_{i=1}^{m} \alpha_i \lambda_i^k q_i = \sum_{i=1}^{m} \alpha_i e^{\lambda_i^* k} q_i$$
$$(15b)$$

where the set $\{\alpha_i\}$ defines the coordinates of $\pi(0)$ for the basis $\{q_i\}$, and λ_i is the eigenvalue corresponding to the left eigenvector q_i, of $\Phi(T,0)$. Given that $\Phi(T,0)$ is of the form given by eq.(6), we may diagonalize the state transition matrix to obtain the final form of eq.(15b), which expresses $\pi(t)$ in terms of the eigenvalues, λ_i^*, of the matrix $\int_0^T Q(\tau)\,d\tau$. Note that $\lambda_i = \exp\{\lambda_i^*\}$ and that $\int_0^T Q(\tau)\,d\tau$ and $\Phi(T,0)$ share the same eigenvectors (since the latter can be expanded to a polynomial function of the former). Therefore, if convergence occurs, it will be exponentially fast to the left eigenvector corresponding to $\lambda_i = 1$ (or equivalently, $\lambda_i^* = 0$). Convergence implies $|\lambda_i| < 1$ (or $\lambda_i^* < 0$) for all other eigenvalues, so that the contribution to $\pi(t)$ due to all other eigenvectors will disappear as $t \to \infty$. Eq.(15b) also explains Theorem 1. If $\pi(0)$ does not equal the left eigenvector corresponding to $\lambda_i = 1$ (or $\lambda_i^* = 0$), then $\pi(t)$ cannot achieve this equilibrium value in finite time (though it can come arbitrarily close).

Should the conditions defined by Theorem 1 be met, then we need only to characterize the state probability trajectory over the interval $t = (0, T)$. More generally, $\pi(kT + \delta)$ (k an integer, $0 < \delta < T$) is given by

$$\pi(kT + \delta) = \pi(0)\{\Phi(T, 0)\}^k \Phi(\delta, 0)$$
$$= \pi(0)e^{k\int_0^T Q(\tau)\,d\tau + \int_0^\delta Q(\tau)\,d\tau} \tag{16}$$

Eq.(16) can be easily obtained from eq.(6) using the fact that $Q(t)$ is periodic, or by application of eqs.(4), (11) and (12) to the more general system. The system can always be solved by determining the definite integrals and then computing the matrix exponential by standard methods [10], or by bypassing the explicit matrix exponential solution and numerically integrating eq.(1a) directly [1], [6]. The form of the solution given by eq.(16) is not amenable to techniques that assume a nonhomogeneous CTMC, such as uniformization [1]. However, from systems/control theory, the method of Floquet [8], [9] allows us to modify a periodic system so as to be able to take advantage of these techniques. This is due to the fact that we may always transform our original periodic dynamic system given by eq.(1a) to a time-invariant one described by

$$\dot{\xi}(t) = \xi(t)R \tag{17}$$

if $Q(t)$ is periodic. This is accomplished by the linear transformation

$$\xi(t) = \pi(t)P(t) \tag{18}$$

The solution to eq.(17) is then simply given by

$$\xi(t) = \xi(0)e^{Rt} \tag{19}$$

R is given as a solution to the following equation:

$$e^{RT} = \Phi(T, 0) \tag{20}$$

For our system, $\Phi(T, 0)$ is given by eq.(6), so that R is then given by

$$R = \frac{1}{T}\int_0^T Q(\tau)\,d\tau \tag{21}$$

Note that R represents the time average of $Q(t)$ over a single period for our original system. (For more general systems, R is given by $1/T$ times the matrix logarithm of $\Phi(T,0)$.) The transformation matrix, $P(t)$, is given by

$$P(t) = \Phi(0, t) e^{Rt} \qquad (22a)$$

where $\Phi(0, t) = \{\Phi(t, 0)\}^{-1}$. As we will later show, $P(t)$ does not need to be explicitly determined when we apply the method of Floquet in our context, thereby avoiding the computation of an additional matrix exponential. Two very important properties of $P(t)$ are that $P^{-1}(t)$ always exists and that both $P(t)$ and $P^{-1}(t)$ are periodic with period T. The periodic nature of $P^{-1}(t)$ is what allows us to obtain a form that is amenable to homogeneous CTMC solution methods. The state transition matrix of our original system is then given by

$$\Phi(t, 0) = P(0)e^{Rt}P^{-1}(t) \qquad (23)$$

Using the relationship $\Phi(t, t) = I$, from eq.(22a) we obtain

$$P(0) = \Phi(0, 0) e^{Rx0} = I \qquad (22b)$$

From eq.(18), we then obtain

$$\xi(0) = \pi(0)P(0) = \pi(0) \qquad (18a)$$

Since $P^{-1}(t)$ always exists, from eq.(22a), it must be given by

$$P^{-1}(t) = e^{-Rt}\Phi(t, 0) \qquad (24)$$

From the periodic nature of $P^{-1}(t)$, we obtain $P^{-1}(kT + \delta) = P^{-1}(\delta)$. Multiplying both sides of eq.(18) by $P^{-1}(t)$ yields

$$\begin{aligned}
\pi(kT + \delta) &= \xi(kT + \delta)P^{-1}(kT + \delta) = \xi(kT + \delta)P^{-1}(kT + \delta) \\
&= \xi(0)e^{R(kT + \delta)}P^{-1}(kT + \delta) \\
&= \pi(0)e^{R(kT +\delta)}e^{-R\delta}\Phi(\delta, 0) \\
&= \pi(0)e^{RkT}\Phi(\delta, 0) \\
&= \xi(kT)\Phi(\delta, 0)
\end{aligned} \qquad (25)$$

After appropriate substitution, eq.(25) of course yields eq.(16), with the important difference that $\xi(kT)$ is defined by an equivalent time-invariant system. For $\delta \neq 0$, we again have the original problem of determining $\Phi(\delta, 0)$. However, we may immediately use R in a homogeneous CTMC solution method, such as uniformization, to compute the transient probabilities for any $t = kT$, i.e.

$$\xi(kT) = \sum_{i=0}^{\infty}\Omega(i)e^{-rkT}\frac{(rkT)^i}{i!} \qquad (26a)$$

where $r \geq \max_i |-r_{ii}|$ is the largest magnitude diagonal element of the matrix R and

$$R^* = \frac{1}{r}R + I \qquad (26b)$$

116

$$\Omega(i) = \Omega(i - 1)R^{\wedge} \tag{26c}$$

$$\Omega(0) = \pi(0) \tag{26d}$$

A large number of time points could be computed simultaneously by computing more than one term of the sum on the left hand side of eq.(26a), for different values of k. Note that eqs.(26b) through (26d) do not depend upon k. Such a parallelized uniformization method may be more efficient for computing transient probabilities for a large number of different values of k than other traditional methods for computing the matrix exponential. These methods require the determination of successive powers of e^{RT} (taking advantage of eq.(12)). For large systems, or for a large number of values for k, this might practically require determining the eigenvalues and eigenvectors of R so that it could be diagonalized, a computationally intensive task. However, the results could also be used to directly compute the matrix exponential. Numerical integration of the original differential equation to obtain these points would require determining a (possibly large) number of intermediate values for t between kT and $(k + 1)T$ (depending upon the required step size). Therefore, a parallelized uniformization method may be very attractive.

Such a parallelized uniformization algorithm can also be used to quickly determine whether the state probabilities converge to a limiting periodic cycle (though a model of a real system probably always will), and if so, to also estimate the rate of convergence. This can be accomplished by examining $\pi(kT)$ for a number of values for k. Given that the realistic systems we are examining have periods on the order of a day or a week, a quick estimate of the convergence rate is very useful.

We will now demonstrate how the results we have developed above may be used to obtain $\pi(t)$ for aperiodic systems, or for $t = kT + \delta$, $0 < \delta < T$, for periodic systems.

Generalization for Arbitrary Time Points and/or Aperiodic Systems

From eq.(25), we have shown that $\pi(kT) = \xi(kT)$, i.e., at time points that are integer multiples of the period T, the values of the transient probabilities before and after the linear transformation are the same. It is very easy to show that this is not true in general for an arbitrary time point $t = kT + \delta$, $0 < \delta < T$. From eqs.(2), (6) and (12), we have

$$\pi(kT + \delta) = \pi(kT)\Phi(kT + \delta, kT)$$
$$= \pi(kT)\Phi(\delta, 0) = \pi(kT) \exp\{\int_0^\delta Q(\tau)\, d\tau\} \tag{27a}$$

From eqs. (2), (7), (21) and (25), we have

$$\xi(kT + \delta) = \xi(kT)e^{R\delta} = \pi(kT) \exp\{\frac{\delta}{T}\int_0^T Q(\tau)\, d\tau\} \tag{27b}$$

Since the above exponents are not, in general, equal for an arbitrary $Q(\tau)$, $\pi(kT + \delta)$ is not, in general, equal to $\xi(kT + \delta)$.

However, assume we are given two different time-varying linear systems defining the probabilistic trajectories $\pi(t)$ and $\pi'(t)$ whose dynamics are of the same form as eqs.(1a) and (1b), with the respective infinitesimal generator matrices $Q(t)$ and $Q'(t)$. Let $Q(t)$ be periodic with period T and let $Q'(t)$ be aperiodic. If both trajectories pass through the same point at time t_0, i.e., $\pi(t_0) = \pi'(t_0)$, and $Q(t) = Q'(t)$, $\forall t$, $t_0 \leq t \leq T'$ (where $T' = kT + \delta$), then it is very easy to show that $\pi(t) = \pi'(t)$, $\forall t$, $t_0 \leq t \leq T'$.

Behavior prior to t_0 is irrelevant, since any behavior after t_0 is entirely defined by the states at t_0 (i.e., $\pi(t_0)$ and $\pi'(t_0)$) and the trajectories of the infinitesimal generator matrices (i.e., $Q(t)$ and $Q'(t)$) for $t > t_0$. Since our dynamic system is causal, future behavior occurring after time T' in no way affects behavior prior to time T'. The periodic and aperiodic systems above are therefore indistinguishable over the interval (t_0, T').

Therefore, given an aperiodic time-varying system for which we wish to compute $\pi(t)$ for an arbitrary time point $t = T'$, we may simply pretend that the system is periodic with period T', and apply the method of Floquet as before. The probability at any arbitrary time point $T' = kT + \delta$ for a periodic system with period T can be computed in a similar manner. In either case, let

$$R = \frac{1}{T'} \int_0^{T'} Q(\tau) \, d\tau \tag{21a}$$

This matrix R may then be used as the infinitesimal generator matrix in any method that assumes a time-invariant system, e.g., uniformization (substitute R as defined above in eqs.(26a) through (26d), ignoring the restriction that t must equal kT), eq.(9) for an M/M/1/K system (substitute the corresponding time averages for λ, μ and ρ), etc.

$\pi(T')$ defined by this time-invariant equivalent system is not an approximation, although standard numerical methods presently available to compute $\pi(T')$ using R can introduce round-off errors, truncation errors (as seen in uniformization), etc. It is easy to show that $\xi(T')$ and $\pi(T')$ are equal, given R as defined above, i.e.

$$\xi(T') = \xi(0)e^{RT'} = \pi(0) \exp\{[\frac{1}{T'} \int_0^{T'} Q(\tau) \, d\tau]T'\}$$

$$= \pi(0)e^{\int_0^{T'} Q(\tau) \, d\tau} = \pi(T') \tag{28}$$

However, we cannot (in general) parallelize this algorithm as before since R must be recomputed for each new time point. As demonstrated previously, $\xi(t) = \pi(t)$ only when t is an integer multiple of the period T' (i.e., $t = kT'$, k an integer). Furthermore, when used in this manner, this equality holds only for $k = 1$, since beyond T', $Q(t)$ is obviously not periodic with period T'.

We will now present a model that we feel captures the actual discrete changes in user population on a LAN. This approach will allow us to convert our original (periodic or aperiodic) nonhomogeneous CTMC to a series of homogeneous CTMCs. With the method above, we may also convert it to a single equivalent homogeneous CTMC for a given time point of interest.

The Model

As previously mentioned, observations of the behavior of actual LAN systems [11] suggest that the number of request sources, and therefore the traffic

intensity, varies in a step-wise manner over the course of a day in a predefined (presumably measurable) pattern. Let $\lambda(t)$ be the overall mean arrival
rate to a given server in our network. This rate changes at (possibly fixed)
known time instances, t_1, t_2, etc.; i.e.,

$$\lambda(t) = \lambda_k \qquad t_{n-1} \leq t < t_n \tag{29}$$

Therefore, we believe our system may be modelled as a homogeneous CTMC
between rate transitions, with $\lambda(t)$ representing the mean arrival rate of a
Poisson process. (McGough et al. [17] consider a system in which the reliability statistics of a flight control system are defined by a CTMC that changes,
depending upon the "mission phase," i.e., take-off, landing, etc.. However, only
mean statistical quantities were approximated.) Numerous CTMC approximate
models of real systems have been constructed using Coxian or phase-type distributions for a variety of more general service time distributions, using special
representations for various arbitration policies, etc. (see, as examples, [18],
[19]). Our arrival processes can be viewed as the result of the merging of a
time-varying number of independent Poisson streams. Each Poisson stream
represents the service requests from an individual user, with an exponentially
distributed think time between requests. Each user could have his own individual request rate (e.g., an expert typist might have a different request rate
from that of a one-finger hunt-and-peck typist). The number of users/requesters
might vary in a precise manner over the course of the day (or week), that
repeated itself each day (or week) and could be easily measured. The model
described above might therefore accurately capture the behavior of many real
systems. The overall arrival rate to a given server, $\lambda(t) \in \mathbb{R}$, can be represented
in the following manner

$$\lambda(t) = \alpha(t)\tilde{\lambda}^T = \sum_{i=1}^{N} \alpha_i(t)\tilde{\lambda}_i \tag{30}$$

where

$$\alpha(t) = [\alpha_1(t), \ldots, \alpha_N(t)]$$

$$\tilde{\lambda} = [\tilde{\lambda}_1, \ldots, \tilde{\lambda}_N]$$

The set $\{\tilde{\lambda}_i\}$ represents the request rates of different populations of users, while
the set $\{\alpha_i(t)\}$ represents the actual number of those user types as a function of
time. Note that at certain times, it may be the case that $\alpha_i(t) = 0$ for a given i.

If state dependent transition rates are required, then $\lambda(t)$ can also be represented as a vector, with each element (corresponding to the transition rate
between a pair of given states) given by eq.(30) using a different value of $\alpha(t)$
for each state. The service rates may also be time- and state-dependent in the
manner described above, i.e., $\mu(t) = \beta(t)\tilde{\mu}^T$.

The infinitesimal generator matrix of our CTMC, $Q(t)$ ($\in \mathbb{R}^{m \times m}$ given m states),
is therefore time-varying, with

$$Q(t) = Q_n \qquad t_{n-1} \leq t < t_n \tag{31}$$

Because our system may be represented by a homogeneous (time-invariant)
CTMC between transitions in arrival rate, we may immediately obtain a simple

explicit expression for the transient behavior of its state probabilities

$$\pi(t) = \begin{cases} \pi(t_0)e^{Q_1(t - t_0)} & t_0 \leq t < t_1 \\ \pi(t_1)e^{Q_2(t - t_1)} & t_1 \leq t < t_2 \\ \pi(t_2)e^{Q_3(t - t_2)} & t_2 \leq t < t_3 \\ \vdots \end{cases}$$

where

$$\pi(t_1) = \pi(t_0)e^{Q_1(t_1 - t_0)}$$
$$\pi(t_2) = \pi(t_1)e^{Q_2(t_2 - t_1)}$$
$$\vdots$$

We may simplify the above equations to

$$\pi(t) = \pi(t_0)e^{Q_1(t_1 - t_0)}e^{Q_2(t_2 - t_1)} \cdots e^{Q_k(t - t_{n-1})} \qquad t_{n-1} \leq t < t_n \qquad (32)$$

We are able to obtain such a simple form because a discontinuity cannot occur for our state variable, $\pi(t)$, at time t_n (i.e., $\pi(t)$ will be the same immediately before and immediately after the change in arrival rate at time t_n). Therefore, the final state occurring just prior to the transition at t_n (for the CTMC defined by the infinitesimal generator matrix Q_n) will serve as the initial state of the dynamic system created immediately after the transition (for the CTMC defined by Q_{n+1}). We do not need to be concerned with the distribution of the time between arrivals when a change in arrival rate occurs (as is the case for an MMPP/G/1 system [20]), because the transition times are fixed and known, and $\pi(t)$ is continuous in time. Eq.(32) is a direct result of the semi-group property.

Any known method for obtaining the individual matrix exponentials could be used. If Q_n defines a simple M/M/1/K system, then given $\pi(t_{n-1})$, we have the following exact formula

$$\pi_j(t) = \sum_{i=0}^{K} \phi_{ij}(t - t_{n-1}, \lambda_n, \mu_n)\pi_i(t_{n-1}) \qquad t_{n-1} \leq t < t_n \qquad (33)$$

where $\pi_i(t)$ represents the ith element of $\pi(t)$ ($\in \mathbb{R}^{1 \times (K+1)}$), and $\phi_{ij}(t, \lambda, \mu)$ is defined by eq.(9). If every Q_n defines an M/M/1/K system, then the transient behavior may be easily and exactly computed by applying eq.(33) recursively.

For a more general Q_n, we could, for example, apply uniformization to recursively solve our time varying system as follows (for $t_{n-1} \leq t < t_n$)

$$\pi(t) = \sum_{i=0}^{\infty} \Omega(i)e^{-q_n(t - t_{n-1})} \frac{q_n^i(t - t_{n-1})^i}{i!} \qquad (34a)$$

$$Q^* = \frac{1}{q_n}Q_n + I \qquad (34b)$$

$$\Omega(i) = \Omega(i - 1)Q^* \qquad (34c)$$

$$\Omega(0) = \pi(t_{n-1}) \qquad\qquad (34d)$$

q_n is once again the largest magnitude diagonal element of the matrix Q_n. Therefore, for $t_{n-1} \leq t < t_n$, n uniformization problems must be solved. This is basically the same result van Dijk obtains in [12], in which he applies the semi-group property of a nonhomogeneous Markov chain, though the above represents a much more computationally tractable form.

By applying our generalized method of Floquet for an arbitrary time point and arbitrary (possibly aperiodic) CTMC, we may avoid computing the intermediate values of $\pi(t_i)$, $0 < i < n - 1$, and obtain $\pi(t)$ (for any $t_{n-1} \leq t < t$) by solving only a single uniformization problem. Let $\pi(0)$ be the initial condition and let

$$
\begin{aligned}
R &= \frac{1}{t} \int_0^t Q(t)\, dt & &\text{in general} \\
&= \sum_{i=1}^{n-1} (t_i - t_{i-1})Q_i + (t - t_{n-1})Q_n & &\text{in the model above}
\end{aligned}
\qquad (21b)
$$

$\pi(t)$ may then be obtained using R obtained above in eqs.(26a) through (26d), substituting t for kT in eq.(26a).

If $Q(t)$ is periodic, then we may once again obtain $\pi(kT + \delta)$ by solving a single uniformization problem using our generalized method of Floquet. We could also obtain $\pi(kT + \delta)$ by first computing $\xi(kT)$ via uniformization, to then be used as the initial condition to obtain $\pi(\delta)$ as described above), $n + 1$ uniformization problems need to be solved, where n is the number of arrival rate transitions within the interval $(0, \delta)$ Without applying Floquet's method, $n + kN$ uniformization problems would need to be solved, where N is the total number of arrival rate transitions within the interval $(0, T)$.

Numerical Example

For our example, we have chosen an M/M/1/5 first-come first-serve queue with arrival rate $\lambda(t)$ and service rate $\mu(t)$. The system can then be defined as a six-state CTMC, where state 0 represents 0 jobs in the system, state 1 represents 1 job in the system, etc. It is assumed that initially there are no jobs present in the system, i.e. $\pi_0(0) = 1$. It is further assumed that $\mu(t) = 0.5$ jobs/unit time, $\forall t$, and that the arrival rate varies in a periodic step-wise fashion, with each period having three steps,

$$
\lambda(t) = \begin{cases}
0.15 \text{ jobs/unit time} & 0 \leq t < .02 \text{ time units,} \\
0.30 \text{ jobs/unit time} & .02 \leq t < .05 \text{ time units,} \\
0.20 \text{ jobs/unit time} & .05 \leq t < .07 \text{ time units.}
\end{cases}
$$

The time point of interest is at the end of the second step after 2 full periods (i.e., after 8 arrival rate transitions). We chose an M/M/1/K queue so that we would have two distinct solution methods to compare, i.e., the closed form expression given by eq.(9), and uniformization. Uniformization was accomplished using the commercially available software tool SHARPE [21].

For each of the two solution methods above, two distinct approaches were also compared. In the first approach, the system was solved at each time step (or arrival rate transition point). The resulting probabilities were then used as the initial conditions for the next time step (8 steps total). In the second approach,

Table 1. Comparison of closed-form solution method. Multiple steps (8) vs. single step.

State	Multiple Steps (8)	Single Step	Percent Difference
0	9.58831483e-01	9.58893462e-01	-6.5e-3
1	4.02838524e-02	4.02231279e-02	0.15
2	8.71841896e-04	8.70605305e-04	0.14
3	1.26824570e-05	1.26645514e-05	0.14
4	1.38825874e-7	1.38624106e-7	0.15
5	1.22657368e-9	1.22470681e-9	0.15

Table 2. Comparison of uniformization solution method. Multiple steps (8) vs. single step.

State	Multiple Steps (8)	Single Step	Percent Difference
0	9.58831483e-01	9.58893462e-01	-6.5e-3
1	4.02838524e-02	4.02231279e-02	0.15
2	8.71841897e-04	8.70605305e-04	0.14
3	1.26824570e-05	1.26645513e-05	0.14
4	1.38825847e-07	1.38624030e-07	0.15
5	1.22656535e-09	1.22468075e-09	0.15

Table 3. Steady-state probabilities at the various time steps.

State	Time Step 1	Time Step 2	Time Step 3
0	7.00510666e-01	4.19575720e-01	6.02467702e-01
1	2.10153204e-01	2.51745435e-01	2.40987085e-01
2	1.89137892e-02	1.51047263e-01	9.63948352e-02
3	6.30459629e-02	9.06283584e-02	3.85579343e-02
4	5.67413677e-03	5.43770151e-02	1.54231737e-02
5	1.70224100e-03	3.26262090e-02	6.16926946e-03

a single equivalent infinitesimal generator matrix over the 8 transitions was computed using eq.(21b), allowing the transient probabilities to be obtained in a single step.

Table 1 summarizes the transient probabilities obtained using eq.(9), for both multiple (i.e., 8) steps and a single step (using R, the time average infinitesimal generator matrix). Table 2 summarizes the same data obtained using uniformization.

Please note that the equilibrium probabilities for time step 2 are quite different from those obtained in Tables 1 and 2. For this particular example, this significant difference indicates that simply examining the equilibrium values at each time step would be inadequate, since the system interarrival rate switches too quickly for equilibrium to ever be achieved. Data taken from an actual LAN [11] suggest that such rapid changes do in fact occur, and typically occur at the most congested periods of the day. In designing a network, analyses of these most heavily congested periods are critical if acceptable performance is to be achieved. Clearly, more attention needs to be given to the transient analysis of such systems whose underlying statistics vary with time.

For comparison, the steady state probabilities for the different stages are given in Table 3.

Summary

A transient model of a nonhomogeneous, possibly periodic CTMC may be very useful in capturing the observed fluctuations in user population size and service time requirements of actual LANs. Modern systems/control theory provides a number of useful results for linear time-varying systems. If the system dynamics are described by a periodic infinitesimal generator, then the state probabilities will also be periodic only when the initial condition state vector is the left eigenvector corresponding to the unity eigenvalue of $\Phi(T, 0)$. Complete characterization of such a system then requires characterization of the behavior over only a single period. For an arbitrary initial condition, the eigenvalues of $\Phi(T, 0)$ can be examined to determine if the state probabilities approach a periodic limit cycle, and if so, at what rate. Floquet's method may be used to obtain a time-invariant representation that is then amenable to such methods as uniformization. Parallelization of the uniformization method may be very efficient over other standard methods for obtaining a large number of sample points, and may be used to determine if and how fast the systems converges to a limit cycle. More importantly, Floquet's method can be extended to obtain transient probabilities for any aperiodic nonhomogeneous system using standard available methods requiring a homogeneous system. This method can then be used to evaluate the model we propose, which we feel accurately captures the behavior of real systems. In this model, user population changes occur in a discrete manner.

References

[1] A. Reibman and K. Trivedi, "Numerical transient analysis of Markov models," *Comput. Opns. Res.*, Vol. 15, No. 1, pp.19-36, 1988.

[2] D. G. Kendall, "On the generalized 'birth-and-death' process," *Ann. Math. Statist.*, Vol. 19, pp. 1-15, 1948.

[3] A. Lamens and R. Consael, "Sur le processus non homogene de naissance et de mort" *Acad. roy. Belg., Bull. classe sci.*, Ser. 5, Vol. 43, pp. 597-605, 1957.

[4] A. Lamens, "Sur le processus non homogene de naissance et de mort a deux variables aleatoires," *Acad. roy. Belg., Bull. classe sci.*, Ser. 5, Vol. 43, pp. 711-719, 1957.

[5] A. T. Bharucha-Reid, *Elements of the Theory of Markov Processes and Their Applications*, pp. 57-128, McGraw-Hill, NY, 1960.

[6] E. L. Leese and D. W. Boyd, "Numerical methods of determining the transient behaviour of queues with variable arrival rates," *Canadian J. Oper. Res.*, Vol. 4, No. 1, pp. 1-13, 1966.

[7] D. Tipper and M. K. Sundareshan, "Numerical methods for modelling computer networks under nonstationary conditions," *IEEE JSAC*, Vol. 8, No. 9, pp.1682-1695, Dec. 1990.

[8] T. E. Fortmann and K. L. Hitz, *An Introduction to Linear Control Systems*, pp.589-599, Marcel Dekker, NY, 1977.

[9] T. Kailath, *Linear Systems*, pp.594-631, Prentice-Hall, NJ, 1980.

[10] C. Moler and C. Van Loan, "Nineteen dubious ways to compute the exponential of a matrix," *SIAM Review*, Vol. 20, No. 4, pp.801-836, Oct. 1978.

[11] K. W. Davis, Measured IBM 4 Mbps token ring data, Sept.-Dec. 1991, personal communication.

[12] N. M. van Dijk, "Uniformization for nonhomogeneous Markov chains," *Op. Res. Letters*, Vol. 12, No. 5, pp. 283-291, Nov. 1992.

[13] L. Kleinrock, *Queueing Systems: Volume 1: Theory*, pp.44-53, John Wiley and Sons, NY, 1975.

[14] P. M. Morse, *Queues, Inventories and Maintenance: The Analysis of Operating Systems with Variable Supply and Demand*, John Wiley and Sons, NY, 1958.

[15] L. Takacs, *Introduction to the Theory of Queues*, Oxford University Press, NY, 1962.

[16] S. Sharma and D. Tipper, "Approximate models for the study of nonstationary queues and their applications to communications networks," *Proc. ICC '93* (submitted).

[17] J. McGough, A. Reibman and K. Trivedi, "Markov reliability models for digital flight control systems," *J. Guidance, Control, Dynamics*, Vol. 12, No. 2, pp. 209-219, Mar.-Apr. 1989.

[18] M. C. Hsueh, R. Iyer and K. Trivedi, "Modeling based on real data: a case study," *IEEE Trans. Computers*, Vol. C-37, No. 4, pp. 478-484, Apr. 1988.

[19] A. Bobbio and K. Trivedi, "Computation of the distribution of the completion time when the work requirement is a {PH} random variable," *Stochastic Models*, Vol. 6, No. 1, pp. 133-150, Apr. 1990.

[20] W. Fischer and K. Meier-Hellstern, "The MMPP cookbook," *Perf. Eval.*, (in press).

[21] R. Sahner and K. Trivedi, "Reliability modelling using SHARPE," *IEEE Trans. Reliability* , Vol. R-36, No. 2, pp. 186-193, June 1987.

Closed Queueing Network Modeling for End-to-End Performance Analysis of ISO LLC and Transport Protocols over Bridged Networks

Takashi Ikegawa

NTT Telecommunication Networks Labs.
1-2356, Take, Yokosuka-shi, 238-03, Japan

ABSTRACT

This paper presents approximate performance analysis method of ISO LLC type 2 protocol and Transport Protocol class 4 over the networks of simple intermediate nodes (bridges) connected by transmission links, such as bridged LANs, frame relaying and ATM cell relaying networks. This analysis method is based on a BCMP-type closed queueing network that model go-back-n retransmissions of data packets lost due to bit errors and buffer overflows at the bridges. The proposed model makes it possible to examine the effect of error recovery mechanism on the end-to-end performance such as throughput and mean end-to-end delay, over the window controlled bridged networks. The approximation accuracy is validated with simulations. Numerical results indicate that approximation accuracy is fairly good, assuming packet length independence, independently occurring bit errors and no buffer overflow. For the case that the overflows happen, we show that a Poisson approximation of arrival process of the overflow queues does not yield good accuracy over a wide range of window sizes because the Poisson process can not properly represent the bursty traffic caused by go-back-n retransmission.

1. Introduction

The use of bridges, which perform only simple functions such as switching and error detection, has been increasingly popular to interconnect high-speed Local Area Networks (LANs) [12]. Recently, there have been emerging wide area networks that provide end-to-end high speed data transfer service by simplification of relaying operation at the intermediate nodes (e.g., Asynchronous Transfer Mode - ATM - cell relaying networks [5] and frame relaying networks [26]). Throughout this paper, an intermediate node which carries out the simple functions such as switching and error detection but is not involved in error-recovery or flow control is denoted as a bridge using LAN terminology [12]. For networks consisting of such intermediate nodes, denoted as bridges connected by transmission lines, namely *bridged networks,* packets dropped due to bit errors or buffer overflows are not recovered within the networks. To support error-free communication applications over the bridged networks, each end-terminal performs a reliable protocol. The existing international standard protocols such as Logical Link Control type 2 protocol (LLC2) [19] and Transport Protocol class 4 (TP4) [18] provide reliable data transmission service. When implementing these protocols, it is necessary to determine values of the protocol parameters (e.g., packet length, window size and timeout value) to achieve satisfactory performance. Thus, this paper presents the performance analysis method useful to gain insights into the interaction among the parameters of the standard protocols over the bridged networks.

For performance analysis of LLC2 and TP4, it is required to model both flow control and error recovery, which are their important protocol functions. The flow control is realized by a window mechanism, i.e., a sender is permitted to transmit up to a fixed number (window size) packets without waiting for acknowledgments. The error recovery mechanism is based on the go-back-n scheme.

One approach for performance analysis of bridged networks where lost packets are recovered is to use an open tandem queueing network with a feedback loop that represents lost packet transmissions [1], [4], [31]. This approach enables us to examine the effect of the error recovery, but it can not treat the behavior of the window mechanism.

The analysis methods that take both the error recovery and the window flow control into account explicitly have been presented in [14] and [21]. In these studies, throughput has been derived utilizing the concept of "virtual transmission time" that contains the duration of error recovery actions and window mechanism proposed in [6]. To apply this concept to the bridged network (unlike a point-to-point link analyzed in [6]), the mean transmission time of a single packet by a sender was approximated by the mean service time of a server (a transmission line or a processor) which becomes bottleneck over the virtual circuit. Limitations of this approach are that the analytical results do not give good estimates when the packet loss occurs frequently at the queues located between the sender and the bottleneck server [14], and it is impossible to evaluate queueing delay of packets that find closed window upon arrival at the sender [6]. Furthermore, networks to which this approach is applicable are limited to those of point-to-point links.

On the other hand, finite population queueing networks (of which closed queueing networks are typical ones) with a feedback loop that represents positive acknowledgement allow us to examine the queueing delay of packets due to closed window (e.g., [28], [29]). Thus, we can evaluate the effect of the window size on the performance measures over a wide range of it by using these models. Another advantage of such models is that this make it possible to analyze the performance of various types of window flow controlled communication networks (e.g., networks with multiple virtual circuits interacting each other [3], [24], [28], multi-access networks such as LANs and satellite communication networks [3], [13], [23], [24], and multi-layered architecture [8], [10], [17]). From advantages above, this paper employs the approach using the closed queueing network to analyze the end-to-end performance of the bridged networks.

In previous work based on finite population queueing network [3], [28], [29] and [30], retransmission of lost packet have been ignored. Several finite population queueing networks that take into account the error recovery mechanism have been proposed in [8], [10], [15], [16], [17] and [27]. However, the error recovery scheme considered in [8], [10], [17] and [27] is an *ideal* selective retransmission scheme rather than go-back-n scheme. Although the LLC2 sender can observe packet loss by two procedures (receipt of negative acknowledgement and timeout), only one procedure was modeled. Recently, in [15] and [16], the author has proposed the closed BCMP-type queueing network to model the error recovery mechanism of LLC2 on a full -duplex point-to-point link, and has shown that approximation accuracy is fairly good. This paper extends that queueing model to the bridged network. In the model proposed here, the buffer overflows that are known to lead to severe performance degradation [7], [25], are taken into account.

In the following, Section 2 describes the bridged network configuration. Section 3 shows a closed queueing network model of a virtual circuit over the bridged network and introduces approximations for analytical tractability. In Section 4, throughput and mean end-to-end delay are obtained using the closed queueing network. In Section 5, we compare analytical results with simulation ones to evaluate the approximation accuracy, and discuss the impacts on the protocol parameters on the end-to-end performance.

2. Network Model

As depicted in Figure 1, we consider the bridged network consisting of N full-duplex links (it is applicable to the bridged networks including the multi-access networks using methods proposed in [13], [23] and [24]) between a source terminal B_0 and B_N. The links L_i and L_{i+1} are interconnected by the bridge B_i. Both the forward line and the feedback one (throughput acknowledgements are transmitted) of the link L_i produce statistically independent bit errors at rate P_i^e.

Each terminal performs LLC2 or TP4. Both LLC2 and TP4 carry out window flow control. In this analysis, this paper employs sliding window flow control, which is window flow control combined with individual acknowledgement scheme, i.e., for each data packet (DT-packet) successfully received by the destination terminal, one positive acknowledging packet (ACK-packet) is transmitted back.

The error recovery of LLC2 allows a source terminal to detect packet loss by the following two procedures: 1) REJ scheme, namely, it can observe the packet loss by receipt of a negative acknowledgement packet (REJ-packet), and 2) timeout scheme, namely, it can do by results of checkpointing procedure carried out after expiration of a timer. It retransmits the lost DT-packets on go-back-n basis upon detection the packet loss.

For TP4, it is impossible for a destination terminal to request retransmission of lost DT-packets, but retransmissions happen only upon timer expiration, i.e., TP4 defines only timeout scheme without the checkpointing function like LLC2>TP4 permits various retransmission strategies [22]. This analysis employs the following retransmission strategy from view point of simplicity of timer and buffer

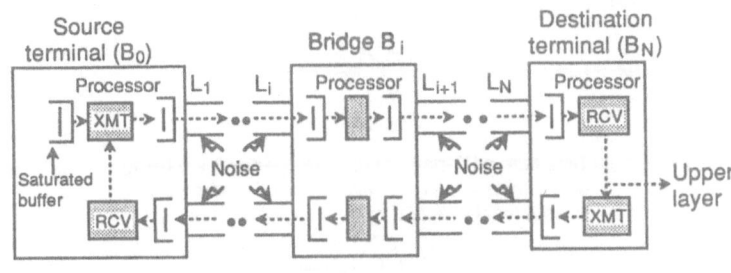

XMT: processor for handling transmission
RCV: processor for handling reception

Figure 1: Network model

managements: all DT-packets waiting for acknowledgement are retransmitted according to the go-back-n basis upon timeout.

In this network model, a buffer pool of a transmitting processor at the source terminal is assumed to always have waiting DT-packets (i.e., heavy traffic condition). However, this condition can easily be relaxed by applying decomposition-aggregation methods [8], [10], [13], [17] and [29].

Packet processing and transmission are carried out on FIFO (First In First Out) discipline. Packet processing times are assumed to be independently and randomly selected from the exponential distribution at each processor. We assume that DT-packets and control packets are distributed exponentially in length with mean values of l_d+l_c and l_c, respectively, where l_d denotes mean user data length, and l_c represents mean control field length (assumed to be equal to mean control packet length).

3. Queueing modeling of the bridged network

A closed queueing network, where the number of circulating packets is equal to window size, can model the sliding window flow controlled network under the heavy traffic condition (see e.g., [28]). This paper presents a closed queueing network which models the error recovery defined in LLC2 and TP4, as shown in Figure 2.

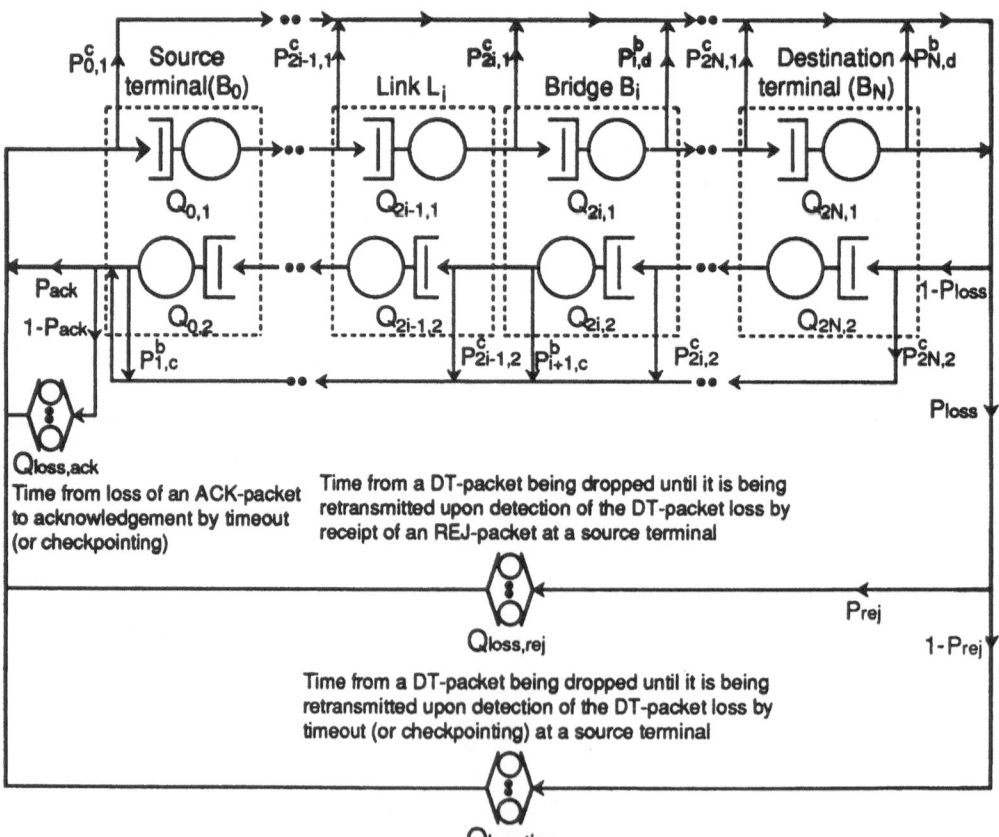

Figure 2: Queueing network model

The forward line and the backward one of link L_i for $i=1,...,N$ are represented as FIFO queues $Q_{2i-1;1}$ and $Q_{2i-1;2}$ whose service times correspond to packet transmission times, respectively. We note that propagation delay can be easily taken into consideration by introducing infinite server (IS) queues (the infinite server queue is defined as the queueing system where the number of servers is sufficient to be always at least one free, which corresponds Type-3 queueing system defined in [2]) whose service times corresponds to the propagation delay [4], [15]. We model two processors of the bridge (or the terminal) B_i for $i=0,...,N$ by FIFO queues ($Q_{2i-1;1}$, $Q_{2i-1;2}$). A DT-packet travels from a transmitting processor (XMT) of a source terminal $Q_{0;1}$ to a receiving processor (RCV) of a destination terminal $Q_{2N;1}$, i.e., forward path. The DT-packet may be dropped because of bit errors or buffer overflow. We denote the probability of a DT-packet dropped due to the bit errors occurring at link L_i by P_{id}^b and denote the probability of a packet being dropped due to the overflow at queue Q_{ij} (that is the probability of arriving one finding a buffer pool fool) by P_{ij}^c The out-of-sequence DT-packet is discarded by the destination terminal because of go-back-n retransmission scheme. The source terminal retransmits the lost DT-packets upon detection of packet loss by an REJ or a timeout scheme for LLC2, and by only timeout scheme for TP4. We introduce IS queues $Q_{loss;rej}$ and $Q_{loss;time}$ represent packet loss detection times by the REJ scheme and by the timeout one respectively. We define P_{loss} as the probability that a transmission of a DT-packet from the XMT of the source terminal will fail (i.e., the probability of a DT-packet being dropped due to bit errors, overflows, or out-of-sequence), and P_{rej} as the probability of a lost DT-packet being retransmitted by an REJ scheme.

The destination terminal sends an ACK-packet upon receipt an in-sequence DT-packet without a bit error. The ACK-packet traverses from an XMT of the destination terminal $Q_{2N;2}$ to an RCV of a source terminal $Q_{0;2}$, i.e., backward path. Similar to the DT-packets, the ACK-packet may be lost because of bit errors or overflows. We denote the probability of a control-packet such as an ACK-packet being dropped due to bit errors occurring at link L_i by P_{ic}^b If all ACK-packets of a whole window are discarded, timeout will be invoked. In this analysis, an IS queue $Q_{loss;ack}$ is introduced to represent this acknowledgement delay due to the timeout.

An exact analysis of actual model is rather difficult. In view of this, we model the virtual circuit over the bridged network using a BCMP-type queueing network [2] (for which the solution is known in an explicit form) by making several approximations as follows:

(A1) Packet lengths are distributed independently at each transmission line (this approximation corresponds to Kleinrock's message independence assumption [20]) (in fact, packet length is preserved at each packet, and lengths of retransmitted DT-packets are equal to one of first transmission),

(A2) Packet loss due to bit errors and overflows occurs independently at each DT-packet (we note that packet loss because of bit errors takes place independently, assuming independent bit errors. However, for overflows, packet loss independence assumption may not hold since go-back-n retransmission traffic reveals burstiness as described in Section 5.)

(A3) Lost of out-of sequence DT-packets happens independently at each DT-packet (in real situation, if a DT-packet is dropped due to out-of-sequence, DT-packets from the next arriving one until an expected one received successfully will be discarded),

(A4) Packet loss detection times are independently and randomly selected from the Cox distribution,

(A5) Effect of control packets except for ACK-packets on waiting time of DT-packets and ACK-packets is ignored,

(A6) We approximate the mean packet loss detection time of an REJ scheme by the mean time of ACK-packets traveling on a backward path when no packet loss occurs,

(A7) Each time a source terminal receives an ACK-packet, timeout will be invoked with the probability of all ACK-packets of a whole window being dropped,

(A8) The elapsed times from the epoch of loss of an ACK-packet to the epoch of acknowledgement by timeout at a source terminal are independently and randomly selected from the Coxian distribution,

(A9) The time required for checkpointing procedure (i.e., the elapsed time from transmission of a polling packet to receipt a corresponding response packet at a source terminal) is negligible compared to timeout value T_{out} by the assumption of T_{out} being large enough.

We make the following approximation regarding of the queues where buffer overflow may happen (denoted as overflow queues), i.e., the queues whose capacity is less than window size:

(B) We approximate an arrival process to the overflow queue by a Poisson process (similar to that of [4], [17], [21], [27] and [31]).

The routing probabilities of the closed queueing network are derived as follows. For P^b_{id} and P^b_{ic} with the independent and exponentially distributed packet length approximation, we can obtain

$$P^b_{id} = \int_0^\infty \{1-(1-P^e_i)^x\} \frac{\exp(-x/l_d+l_c)}{l_d+l_c} dx = 1 - \frac{1}{1-(l_d+l_c)\log(1-P^e_i)} \tag{1}$$

$$P^b_{ic} = 1 - \frac{1}{1-l_c\log(1-P^e_i)} \tag{2}$$

From the assumptions of independence and exponential service time, and from the approximation (B), the overflow queue can be modeled as a finite M/M/1 queue. Hence the values of P^c_{ij} can be derived explicitly when arrival rate and service rate are given (the arrival rate is calculated by using iterative algorithm described in the following section). Note that the value of P^c_{ij} is equal to zero for $Q_{ij}C$ where C is defined as set of overflow queues.

With the approximation of (A3), P_{loss} and P_{rej} become fixed, and given as follows (Detailed derivation can be found in [15] and [16]):

$$P_{loss} = \frac{P_dW}{P_dW+1-P_d} \tag{3}$$

$$P_{rej} = \begin{cases} \dfrac{(1-P_c)(1-P_d)(1-P_d^{W-1})}{1-P_d^W} & \text{for LLC2} \\ 0 & \text{for TP4} \end{cases} \tag{4}$$

where

W: window size

P_d : the probability of a DT-packet being dropped because of bit errors or overflows on a forward path, which can be expressed as

$$P_d = 1 - \prod_{i=1}^{N} (1 - P_{id}^b) \prod_{j=0}^{2N} (1 - P_{j1}^c) \tag{5}$$

P_c : the probability of a control packet being dropped on a backward path, which is given by

$$P_c = 1 - \prod_{i=1}^{N} (1 - P_{i2}^b) \prod_{j=0}^{2N} (1 - P_{j2}^c) \tag{6}$$

Letting P_{ack} be the probability of timeout occurring after receipt of an ACK-packet at a source terminal, P_{ack} is expressed as $1 - P_c^W$ from the approximation of (A7).

We denote the mean service time of Q_{ij} as τ_{ij}. The value of $\tau_{ij}(i=0,...2N, j=1,2)$ is determined when the corresponding transmission link capacity or mean processing time is given (note that service times of servers on the forward path are of only DT-packets, and ones on the backward path are of only ACK-packets from the approximation of (A5)). The approximation (A6) dictates that $\tau_{loss;time}$ can be expressed as $\sum_{i=0}^{2N} t'_{i2}$ where t'_{i2} represents the mean queueing delay including packet service time at Q_{ij} when no packet loss occurs. Under the approximation (A9), $\tau_{loss;time}$ and $\tau_{loss;ack}$ are given in the following form:

$$\tau_{loss;time} = \tau_{loss;ack} = \begin{cases} T_{out} / \{1-P_{ch}\} & \text{for LLC2} \\ T_{out} & \text{for TP4} \end{cases} \tag{7}$$

where, for LLC2, the form results from the equality $T_{out} \sum_{k=1}^{\infty} P_{ch}^{k-1}(1-P_{ch})$ with P_{ch} defined as the probability that a checkpointing procedure fails (i.e., the probability that each packet of a polling packet and the corresponding response packet is lost) which is given as

$$P_{ch} = 1 - (1-P_a)(1-P_c) \tag{8}$$

where P_a is defined as the probability of a control packet being dropped on a forward path, which is given as follows:

$$P_a = 1 - \prod_{i=1}^{N} (1 - P_{i1}^b) \prod_{j=0}^{2N} (1 - P_{j1}^c) \tag{9}$$

4. Performance measures

In this section, we derive throughput T_p, i.e. mean number of bits received correctly at a destination terminal, and mean end-to-end delay T_{delay}, i.e. , mean elapsed time that begins when a source terminal transmits a DT-packet and ends when a destination terminal receives the DT-packet correctly. We define the following notations:

σ_{ij}: mean relative arrival rate of Q_{ij}. The values of σ_{ij} can be obtained by solving flow balance equations under the condition $\sigma_{ij}=1$, if the routing probabilities of the queueing network are given (cf. [11]. pp. 64).

131

t_{ij} mean queueing delay (including packet service time) of Q_{ij}

λ: overall throughput of the queueing network expressed as

$$\lambda = \frac{W}{\sum_{(i,j)} \sigma_{ij} t_{ij}} \tag{10}$$

A. *Throughput*

The mean number of DT-packets received by a destination terminal correctly per unit time is given by $\lambda\sigma_{2N;2}$. that is, the mean number of ACK-packets transmitted by the destination per unit time. Therefore, the throughput T_p is obtained by:

$$T_p = \lambda\sigma_{2N;2} / d \tag{11}$$

B. *Mean end-to-end delay*

The mean number of transmission times of a DT-packet before it is successfully received at the destination terminal is given by $\sigma_{01}/\sigma_{2N;2}$. Furthermore, the mean number of times a DT-packet must be retransmitted by an REJ scheme and by a timeout before it is successfully received at the destination terminal are obtained as $\sigma_{loss;rej}/\sigma_{2N;2}$ and $\sigma_{loss;time}/\sigma_{2N;2}$, respectively. The mean delay of a DT-packet traversing from an XMT of a source terminal to an RCV of a destination terminal is given by $\sum_{i=0}^{2N} t_i$ The delays from a DT-packet being dropped until it is being retransmitted by an REJ scheme and by a timeout scheme are approximately given by $t_{loss;rej}$ and $t_{loss;time}$ respectively. Consequently, the mean end to end delay T_{delay} can be written in the form:

$$T_{delay} = \frac{\sigma_{01} \sum_{i=0}^{2N} t_{i1} + \sigma_{loss;rej} t_{loss;rej} + \sigma_{loss;time} t_{loss;time}}{\sigma_{2N;2}} \tag{12}$$

C. *Iterative solution algorithm*

To actually calculate the quantities required for the performance analysis (i.e. $\{t_{ij}\}$, $\{\sigma_{ij}\}$, and λ), we must know $\{P^c_{ij}\}$. However, the value of P^c_{ij} where $Q_{ij} \in C$ is dependent on the arrival rate to the overflow queue which is a function of λ. The following iterative algorithm is used to solve the queueing network when $C \neq \emptyset$.

(Step 1) Assign an initial value to $\lambda = \lambda_1$, e.g. set λ to one when no overflow takes place.

(Step 2) - Calculate the mean arrival rate to each overflow queue with $\lambda = \lambda_1$, and derive P^c_{ij} by solving the finite M/M/1 queue for $Q_{ij} \in C$.

- Calculate $\{\sigma_{ij}\}$ by solving flow equilibrium equations under the condition $\sigma_{11} = 1$ of the closed queueing network for the obtained values of P^c_{ij}

- Derive $\{t_{ij}\}$ using $\{\sigma_{ij}\}$

(Step 3) Calculate λ_2 by substituting the values of the $\{\sigma_{ij}\}$ and the $\{t_{ij}\}$ into equation (10). If $|\lambda_1 - \lambda_2| \leq \varepsilon$ (e.g. $\varepsilon = 10^{-6}$), then stop. Otherwise, let $\lambda_1 = \lambda_2$ and go to (Step 2).

5. Numerical Results

In this section, numerical results are presented and compared with the simulation results to evaluate the effect of approximations made in the analysis. The following parameter values are used for the numerical results.

- We consider the bridge network consisting of three links, i.e. N=3, with transmission capacities of L_1=10 Mbps, L_2=1.544 Mbps, L_3=10 Mbps.
- The mean processing times of packets are assumed to be 2 msec, both at the source and the destination terminals. Furthermore, the mean packet processing time at a bridge is assumed to be equal to 1 msec.
- Timeout value is assumed to be constant and equal to 250 msec.
- The mean control field length (equal to mean control packet length) is assumed to be equal to 25 bytes.

First, we evaluate the approximation accuracy under the overflow condition (i.e. all buffer capacities are infinite). Figures 3 and 4 show that the throughput and the mean end-to-end delay as a function of the mean length of the user data l_d, respectively. In these examples, window size is assumed to be equal to seven, bit error rates L_1 and L_3 are negligible while two values of L_2, 10^{-5} and 10^{-6} are considered.

In Figures 5 and 6, the throughput and the mean end-to-end delay are respectively are varied against DT-packet bit error probability due to bit errors at link L_2, P_{2d}^b. We have l_d=256 bytes, $P_1^e=P_3^e=0$ and P_2^e varied. To verify the packet length independence assumption (PLIA), we simulate for two cases: with and without PLIA. In the latter, the packet length, after being chosen from an exponential distribution at the first transmission, is preserved in downstream nodes.

As shown in Figures 5 and 6, relative approximation errors are in the acceptable range for the PLIA holding. The relative approximation error is defined as the absolute value of the difference between the analytical results and the simulation is divided by the analytical one. However, as shown in these figures, the errors

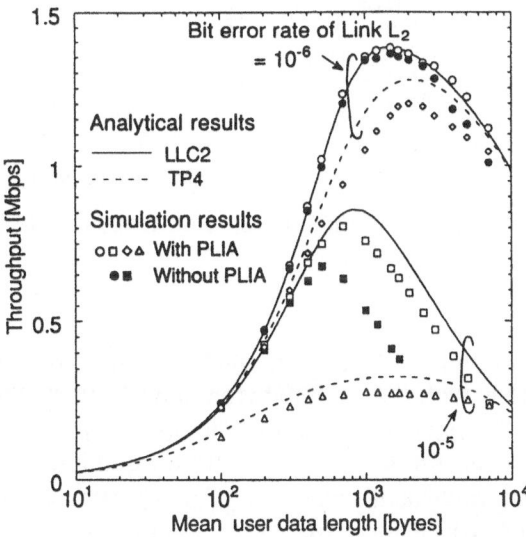

Figure 3: Throughput versus mean user data length for different bit error rates of link L_2

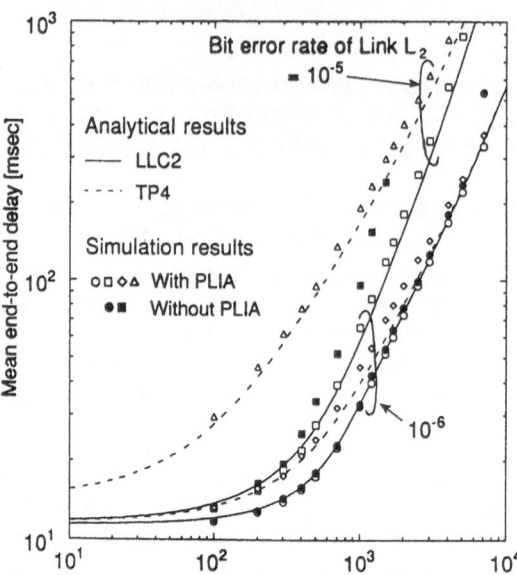

Figure 4: Mean end-to-end delay versus mean user data length for different bit error rates of link L_2

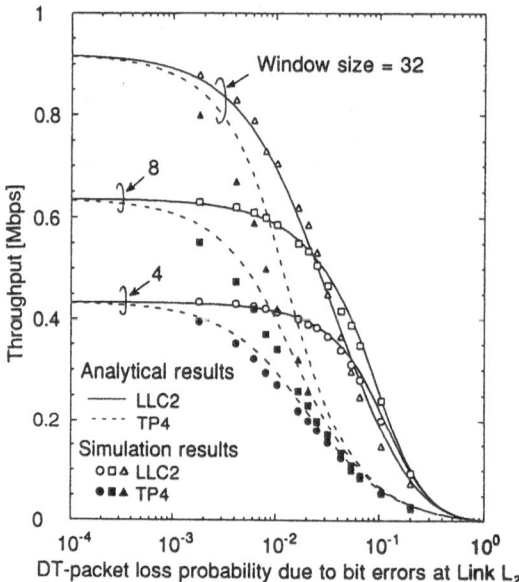

Figure 5: Throughput versus DT-packet loss probability due to bit errors at link L_2 for different window sizes

are noticeably large when the PLIA is not used over the range when DT-packet is long and bit error rate is high (i.e. $l_d \geq 800$ bytes and $P_2^e = 10^{-5}$). The main reason for this is that the actual DT-packet loss probability due to bit errors when PLIA is not used is much larger than the one obtained from equation (1) (e.g., for $l_d = 1000$ bytes and $P_2^e = 10^{-5}$, an analytical result of P_{2d}^b is 7.6×10^{-2} is used to approximate the actual value of 9.3×10^{-2} without the PLIA). On the other hand, when the PLIA is assumed to hold, then the approximations (A2) to (A9) are observed to be valid.

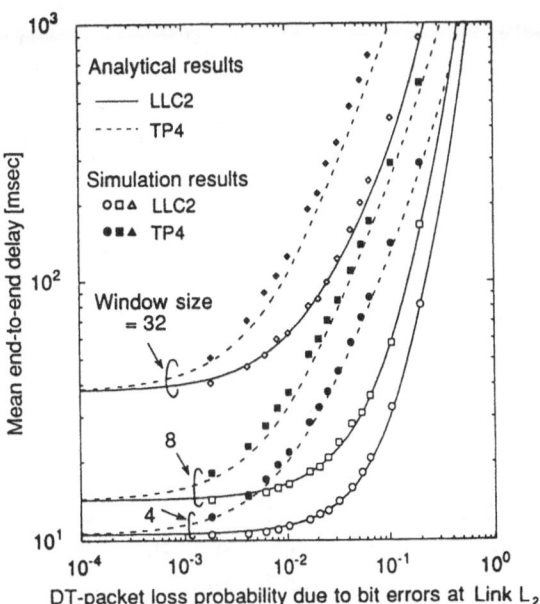

Figure 6: Mean end-to-end delay versus DT-packet loss probability due to bit errors at link L_2 for different window sizes

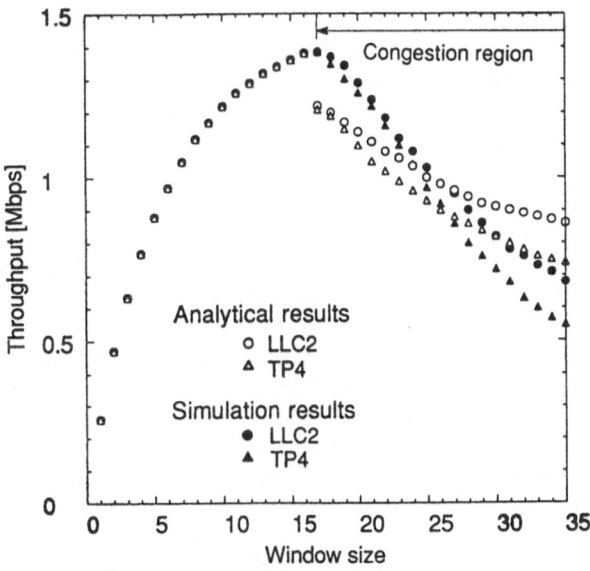

Figure 7: Throughput versus window size

Next, the accuracy of a Poisson approximation for the arrival process at the overflow queues is examined. Here, we assume the all buffer size at the bridge B_1 (including a server) are 16 [packets] but all other buffets are infinite, and assume the bit error free condition for l_d =512 bytes. We show the throughout and the mean end-to-end delay as a function of window size in Figure 7 and 8 , respectively. These figures show that the approximation error is acceptable over the region of the moderate congestion (around window size equal to 25).

However, on the starting point of the congestion, the error is rather large because the output process of each queue of the BCMP-type closed queueing network is

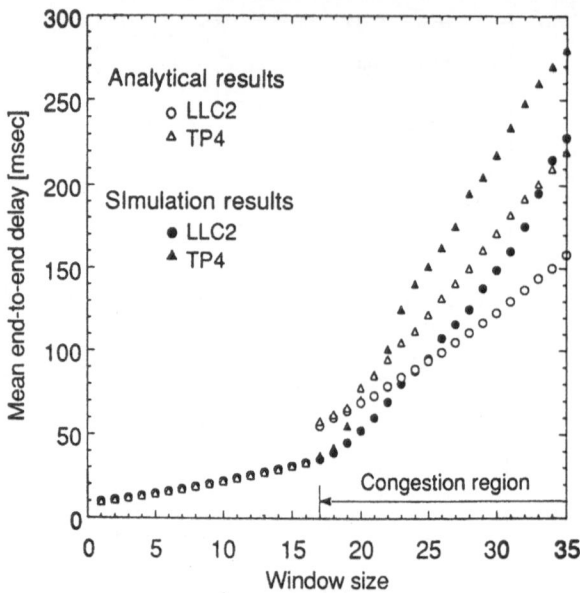

Figure 8: Mean end-to-end delay versus window size

no longer Poisson ([11] pp. 146-167). Furthermore, under heavy congestion condition, the analytical method noticeably underestimates the mean end-to-end delay. This reason is explained as follows. Actually, the retransmitted packets arrive at the queues in *bursts* because the go-back-n scheme causes a source terminal to retransmit the *all* unacknowledged DT-packets upon detection of packet loss, but the Poisson process can not properly represent this bursty traffic.

Figure 3 shows that the optimum length which gives the maximum throughput exists (the reason is the same as the study of performance analysis for a point-to-point link [6]). As shown in Figure 5, the throughput approaches the maximum throughput of the bottleneck queue (that corresponds to a forward line of link #2:FIFO queue Q_{31} in this case) for the larger window size when no packet loss happens or loss probability is small enough. On the other hand, when packet loss occurs frequently, the throughput drops more sharply for the larger window size. The reason of latter effect is explained as follows: the number of retransmissions of packet successfully received but discarded increases for the larger window size because go-back-n scheme causes the retransmission of full window packets upon detection packet loss.

As shown in the figures from Figure 3 to 8, LLC2 exhibits significantly better performance than TP4, because of its more efficient retransmission strategy using an REJ scheme, as indicated in [22].

6. Conclusion

This paper proposed the approximate closed queueing network that models the bridged network where the packets dropped due to bit errors and buffer overflows are recovered by LLC2 or TP4 protocol. This model makes it possible to examine the effect of error recovery mechanism on the performance over the window flow controlled bridged networks. Compared performance measures derived the analysis with simulation results, the accuracy is reasonably good, assuming packet length independence, independently occurring bit errors, and no buffer overflow. However, when the buffer overflows happen, this analysis does not yield good estimation over a wide range of window sizes because the arrival process to the overflow queues cannot be approximated by a Poisson process. The arrival process approximation of the queue that gives better accuracy is an important open issue for end-to-end performance analysis of the bridged networks.

The methodology using the closed queueing network is rich in that it enable to apply the window flow controlled bridged networks consisting of combinations of many types of networks (including point-to-point link, and multi-access networks such as LANs). Another potential application of the methodology is performance evaluation of dynamic window flow control mechanism for bridged networks [7], [25].

Recently, the protocols suited to high bandwidth, low error-rate communication networks (such as bridged networks), called light weight transport protocols, have been studied (see e.g., [9]). Almost all the light-weight transport protocols employ the window flow control with an automatic repeat request of selective retransmission. The next step of this research is to model the *real* error recovery procedures of the selective retransmission using closed queueing networks, and to compare the performance of light-weight protocols with one of the existing protocols such as LLC2 and TP4.

Acknowledgement

The author would like to thank Mr. Kensaku Kinoshita, Executive Manager, and Mr. Takafumi Saito, Research Group Leader, of NTT Telecommunication Networks Labs., for their encouragements during this research, and Mr. Jyun-ichi Kajitani of Waseda University for his assistance in simulation programming. The author is also indebted to Mr. Teruyuki Kubo for reviewing the manuscript.

References

[1] J.J. Bae, T. Suda, and N. Watanabe, *Evaluation of the effects of protocol processing overhead in error recovery schemes for a high speed packet switching network: link by link versus edge to edge schemes*, IEEE JSAC, 9-9, 1496-1509, 1991
[2] F. Baskett, K.M. Chandy, R.R. Muntz, and J. Palacios, *Open, closed, and mixed networks with different classes of customers*, J. ACM, 22, 248-260, 1975
[3] B. Berg and R.H. Deng, *End to end performance of interconnected LANs*, Computer Communications, 14-2, 105-112, 1991
[4] A. Bhargava, J.F. Kurose, D. Towsley, and G. Vanleemput, *Performance comparison of error control schemes in high speed communication networks*, IEEE JSAC, 6-9, 1565-1575, 1988
[5] J-Y. LeBoudec, *The Asynchronous Transfer Mode: a tutorial*, Computer Networks and ISDN Systems, 24, 279-309, 1992
[6] W. Bux, K. Kümmerle, and H.L. Truong, *Balanced HDLC procedures: a performance analysis*, IEEE Trans. Comm., 28-11, 1889-1898, 1980
[7] W. Bux and D. Grillo, *Flow control in local area networks of interconnected token rings*, IEEE Trans. Comm., 33-10, 1058-1066, 1985
[8] A.E. Conway, *Performance modeling of multilayered OSI communication architectures*, ICC'89, 651-657, 1989
[9] W.A. Doeringer, D. Dykeman, M. Kaiserwerth, B.W. Meister, H. Rudin, and R. Williamson, *A survey of light weight transport protocols for high speed networks*, IEEE Trans. Comm., 38-11, 2025-2039, 1990
[10] S. Fdida, H.G. Perros, and A. Wilk, *Semaphore queues: modeling multilayered window flow control mechanisms*, IEEE Trans. Comm., 38-3, 309-317, 1990

[11] E. Gelenbe and G. Pujolle, *Introduction to queueing networks,* John Wiley & Sons, 1987

[12] Special issue on bridges and routers, IEEE Network Magazine, 2-1, 1988

[13] O. Gihr and P.J. Kuehn, *Comparison of communication services with connection oriented and connectionless data transmission,* Computer Networking and Performance Evaluation, 173-186, North Holland, 1986

[14] T. Ikegawa and K. Motomura, *Throughput analysis of data link protocols for bridged networks,* Trans. IEICE, J74-B-I, no. 12, 1029-1041, 1991 (in Japanese)

[15] T. Ikegawa, *Queueing network modeling for performance analysis of HDLC based protocols,* Trans. IEICE, J75-B-I, no. 11, 723-733, 1992 (in Japanese)

[16] T. Ikegawa, *Performance analysis of window flow controlled networks with go-back-n error recovery: closed BCMP-type queueing network approach,* submitted

[17] H. Inai, T. Nishida, T. Yokohira, and H. Miyahara, *End-to-end performance modeling for layered communication protocol,* INFOCOM'90, 442-449, 1990

[18] ISO, *Connection oriented transport protocol specification,* ISO 8073, 1984

[19] ISO, *Logical link control,* ISO 8802/2 and ANSI/IEEE Sd. 802.2, 1985

[20] L. Kleinrock, *Queueing systems: vol. 2,* John Wiley & Sons, 1976

[21] D.T.D. Luan and D.M. Lucantoni, *Throughput analysis of a window based flow control subject to bandwidth management,* INFOCOM'88, 4C3.1-4C3.7, 1988

[22] B.W. Meister, *A performance study of the ISO transport protocol,* IEEE Trans. Comp., 40-3, 253-262, 1991

[23] L.C. Mitchell and D.A. Lide, *End to end performance modeling of local area networks,* IEEE JSAC, 4-6, 975-985, 1986

[24] M. Murata and H. Takagi, *Two layer modeling for local area networks,* IEEE Trans. Comm., 36-9, 1022-1034, 1988

[25] M. Nassehi, *Window flow control in frame relay networks,* GLOBECOM'88, 54.1.1.-54.1.7., 1988

[26] M. Rahnema, *Frame relaying and the fast packet switching concepts and issues,* IEEE Network Magazine, 18-23, 1991

[27] K.K. Ramakrishnan, *Analysis of a dynamic window congestion control protocol in heterogeneous environments including satellite links,* Proc. Computer Network Symp., 94-101, 1986

[28] M. Reiser, *A queueing network analysis of computer communications networks with window flow control,* IEEE Trans. Comm., 27-8, 1199-1209, 1979

[29] M. Reiser, *Admission delays on virtual routes with window flow control,* Proc. Perf. Data Comm. Sys. Appl., Pujolle (Ed.), North Holland, 67-76, 1981

[30] M. Schwartz, *Performance analysis of the SNA virtual route pacing control,* IEEE Trans. Comm., 30-1, 172-184, 1982

[31] A. Wolisz and R.P-Zeletin, *Modeling end to end protocols over interconnected heterogeneous networks,* Comp. Comm., 15-1, 11-22, 1992

A PREVIEW OF APPN HIGH PERFORMANCE ROUTING

James P. Gray and Marcia L. Peters[1]

IBM Corporation
SNA Studies
[1]Networking Systems Architecture
Internal Mail Address C74/673
P.O. Box 12195
Research Triangle Park, NC 27709

Abstract: After a brief review of APPN—Advanced Peer-to-Peer Networking—and a survey of existing routing techniques, a new SNA approach to routing called HPR—APPN High Performance Routing—is introduced.   Topics covered in this overview include HPR function placement within the OSI layered model, priority scheduling for multilink transmission groups, Automatic Network Routing, Rapid Transport Protocol, Adaptive Rate-Based congestion control, the relationship of effective congestion control algorithms to throughput and response time, and HPR's selection of frame relay as a preferred data link control.

Trademarks: IBM, Operating System/2, OS/2, VTAM, AS/400, APPN, and Advanced Peer-to-Peer Networking are trademarks of International Business Machines Corporation.

INTRODUCTION

What is APPN?: APPN—Advanced Peer-to-Peer Networking—is an extension of SNA—Systems Network Architecture. APPN was first announced in 1987. At the time of writing, APPN was available on the following IBM products:

- AS/400 [21] [6] [7]
- 3174 Establishment Controller [11]
- OS/2 Communications Manager [25] [9]
- System/36 [7]
- DPPX/370
- System/390 mainframe and 3745 front end processor (VTAM 4.1, NCP 6.2)
- 6611 Network Processor router

Availability on the RISC System/6000 via AIX SNA Services is planned for 1993. A number of other companies also offer APPN in their products, including Brixton, InSession Inc., and Systems Strategies Inc.   Additional vendors have demonstrated APPN prototypes or are expected to offer APPN in their products, including: Advanced Computer Communications, Apple Computers, 3Com, Cabletron Systems, Cisco Systems, CrossComm, Data Connection Ltd., Network

Equipment Technologies, Network Systems Corporation, Novell, Retix, Siemens-Nixdorf, Ungermann-Bass Access-One, Unisys, and Wellfleet Communications. IBM opened the APPN End Node protocols in 1991 by publication of the SNA Type 2.1 Node Reference [23]. In March 1993 IBM opened the APPN Network Node protocols by source code licensing, technology licensing, and publication [24].

What is APPN/HPR?: APPN/High Performance Routing, which IBM called *APPN+* when revealing its future networking directions in March, 1992, is a further extension of SNA to take advantage of fast links with low error rates. It replaces ISR (intermediate session routing)—the routing technique in current APPN products—with HPR (high performance routing). One of its key benefits is the ability to nondisruptively reroute sessions around failed nodes and links. APPN/HPR can improve intermediate routing performance by 3 to 10 times, greatly reduce network node storage requirements, and augment existing LU-based flow control with an advanced congestion avoidance algorithm called ARB—Adaptive Rate-Based congestion control. In late 1992, IBM said it intends to make APPN/HPR technology available in a future release of its network node source code license, then in March 1993, IBM announced plans to publish a draft specification of the HPR formats and protocols in the third quarter.

APPN'S CLIENT-SERVER MODEL

The term *peer-to-peer* is misleading because it captures only one of two equally important aspects of APPN. Equally descriptive terms could have been *distributed* or *client/server*. *Distributed* means decentralized. APPN is not locked into the strict hierarchical topology required by its predecessor, SNA subarea architecture: APPN architecture supports a variety of physical topologies such as star, hub, mesh, and hierarchical, over a variety of transmission media including most popular LAN and WAN media. *Distributed* says that important data and important applications may reside on any computer in the network. In other words, computers other than mainframes are running significant applications. Such applications are network-centric rather that mainframe-centric. The term *peer* accurately characterizes this aspect of APPN. *Client-server* says that in application design, function placement is flexible [17]. It recognizes that computers differ in terms of physical location and physical security, the level of attended support (e.g. operator intervention, regular backup, and so forth), connectivity to other computers, permanent storage media (disk, tape, and the like), the amount of RAM for running large applications, support for multitasking, and computing capacity (MIPs, FLOPs). Client-server principles encourage flexible application design so that smaller computers can take advantage of the capabilities of larger computers. It is the client-server aspect of APPN that the term *peer* does not suggest.

With client-server concepts in mind, and to support emerging networked applications for distributed computing, APPN defines two main node types: the end node (EN) and the network node (NN). APPN also interoperates well with a pre-APPN node type called the Low-Entry Networking end node (LEN EN), a subset of the APPN EN. LEN was a precursor to APPN; APPN is the strategic base for enhancements to SNA. (All three of these node types are classified as SNA Type 2.1 nodes.) This general architectural structure makes sense. The control plane in an NN requires more physical resources (CPU, RAM, disk) to support the network control applications and services that it provides. By offloading most of these functions to an NN, an APPN EN can be relatively small and inexpensive, and/or have more of its resources left for applications, and still partake of APPN's automation and dynamics. An NN may also have specialized routing hardware and attachments to many WAN links.

A network node is a **server**, in the sense of providing network services to its end node clients. The services are directory services and route selection services. An brief overview will illustrate the respective client and server roles of ENs and NNs.

Directory services means locating a partner application: determining what computer the application currently resides on. This avoids the EN having that information predefined. This also gives network administrators flexibility in moving an application to the computer best able to support it. With APPN, moving an application requires only local definition changes at the computer where the application used to be, and at the computer it moves to. All other computers learn this information dynamically, as needed.

Route selection services means picking the best route for the session, based on a user-specified class of service and transmission priority, what possible routes are currently available through the network, and the destination computer's location. Route selection also avoids congested nodes, and randomizes among equivalent routes, in order to distribute the load.

These network control applications of APPN have been described extensively in the literature [30], [31], [10], [8], [12], [36], [20]. Less often discussed are its lower layers.

A network node is also a **router**: it forwards traffic that crosses it on the way to somewhere else. The second half of this article discusses APPN routing techniques in more detail.

First though, let's examine the differences in these APPN node types using a layered architecture model. We will cover differences in network control applications (in the uppermost, or transaction, layer) and differences in the lower layers (Layers 2, 3 and 4—the data link, network and transport layers). With this foundation we can examine APPN high performance routing, describe its benefits, show how it complements existing APPN intermediate session routing functions, and consider what kinds of networking products it is appropriate for.

Figure 1 shows APPN's functions divided approximately into horizontal strata, according to the OSI networking model, like a layer cake. [18], [28] In addition, one can imagine vertical slices through these layers, like a slice of cake, containing some cake from each of the layers. There may be chunks of fruit in the cake, so that each slice is a little different. One of these slices is the *user plane*, the piece of system software on a computer directly supporting end user application programs running on that computer. APPC—Advanced Program to Program Communication—also called LU 6.2 or Logical Unit type 6.2—is an example of a user plane. Another slice is the *systems management plane* (shown behind the user plane). Systems management is often structured according to client-server principles, with a client component in an end-user's computer being called an *agent* or *entry point*, and a server component in a specialized (but not necessarily centralized) computer being called a *manager* or a *focal point*.[2] The slice that this paper focuses on is the *control plane*, sometimes called the *control point* in an SNA node. Its job is to support the user and management planes, automating such chores as the distribution of routing information and directories. Bear in mind that while some networking architectures (LAN architectures, in particular) combine the control and management planes, in APPN they are distinct. Discussion of the management plane is outside the scope of this paper.

The next section introduces the APPN node types.

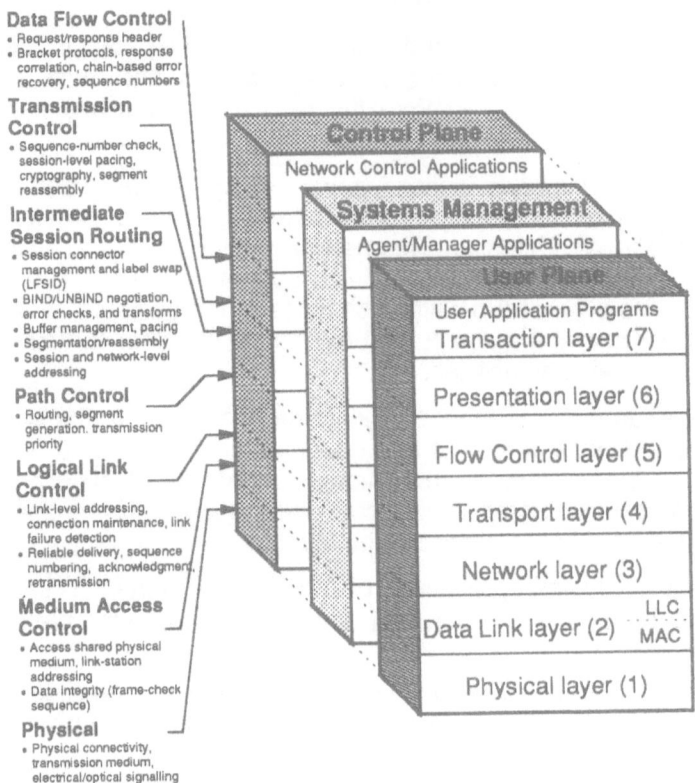

Data Flow Control
- Request/response header
- Bracket protocols, response correlation, chain-based error recovery, sequence numbers

Transmission Control
- Sequence-number check, session-level pacing, cryptography, segment reassembly

Intermediate Session Routing
- Session connector management and label swap (LFSID)
- BIND/UNBIND negotiation, error checks, and transforms
- Buffer management, pacing
- Segmentation/reassembly
- Session and network-level addressing

Path Control
- Routing, segment generation, transmission priority

Logical Link Control
- Link-level addressing, connection maintenance, link failure detection
- Reliable delivery, sequence numbering, acknowledgment, retransmission

Medium Access Control
- Access shared physical medium, link-station addressing
- Data integrity (frame-check sequence)

Physical
- Physical connectivity, transmission medium, electrical/optical signalling

Control Plane
Network Control Applications

Systems Management
Agent/Manager Applications

User Plane
User Application Programs
Transaction layer (7)
Presentation layer (6)
Flow Control layer (5)
Transport layer (4)
Network layer (3)
Data Link layer (2) LLC / MAC
Physical layer (1)

Figure 1. Layered Model Showing SNA Function Placement

LEN End Node: A LEN end node has no network control transaction programs at all, and no client support for requesting the services of a network node: it has no APPN control plane. The routing function in a LEN end node is simply to transmit to an adjacent node using a predefined link and any required link signalling information (such as the link station address on an SDLC link, the MAC and SAP addresses on a LAN, or a selected adapter and telephone dial digits for a switched connection). Without default routing (explained in Figure 2), a LEN end node must have definitions for the locations of all partner applications. One consequence is that if a network administrator decides to relocate an application to a different computer, she needs to distribute updated definitions to **every** LEN EN that accesses that application. (This is one of the things APPN was invented to fix!) The left panel of Figure 2 illustrates such a configuration and the definitions required at LEN EN *A*.

Alternatively, a LEN EN may define all partner applications as residing on an adjacent NN (whether they actually reside there or not). This is termed "default" routing and is shown in the center and right panels of Figure 2. In this case the LEN EN acts as a passive client, accessing the services of a network node server indirectly, by simply attempting to route data to it (sending it a BIND, in SNA parlance). While this method relieves the LEN EN of predefining individual network addresses for all its partner applications (a single definition "all partners reside on the NN" will suffice), the resulting sessions always traverse the network node, not always an efficient route (especially where direct mesh connectivity exists between every pair of computers, as on a LAN). On the other hand, default routing may be quite acceptable if the LEN EN is a portable computer with

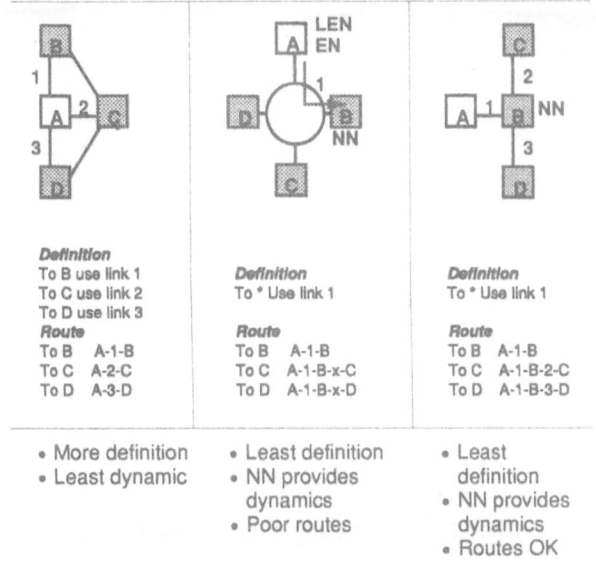

Definition	Definition	Definition
To B use link 1	To * Use link 1	To * Use link 1
To C use link 2		
To D use link 3		
Route	**Route**	**Route**
To B A-1-B	To B A-1-B	To B A-1-B
To C A-2-C	To C A-1-B-x-C	To C A-1-B-2-C
To D A-3-D	To D A-1-B-x-D	To D A-1-B-3-D

• More definition	• Least definition	• Least definition
• Least dynamic	• NN provides dynamics	• NN provides dynamics
	• Poor routes	• Routes OK

Figure 2. LEN End Node Explicit and Default Routing

a single dial-up line to access network computing resources. In such a configuration, all connections would traverse the NN anyway, as the right panel of Figure 2 illustrates.

APPN End Node: An APPN end node adds a small control plane, with client transaction programs (at the transaction layer—layer 7) to register its applications (Logical Units, or LUs, in SNA lingo) to a network node and request services like locating destination applications and selecting routes. [23] Figure 3 illustrates both APPN and LEN end nodes (the latter being a user plane without an APPN control plane).

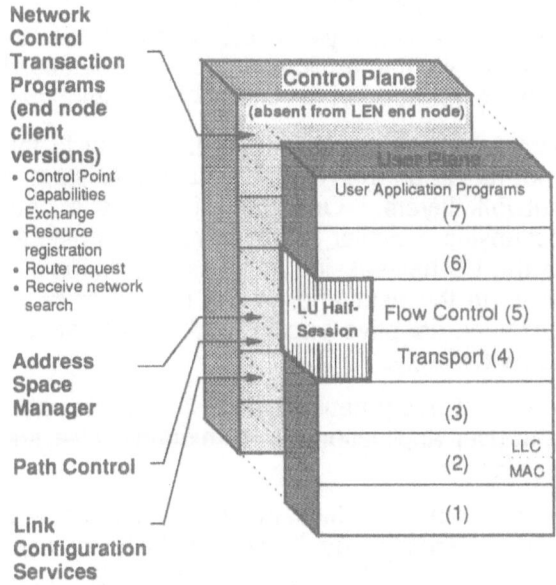

Figure 3. An APPN End Node

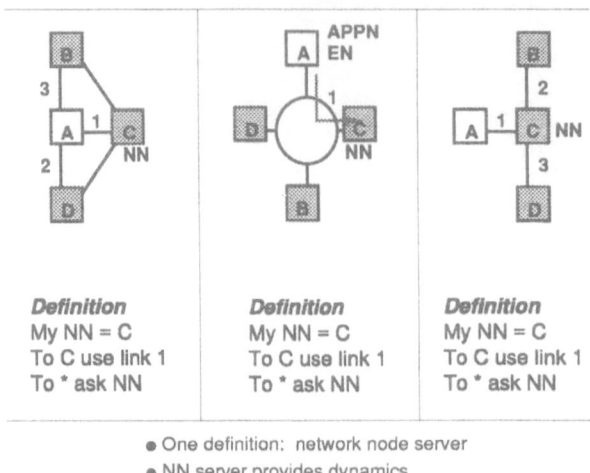

Definition
My NN = C
To C use link 1
To * ask NN

Definition
My NN = C
To C use link 1
To * ask NN

Definition
My NN = C
To C use link 1
To * ask NN

- One definition: network node server
- NN server provides dynamics

Figure 4. An APPN EN Knowing Only its Network Node Server

Unlike the LEN EN which interacts with a network node server passively if at all, an APPN EN actively requests NN services. Some of the benefits over the LEN EN are better dynamics, less definition, and better routes. An APPN EN uses the route provided by its network node server. A different route may be provided every time a new session is set up, and the route provided does not necessarily traverse the NN. This point is illustrated below. In Figure 2, *A* was a LEN EN; in Figure 4, *A* is an APPN EN.

The routing function in an APPN EN is still minimal: an EN can only be the endpoint of a session, never an intermediate node of someone else's session. The transport layer of an APPN EN is enhanced by its support for adaptive pacing. The network layer is the same as a LEN EN.

APPN Network Node: A network node adds specialized network control transaction programs at the transaction layer in the control plane, to manage distributed directories and maintain the replicated topology database used for route computation, as well as server support for end node clients. [24]

It also adds *intermediate session routing*—ISR—the ability to forward packets for applications that do not reside on the NN itself. ISR consists of enhancements at the transport and network layers. One part of ISR is a component called a *session connector*, occupying a similar position and performing similar functions in the protocol stack as the LU half-session in a session endpoint node. (Compare the shaded components in the user and control planes of Figure 5.) ISR functions include error recovery, adaptive pacing, and the adjustment of packet sizes via segmentation and reassembly.

As Figure 5 shows, not every network node has a user plane. A router that does not host any end-user applications is an example of a specialized NN that does not need a user plane.

Many people—even some of IBM's marketing literature—describe SNA as "non-routable." This is not strictly true. The capability for intermediate session routing previously existed in SNA Type 5 and Type 4 nodes such as VTAM and NCP, using FID4 transport (layer 4) over explicit routes and virtual routes (ERs and VRs—the SNA path control network—at layer 3). The backbone was phys-

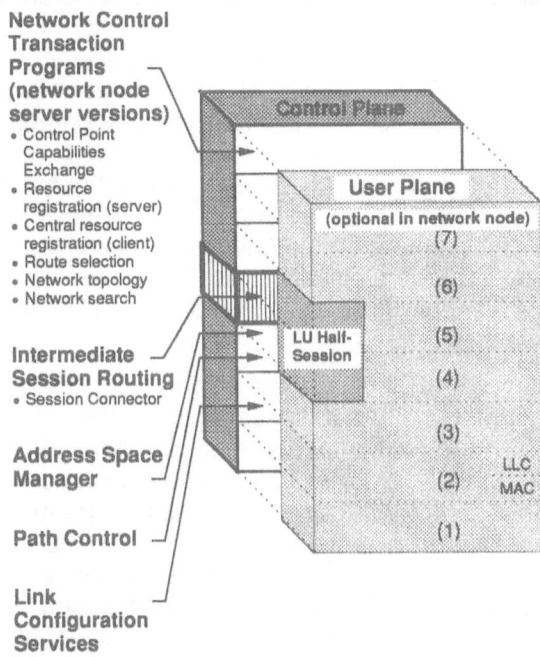

Network Control
Transaction
Programs
(network node
server versions)
• Control Point
  Capabilities
  Exchange
• Resource
  registration (server)
• Central resource
  registration (client)
• Route selection
• Network topology
• Network search

Intermediate
Session Routing
• Session Connector

Address Space
Manager

Path Control

Link
Configuration
Services

Figure 5. An APPN Network Node's Extended Control Plane

ically structured as a mesh, and the peripheral network, using FID2 transport, was strictly hierarchical. Setting up the SNA "routing tables"—ER and VR path definitions—was a laborious manual process or required the use of a tool like NetDA (Network Design Aid). APPN enhances the FID2 transport that first emerged in subarea SNA's peripheral network, and automates the maintenance of routing information and directories. **APPN is native routing technology for SNA.**

Let's examine intermediate session routing in more detail, to understand how High Performance Routing differs from it.

EXISTING ROUTING TECHNIQUES

The literature—and vendors' product lines—are filled with a variety of routing techniques, including routing by network address, label swapping, and source routing. And these routing techniques are often supplemented by network control algorithms to dynamically distribute routing tables or maintain a topology database. They may also be supplemented by discovery or address resolution protocols to dynamically map the name of a desired communication partner to a network-layer address or routing information. All the routing techniques discussed below are suitable, in general, for implementation in either hardware or software, while network control algorithms and address resolution protocols are frequently implemented in software.

Routing by network address is the technique used in Internet Protocol (IP). A single 32-bit routing label that must be unique within the scope of an entire inter-network represents the final destination, and serves as an index into a routing table specifying the next hop. The next hop taken depends on the current state of the routing table at the node processing the packet [16]. Several algorithms exist to distribute IP routing tables, some standard and some proprietary. One of the mostly widely used standards, the Routing Information Protocol (RIP), distributes the entire IP routing table periodically at timed intervals. This type of table distribution is called a *path status* algorithm. As individual IP subnetworks become larger, the amount of administrative traffic generated by these regular routing table updates grows exponentially, placing an upper bound on the size of an individual IP subnetwork. IGRP, Cisco's proprietary routing algorithm, also uses a path status algorithm to distribute its routing table updates and, consequently, also places an upper bound on the size of a subnetwork. Within a single IP subnet, it is necessary that all routers support the same algorithm. Hence the focus on standards rather than on proprietary techniques. A relatively new standard algorithm for TCP/IP, Open Shortest Path First (OSPF), is gaining in support among router vendors [16] and uses a more efficient *link-state* type of algorithm (defined below).

Label swapping is a technique used in current APPN (APPN/ISR) and, interestingly, also in the CCITT high-speed recommendation for Asynchronous Transfer Mode (ATM). A packet bears a single network-layer routing label, representing the next hop. A router or high-speed switch substitutes a new label before transmitting the packet. In connection-oriented protocols, the label-swap tables are generally set up in intermediate nodes when the connection is established, based either on predefined information or on a topology database reflecting the state of the network at the time of connection establishment. The APPN topology database updates are only distributed when information changes. This type of algorithm is called a *link state* algorithm. Link-state algorithms generate much less administrative traffic than path status algorithms, removing one barrier to the growth of larger individual subnetworks. TCP/IP's OSPF is also a link state algorithm.

Source routing is a third routing technique, commonly seen in LAN bridging. A source-routing variant called Packet Transfer Mode (PTM) is currently being applied in high-speed trials such as the Aurora test bed. In source routing a list of routing labels, representing the entire route, prefixes the packet. The route is determined in advance, usually based on a discovery protocol or a topology database. Some argue that source routing is the most efficient of these three techniques, requiring the least processing at intermediate nodes, yielding the maximum throughput.

Link-Sharing—a Fact of Life: Whatever routing technique a protocol uses, most state-of-the-art protocols (APPN included) provide a means for different routing stacks to share the transmission medium. A medium access control (MAC) sublayer provides a graceful and standard way to share the link. Examples of media with a distinct MAC sublayer are token ring (802.5) [22], Ethernet, 802.3. and FDDI. The Point to Point Protocol for HDLC (PPP—RFC 1330 [4], [34]); and Frame Relay (Multiprotocol Encapsulation—RFC 1294 [5] and the Frame Relay Forum Implementation Agreements) are not strictly MAC-layer technologies but permit similar link sharing. If the medium has a MAC sublayer, logical channels can be established between paired adjacent link stations on the basis of predefinition, discovery, or a capabilities exchange protocol peculiar to that medium. It is likely that a MAC sublayer or similar standard will also be defined for any new transmission technologies that may emerge in the future.

Traditional SNA Approaches to Routing and Error Recovery: Traditional SNA—subarea SNA and APPN/ISR—has been fundamentally connection-oriented in its approach to routing. In general, traditional SNA attempts to provide high reliability over a variety of links (including error-prone links with poor characteristics). Error recovery is performed at the data link layer (layer 2) with connection-oriented data link controls like SDLC, X.25 ELLC, or LAN logical link control type 2 (IEEE standard 802.2). [26] [32] Recovery is also performed at the

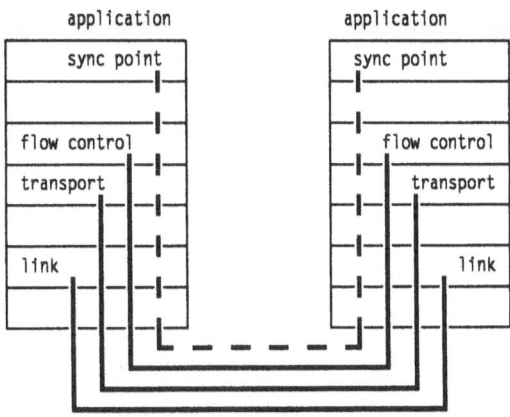

Figure 6. Four Levels of Error Recovery in Traditional SNA

transport layer—layer 4 (in reassembling segmented data, the transmission control component of the LU's half-session ensures that no packets are missing or out of order) and at the data flow control layer—layer 5 (the half-session enforces chains as the unit of application-level error recovery). Figure 6 illustrates these three levels of error recovery. If any of these protocols are violated, due to failure of the underlying transmission facilities, unrecovered packet losses in the network due to congestion at intermediate nodes, inability of the receiving application to buffer all the received data, or transparent rerouting by a subnetwork that causes some packets to arrive out of order, traditional SNA deactivates the session. With its key design point to support business-critical applications like finance, order entry, inventory control, and credit authorization, traditional SNA has industrial-strength algorithms to detect and prevent the occurrence of these error conditions. In addition, sync point (an APPC checkpointing service) ensures the integrity of distributed databases. In case of session, database, or processor failure, all applications and databases assume a known state and a distributed transaction can resume exactly where it left off once communication is reestablished.

Routing in APPN/ISR: APPN/ISR determines the route for a session when the session is set up; the route remains in effect for its duration. The route is chosen based on a user-specified class of service, a transmission priority, the destination, and the available routes, with randomization if more than one route is acceptable. Thus different sessions between the same pair of LUs can have different routes based on user-specified parameters. Every session has a unique identifier (an "FQPCID"), assigned by the origin node, that refers to the session at every node it traverses, throughout its lifetime. This identifier is used for network management and by the transaction programs in the control plane during session setup, but not for routing.

The routing field or network-layer header in APPN/ISR has a single routing label 17 bits long called a *Local-Form Session Identifier* (LFSID), defined by the SNA FID2 transmission header format. It needs to be unique only on a given link. Because it uses FID2, APPN/ISR routing can coexist with pre-APPN traffic, such as 3270 terminal traffic, using the same link. An APPN route is a series of *session stages*, end to end, each with its own LFSID routing label. In a network node, the LFSID indexes a "routing table." This table is distinct from the topology database. Each entry in this table is called a *session connector* and is created at the time of session establishment. As an NN forwards a packet from an inbound link to an outbound link, it replaces the LFSID in the packet header with the LFSID from the session connector. APPN routing can therefore be clas-

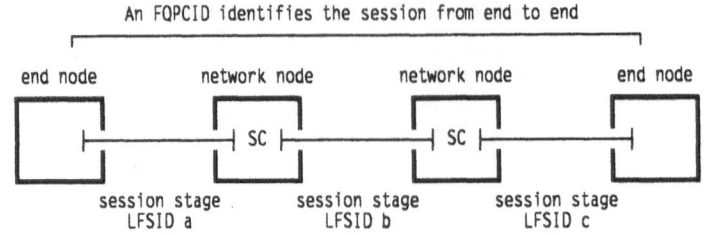

An FQPCID identifies the session from end to end

Figure 7. Session Stages Interconnected by Session Connectors

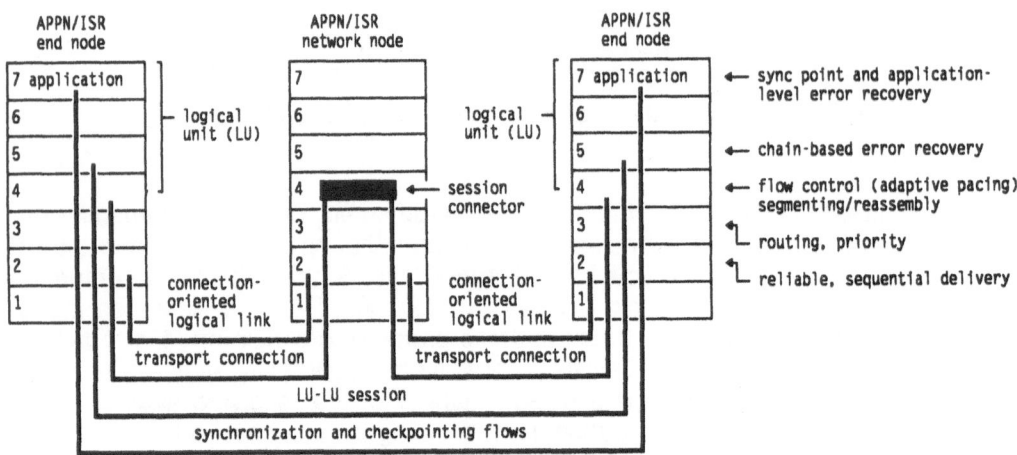

Figure 8. APPN Intermediate Session Routing

sified as *label-swap routing*. There can be many SNA sessions at once on a logical link, each with a distinct LFSID. APPN nodes usually select LFSIDs dynamically, during session set-up. (Pre-APPN nodes may have LFSIDs preassigned.)

Because of APPN's original design point to support business-critical applications over good-to-poor links, in additional to label-swapping, ISR performs additional, transport-layer functions at intermediate nodes. These other functions include segmenting and reassembly, pacing, and priority queuing for transmission. Figure 8 illustrates function placement in APPN intermediate session routing. Let's examine each of these functions more detail.

Segmenting and Reassembly: Different links in a network may support different maximum packet sizes, for reasons of link speed, transmission delay, data link control timing requirements, fairness, and node buffer capacities. [37] This point is illustrated by one of the graphs from "Data Link Control and Contemporary Data Links" by Traynham and Steen (IBM technical report 29.0168, June 1977), reproduced in Figure 9.

In an architecture like APPN that embraces a variety of LAN and WAN transmission media of varying speeds and characteristics, it is not reasonable to expect every node to agree on the same "best" packet size. As a general strategy to maximize performance, APPN/ISR sends the largest packet size allowable on each link. When necessary, segmentation and reassembly are done at intermediate nodes. This is one of the functions performed by APPN intermediate session routing. Any changes to ISR must address the issue of packet size in some manner.

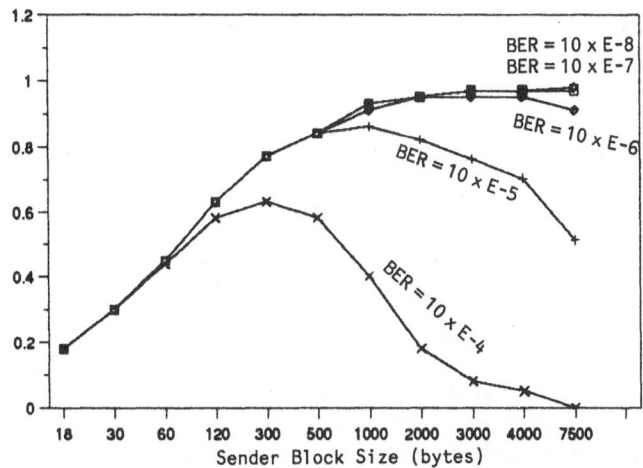

Figure 9. Throughput Efficiency as a Function of Bit Error Rate and Packet Size

Adaptive Pacing: Pacing is a flow control and congestion control technique to adjust the sender's transmission rate according to the capacity of the receiving node's buffers. Pacing is another function performed by APPN intermediate session routing. In APPN/ISR, pacing occurs independently on each session stage or BIND hop. APPN nodes support both fixed and adaptive pacing. Adaptive pacing is preferred, while fixed pacing permits interoperation with older SNA nodes. Because each APPN session stage is independently paced, every node (nodes supporting applications, as well as routers) can adapt the pacing for the traffic it handles in accordance with its own local congestion conditions. This is the basis for global flow control and congestion management in APPN/ISR. Any changes to ISR must improve upon this already-superior existing function.

With fixed pacing, a predetermined number of packets can be sent before the sender has to wait for the receiver to give permission to send more. Adaptive pacing is a more powerful and flexible scheme wherein a sender can send only a limited number, or *window*, of packets per explicit grant of permission to proceed. The window size is changed dynamically, based on conditions at the receiver. This lets a receiving node manage the rate at which it receives data into its buffers. Adaptive pacing provides a node supporting many sessions, or unpredictable bursts of traffic, a dynamic way to allocate resources to a session that has a burst of activity, and to reclaim unused resources from sessions that have no activity (rather than predefining a buffer pool of a particular size for every active session). Thus we see that adaptive pacing allows the receiving node to use its available buffer resources efficiently. It can also prevent potential protocol deadlocks.

If a node is running low on buffer resources, it uses pacing to tell the upstream node to slow down. If that node becomes congested, it in turn may tell its upstream node to slow down. When a node is not congested, it gives the upstream node permission to send faster. When a receiving node gives a sender permission to send a certain window size, the receiver has reserved sufficient buffers in advance, guaranteeing that data, once sent, will not be lost due to congestion. In practice, many products use statistical or demand buffering schemes, which are acceptable as long as confirmed buffers are available when needed.

A separate instance of adaptive session-level pacing exists for each session running to or through a node, to manage the flow of data on one LU-LU session.

Adaptive session-level pacing also applies to the sessions between APPN control points. Adaptive session-level pacing occurs independently on each session stage. Pacing is done by the half-session component of the LU in a node containing a session endpoint, and by the session connector component in an intermediate node. This will become important when we examine how HPR can replace the ISR function (including the session connector) in network nodes, and how HPR can supplement the equivalent component—the LU half-session—in session endpoint nodes.

Priority Queuing for Transmission: Transmission priority permits more important data to pass less important data at queuing points in the network. Priority is another function performed by APPN intermediate session routing. Any changes to ISR must also be equal to or better than the existing support in this area. APPN has four priority levels: a *network* priority, and three session-level priorities: *high*, *medium*, and *low*. Network priority is the highest and is reserved for network control traffic such as pacing messages, topology database distribution, and session establishment. The other three priority levels are for user traffic. A user selects a priority level indirectly, by specifying a mode name defining a session's characteristics. The mode name maps to a class of service definition, which in turn specifies the priority level associated with that class of service. The transmission priority selected for a session is carried in the session activation request (BIND) at session establishment, allowing every node along the path to assign the same priority value, to be used in routing. A transmission priority applies to the session for its lifetime, at every node it traverses. Both ENs and NNs support transmission priority.

Transmission priority is implemented by the path control component in APPN. One function of path control is to direct traffic to the right outbound link. Path control can also multiplex different sessions on a single link. Another function of path control is to ensure that higher-priority data is transmitted before lower-priority data. This is generally implemented as four different queues into which message units are placed, depending on the priority associated with the corresponding session. After the DLC finishes transmitting the current message, path control picks the next message for transmission, selecting from the highest priority queue having a message unit waiting. To ensure that lower-priority data is not preempted indefinitely, an aging mechanism is also used.

Many people have observed that some protocols duplicate certain functions at layers 2 (data link control) and 4 (transport), leading to difficulties and ambivalence in discussions of the subject (especially at meetings of international standards bodies!). Furthermore, the current 7-layer model (see Figure 1), mirrored in the organization of standards bodies, does not adequately describe a new class of protocols that are so versatile they can act either as a transport (layer 4) or as a virtual link (layer 2). [19][27][35] The current paradigm is ripe for an overhaul.

Logical link control (LLC—the upper half of layer 2) was originally created to permit the coexistence of both connection-oriented and connectionless service, between multiple link stations, on the same LAN segment. With the extension of this technology by bridging, and then remote bridging which introduced longer and variable end-to-end delays, came sensitivities to the LLC timeout values which are used to detect link or link-station failure. Nevertheless, the principle of extending a data link layer connection across multiple hops is now firmly established. The advantage of data forwarding at a low layer is performance. The disadvantage is that higher layer protocols (like APPN) that select routes based on APPN link characteristics can't see the actual characteristics of the links interconnecting remote bridges. A bridged link appears as a single link in the APPN topology

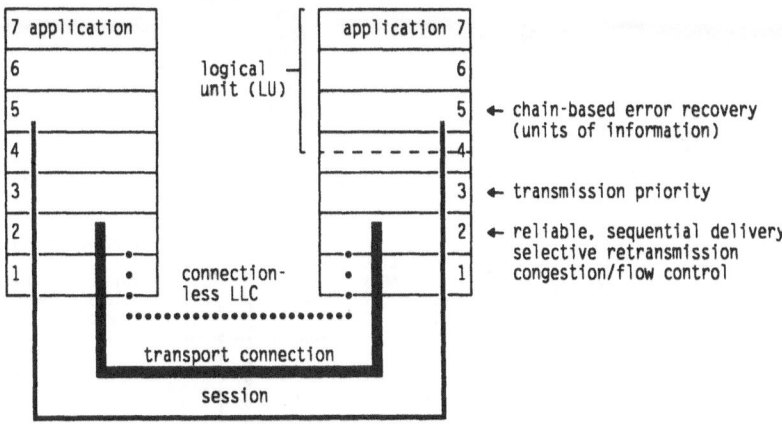

Figure 10. A Transport-Oriented Logical Link Control

database, with a single set of link characteristics hiding the complexity of multiple hops. The inability to know the true hop count, or to distinguish between a slow link and a fast one, can lead to poor route choices.

One solution to this problem is to replace pairs of remote bridges by pairs of APPN/ISR network nodes. However, this only solves part of the problem, since the ISR functions of pacing and segmentation/reassembly, absent from bridges, would reintroduce delays, and current bridging standards lack support for priority.

A solution was needed to better integrate the bridging concept into APPN. With good links, some of traditional SNA's error recovery is redundant. One possible way to reduce overhead is to omit error recovery at the data link layer, replacing the usual connection-oriented LLC with a connectionless LLC. The transport is replaced by APPN's versatile new transport protocol, RTP—Rapid Transport Protocol, which efficiently provides any needed error recovery over multiple hops. One drawback to this approach is the placement of transmission priority (path control) **below** the transport.

An better function placement is shown in Figure 10. The new transport becomes part of the LLC sublayer in layer 2, acting as an enhanced logical link. This is true when the logical link comprises not only a single hop, but multiple hops. In this paper we'll call this versatile new class of protocols *transport-oriented logical link controls*. Thus, a transport-oriented LLC like RTP, spanning multiple links and nodes, if it meets the needs of upper layer components to which it provides service, can replace one or more APPN/ISR session stages, acting as a "virtual link." APPN+ takes advantage of this principle (see Figure 11).

The function placement of Figure 10 is better for several reasons. First, it places transmission priority **above** the new transport—important for a node with more than one link. Second, it permits the existing transport (such as an LU half-session—see Figure 3) to be kept intact for the other key services it provides, and because of its integration and packaging into system software (like APPC).

The first property, the coupling of priority scheduling to the rest of the protocol stack, can be understood as follows. In subarea and APPN SNA, the users declare the class of service (COS) needed for their traffic at the transaction program API or user logon API by giving a mode name. The mode name is mapped to a class of service, which in turn specifies both the route through the network (either by listing valid choices in the subarea case, or by defining the parameters of the route selection algorithm in the APPN case) and the trans-

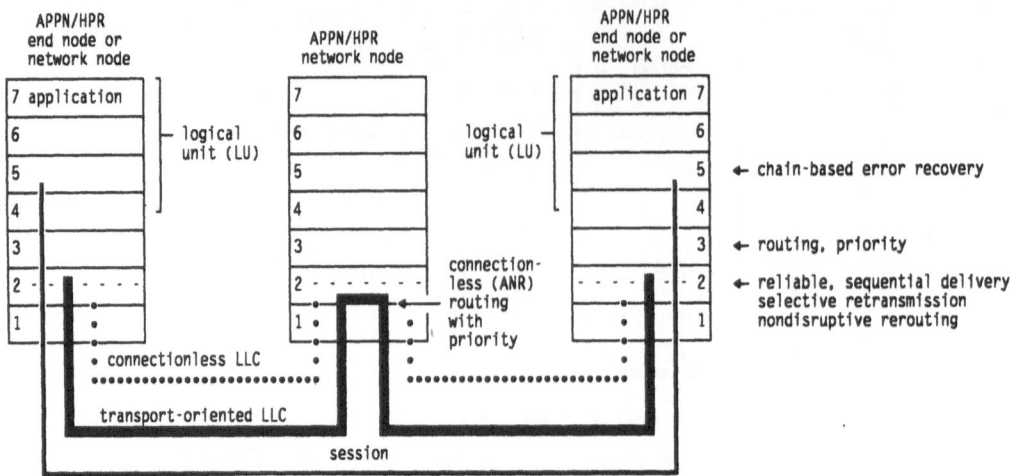

Figure 11. A Multiple-Hop Transport-Oriented Logical Link Control

mission priority to be used for the traffic. In subarea, the route is fixed by the ER routing tables and the VR-to-ER mapping, while the priority is carried in the transmission header of each packet. In APPN, the route is fixed by the route selection control vector on the session initiation request (BIND) and the priority is carried in the BIND and saved at each intermediate session routing point, where it is used on the fly for each packet. In both subarea and APPN, the actual priority queueing is done at the top of the DLC component: when the line is finished transmitting the current frame (for example, at the end-of-frame interrupt from the hardware), the highest priority message is taken off the DLC transmit queue (or, in the case of multilink transmission groups, the MLTG transmit queue). An aging algorithm ensures that even low priority traffic gets through under heavy loads.

HPR needs to preserve the relations above, so needs to have priority scheduling queues at the end-of-frame events even on connectionless DLCs. HPR ties the session class-of-service to these queues by encoding priority bits into the HPR headers. If this were not done (e.g., as it cannot be done in networks that lack priority link queuing, or adequate coupling of it to user COS) then HPR would not have preserved the COS semantics at the user APIs.

One may well ask why, if the new transport is so much *like* an LLC, IBM did not choose an existing standard (on a LAN), with traditional bridging (to get across the WAN). The answer is that traditional LLCs are not up to the task. They are restricted to particular media and are not optimized for multiple hops. A new class of transport protocols was needed. APPN/HPR is not limited to LANs: it can run over any transmission medium that supports an unacknowledged (or connectionless) type of service, for example: LAN LLC, LAP-D, LAP-E, or SDLC (using unnumbered information—UI—frames) or X.25 (using QLLC). Furthermore, RTP provides advanced functions like selective retransmission and adaptive rate-based congestion control that no existing standard supports. Another reason is that even when a particular bridging technology supports non-disruptive rerouting at layer 2 (and many new ones, like frame relay or Data Link Switching [33], do), its selection of a new route is not integrated into APPN and is not based on class of service. An advantage for APPN/HPR, as compared with bridging or TCP/IP to span the wide area network, is that its route selection includes awareness of link characteristics (speed, delay, cost, security) and node characteristics (route addition resistance, congestion).

Introduction to APPN High Performance Routing: APPN/HPR augments APPN/ISR's layer 3-4 transport and network functions with two new elements: Rapid Transport Protocol (RTP) and Automatic Network Routing (ANR), shifting the locus of APPN routing from layer 4 down to layer 2. [3] [15] [14]. Each is described more fully below.

APPN/HPR also includes an ISR/HPR boundary function (not shown) to adapt an HPR-capable part of the network to an ISR-only part of the network, plugging one side of an ISR session connector to the end of an RTP transport-oriented logical link.

Because APPN/HPR is completely integrated into SNA and does not change APPN's control plane at all, any node can be upgraded to the HPR level of function transparently, continuing to interoperate with adjacent nodes still at the ISR level of function. As soon as two or more adjacent nodes are HPR-capable, initial benefits of HPR—non-disruptive rerouting, adaptive rate-based congestion control, selective retransmission, fair multiprotocol transport—start to be realized. As soon as two or more HPR-capable links exist back-to-back, such that high performance routing replaces intermediate session routing in at least one intermediate node (shown in Figure 11), further HPR benefits are experienced—fast routing *with* priority and reduced intermediate node storage.

HPR provides a connection-oriented transport (RTP), end to end, over connectionless source routing (ANR), over the minimal data link control. RTP acts as a "virtual link." The amount of function that RTP demands of the underlying DLC depends on the quality of the transmission medium. On high quality links, the DLC is not asked to provide reliable delivery, sequence numbering, retransmission, or acknowledgment. It merely provides a frame check sequence: errored packets are simply discarded. Links with bit error rates on the order of 10^{-7} or better are good candidates to use a connectionless DLC under RTP. Such links typically use digital transmission over fiber media. On high-error-rate links a connection-oriented DLC with error recovery may be used under RTP. In either case the benefits of HPR are significant.

An HPR path can also include multilink transmission groups—essentially, a bundle of links between adjacent nodes that are treated as a single "fat" pipe. Benefits of MLTG include high availability (if one link of a multilink group fails, the MLTG remains operational) and bandwidth on demand (additional switched links can be dialled up to augment the bandwidth an existing link). Long a feature of subarea SNA networking, MLTG can easily be added to HPR, without the performance and storage penalty of refifoing disordered packets at each MLTG hop. Reordering only needs to be performed at the endpoints of the RTP logical link, a task RTP was designed to accomodate.

RTP insulates the upper layers—the LU—and the user from any awareness of path switching, multipath routing, network-related congestion control activities, retransmissions, acknowledgments, packet resequencing, multiplexing, and so forth. Thus a user's investment in existing SNA applications is completely preserved.

ANR—Automatic Network Routing: The functions of ANR used by APPN/HPR are the following:

- Source routing with locally specified labels
- Connectionless, stateless, fast routing
- Discarding incoming packets in the event of congestion
- Servicing the outbound transmission link based on priority.

153

There's not much more to say—ANR is elegant and simple. [3], [15], [14]. ANR functions are done at every node along the path of an RTP transport-oriented logical link.

RTP—Rapid Transport Protocol: The functions of RTP used by APPN/HPR are the following:

- Connection awareness (of each individual session using the RTP logical link, the session partners of that session, the transmission priority, the current ANR route for the network connection, and if a path switch has occurred, all previous routes used by the logical link over its lifetime)
- Optional reliable delivery (sequence numbering, acknowledgment, and selective retransmission)
- Reordering, if needed (may be needed after a path switch, or if the route contains one or more multilink transmission groups)
- Determining the smallest maximum packet size along the path of the RTP logical link and ensuring (through segmentation) that all message units offered to the link are the proper size (this function eliminates the need for segmentation and reassembly at intermediate nodes)
- Flow control and congestion control/avoidance (Adaptive Rate-Based — ARB)
- Providing an interface to the ANR sublayer (below)
- Providing an interface to SNA path control (above)
- Non-disruptive rerouting
- Multiplexing (more than one session having the same class of service and transmission priority requirements can share a single transport-oriented logical link).

RTP is connection oriented. These RTP functions are done only at the endpoints of an transport-oriented logical link, not at any intermediate nodes (as shown in Figure 11).

At this point, the reader may be struck by apparent similarities between HPR and TCP/IP. One of the features that sets them apart, however, is HPR's congestion control mechanism, described in the following section.

ARB—Adaptive Rate-Based Congestion Control: A new technique for flow control and congestion control was needed for APPN+, to compensate for the loss of adaptive pacing function in intermediate nodes. This new function is called ARB—adaptive rate-based congestion control.

The function of ARB is to regulate the input traffic (load) offered to the RTP logical link in the face of changing network conditions. It is preventive, rather than reactive. When the network approaches congestion (increasing delay, decreasing throughput), ARB reduces the input traffic rate until the network's capacity is restored. When possible, ARB increases the sending rate without exceeding the rate that the receiving endpoint can handle.

ARB uses a closed-loop feedback mechanism based on information exchanged periodically between RTP components at the **endpoints** of an RTP logical link. (No ARB function is performed in intermediate nodes.) The feedback consists of information about two rates: the rate at which RTP accepts data arriving from the network, and the rate at which the RTP hands off the data to a recipient (such as the SNA path control component). Based on this feedback, the sender predicts when congestion is likely to develop, and takes steps to prevent it. If congestion does occur, the sender takes stronger measures to bring the network back to normal.

Because ARB addresses fairness at the data link layer, among different RTP logical links (which may be carrying multiple SNA sessions), APPN's existing

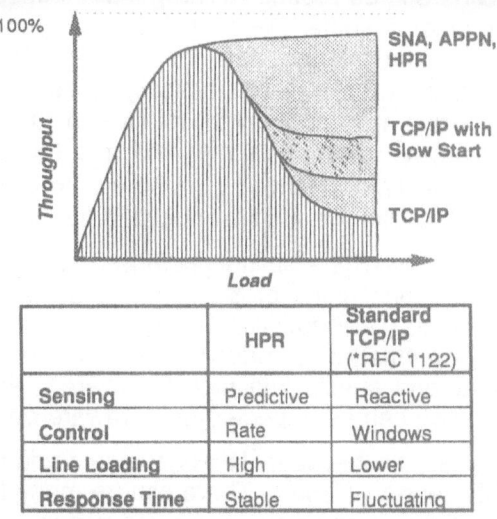

	HPR	Standard TCP/IP (*RFC 1122)
Sensing	Predictive	Reactive
Control	Rate	Windows
Line Loading	High	Lower
Response Time	Stable	Fluctuating

Figure 12. Relationship of Effective Congestion Control to Throughput and Response Time

adaptive session-level pacing algorithm continues to be used in networks with APPN/HPR, to provide fairness among SNA sessions at the data flow control layer (layer 5) in the LU half-session.

Analysis and simulation convinced researchers Rong-Feng Chang et al. [13] that ARB is superior to the "slow-start" congestion control algorithm of standard TCP/IP architecture [1], introduced after the Internet experienced a series of congestion collapses in October of 1986 [29]. In particular, ARB allows high link utilization rates (on the order of 80—90%) and is preventive, rather than reactive. The same study suggested that "slow-start" causes expensive link under-utilization, with lower design loadings (which should be a matter of concern to network administrators and people with budgets). The study also concluded that TCP slow-start exhibited unfairness and bias against certain kinds of traffic due to wide oscillations in packet delay and throughput (these should be of great concern to network users). Additional deficiencies of TCP slow-start cited in [13] included periodic packet losses, systematic discrimination against particular connections, bias against connections with long round-trip times, and bias against bursty traffic. It is worthwhile to note that many IP router vendors address these *TCP architecture* defiencies with proprietary extensions or product-internal enhancements. One informal mechanism used by several vendors is to give routing priority to short packets, which are assumed to be acknowledgments.

Figure 12 illustrates the relationship of effective congestion control algorithms to throughput and response time, which translate directly into cost savings and user satisfaction.

HIGH PERFORMANCE ROUTING VS. INTERMEDIATE SESSION ROUTING

Like APPN/ISR, APPN/HPR is connection-oriented, with the entire route for a session determined in advance.

Unlike ISR, which uses label-swapping, the entire route prefixes each HPR packet. The route is encoded as an arbitrary-length string of routing labels. This technique is classified as *source routing*. (The general concept of source routing may be familiar, being used in some LAN MAC-layer bridging protocols [22]). The particular source routing technique used for APPN/HPR is ANR—automatic network routing. The labels vary in length, typically 1-2 bytes. Each routing label has local significance only: it is meaningful relative to the node processing the label. The first label always represents the next hop (or, at the last node, the terminus of the RTP logical link). Routing consists of stripping off the first routing label and transmitting the packet on the link indicated by the stripped label.

Because of ANR's simplicity, existing platforms that implement it may well realize significant performance gains of 3—10 times on their existing hardware. That is, on many platforms, HPR function might be deliverable in the form of a software upgrade. (By contrast, other advanced routing techniques such as ATM often require expensive and specialized new hardware.) Performance gains at the upper end of this range can be expected on hardware optimized for ANR.

A single RTP logical link between HPR-capable endpoints replaces two or more back-to-back ISR session stages. There can be many RTP logical links at a time on a transmission link; intermediate nodes are unaware of, and unconcerned with, these individual connections: they only see a stream of packets prefixed with routing labels representing their outbound links.

Many SNA sessions can share a single RTP logical link, provided they have the same class of service and transmission priority.

If a path fails, the RTP component at the endpoint of the logical link obtains a new path, still based on the desired class of service and transmission priority, and keeps going—all without the session's awareness. If packets get lost or out of sequence during the switch, RTP takes care of it transparently. As a result of path switch, it's even possible for the forward and reverse directions of data flow to follow different routes through the network.

Frame Relay—A High Speed "SDLC" for HPR: While SNA absorbs and runs over many different types of links, from S/390 channels to the synchronous framing mode of SDLC, SDLC has always played an important role in both subarea SNA and in APPN. In some sense, it has been the template DLC, the one on which the others were modeled. This shows up in the presence of XID exchanges on other DLCs, when XID is the name of a control frame within SDLC. SDLC, however, has the disadvantage of being a single access link, as contrasted with 802.x LANs and frame relay, both of which support link connections to multiple partners through a single hardware connection. Multiple access helps to reduce costs of ports and access lines into private or public carrier networks. And, while 802.x MACs have difficulty in working directly over WAN distances and at commonly available carrier line speeds, frame relay is well-adapted to use in WAN configurations.

In light of its benefits, frame relay has been adopted as one of the two preferred DLCs for HPR: preferred in the sense that products are encouraged to implement both frame relay and 802.x LAN connections for HPR support. This encouragement stops short of a mandatory architecture requirement because of the wide diversity of products and market niches in the world, but certainly the general-purpose networking products such as the IBM 6611, 3745, 3174, 3172, AS/400, and CM/2 products are expected to support both FR and 802.2 LANs for HPR connections.

HPR's support of FR includes both "through a carrier" and "null carrier" configurations. Since FR uses HDLC framing, it runs on existing SDLC adapters as a software or microcode upgrade; being configurable as null network connections allows it to be used wherever point-point full-duplex SDLC lines are now used, without changing line provisioning in any way. Multidrop SDLC lines pose another problem: some can be converted to multiple-connection FR services through a carrier network (the carrier's network becomes the "multidrop" in a certain sense, but with added function as compared to multidrop since the multidrop polling is removed); some can be reconfigured to use APPN/HPR routing services, perhaps even with line cost savings; some will have to remain as SDLC multidrop until we can support multidrop FR configurations (while multidrop is not part of the FR standard, it is easy enough to add, and we intend to do so).

Areas for Future Study: It should also be possible to have multi-path RTP logical links—so that, by choice, a session uses two or more different ANR paths at once. This is sometimes called *bifurcated* routing. This would be advantageous in several ways. It could further reduce the impact of network failures (provided that at least one path remains viable). It could provide a way to aggregate the physical capacity of several links, each with inadequate bandwidth by itself, into a trunk with sufficient bandwidth. It could allow network providers to take advantage of costly parallel capacity, installed for reasons of high availability.

While such an option adds complexity to HPR's non-disruptive rerouting logic and to network management, it appears to be a fruitful avenue for further research. Another possibility is to couple this function with awareness of actual bandwidth requirements of connections and the knowledge of link utilization in a sophisticated high-speed network.

Also for a future study is a thorough survey of the existing literature on congestion control and analysis of trends concerning connection-oriented and connectionless transports.

Conclusions and Implementation Recommendations: We've shown how APPN/ISR and APPN/HPR can work together and discussed HPR routing in some detail. What sort of a product can benefit from implementing HPR? An obvious candidate is a router. A router typically supports several high-speed WAN link attachments and has the capacity to switch large numbers of packets. This appears to be natural fit with HPR's design points.

Can a computer that supports user applications (typically, an end node) also benefit from implementing HPR? We believe the answer is a resounding **yes**. Consider a typical installation with many individual workstations attached to a LAN—possibly quite a large, bridged LAN, shown in Figure 13. Several routers (3 and 4, 10 and 11) provide connectivity between stations on the LAN (1 and 13) and a WAN backbone made up of additional routers and packet switches (not shown). So far, we have shown HPR's benefit when a WAN link fails (if any of links 5—9 fails, the HPR-enabled routers can switch to a new route across the WAN). But one class of failures for which few solutions exist today is failure of a WAN access node such as routers 3, 4, 10 or 11. The typical user (1) is in session with applications both on the local LAN (2) and across the WAN, perhaps

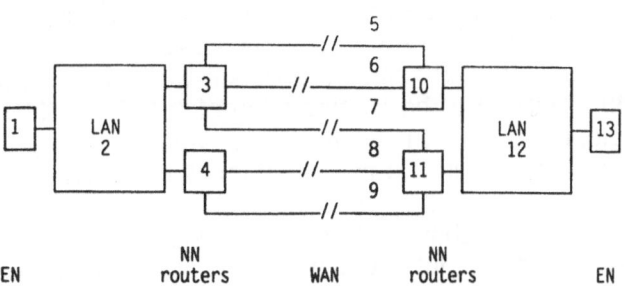

Figure 13. Benefits of HPR for Both End Nodes and Network Nodes

to a remote LAN (12, 13). Today, if the router or LAN segment or bridge through which a session is routed goes down or experiences problems, the session often fails. If the user's computer (1 or 13) includes HPR function, path switching will likely be successful if a local path to an alternate NN exists. Performance and storage benefits are also realized in network nodes 3, 4, 10, and 11 if end nodes 1, 13 support HPR.

SUMMARY

APPN/HPR, also known as APPN+, is a promising new technology for network nodes and end nodes that transparently extends SNA, replacing the intermediate session routing function in selected nodes with the faster high performance routing. APPN/HPR includes a new connection-oriented transport layer protocol, Rapid Transport Protocol (RTP), one of a new class of transport protocols that can also serve as a logical link (with priority) over multiple hops. APPN/HPR also includes a new type of connectionless source routing called Automatic Network Routing (ANR). APPN/HPR provides nondisruptive rerouting based on class of service, fast packet switching, minimal intermediate node storage, a new adaptive rate-based congestion prevention algorithm (ARB), and a drop-in software migration from existing SNA networks based on an ISR/HPR boundary function, for seamless interoperation with current SNA products and protocols.

ACKNOWLEDGMENTS: The authors are indebted to their colleagues at IBM Networking Systems, Ray Bird and Lap Huynh, for creating the HPR architecture specification, and to Dr. Raif Onvural for his kind encouragement, without which this article would not have been written. Thanks are also due to Barry Groner, Jane Munn, and Rick McGee, managers who supported early publication of this material.

References

[1]   R. Braden (editor), Requirements for Internet Hosts Communication Layer RFC 1122, 1989.

[2]   Michael O. Allen and Sandra L. Benedict, "SNA Management Services Architecture for APPN Networks," *IBM Systems Journal*, vol. 31, no. 2, pp. 336-352, 1992. Describes network management architecture for APPN.

[3]   B. Awerbuch, Israel Cidon, Inder S. Gopal, Marc Kaplan, and Shay Kutten, "Distributed Control for PARIS," *Proceedings of 9th ACM Symposium on Principles of Distributed Computing*, Quebec, Canada: ACM, August 1990.

[4]   F. Baker, The Point-to-Point Protocol Extensions for Bridging RFC 1220, 1991.

[5]   T. Bradley, C. Brown, and A. Malis, Multiprotocol Interconnect Over Frame Relay, 1992.

[6]   IBM Corp., AS/400 Distributed Systems Implementation Guide, GG22-9458. Discusses the decision criteria that must be considered when choosing a topology for an AS/400 APPN network

[7]   IBM Corp., S3/X and AS/400 APPN Nodes Using the SNA/LEN Subarea, GG24-3288. Describes the incorporation of a S/370 SNA subarea into a network comprising APPN network nodes. Intended for systems programmers and systems engineers in the intermediate systems and VTAM/NCP areas.

[8]   IBM Corp., APPN/Subarea Networking Design and Interconnection, GG24-3364. A guide for planning interconnection of APPN and SNA subarea networks.

[9]   IBM Corp., Networking Services/2 Installation, Customization, and Opera-
        tion, GG24-3662. Provides planning information for IBM SAA Networking
        Services/2.   Contains an extended example on connecting Networking
        Services/2 and AS/400, with their respective configurations.

[10]   IBM Corp., APPN Architecture and Product Implementations Tutorial,
        GG24-3669. Tutorial on APPN, with an overview of various product imple-
        mentations.

[11]   IBM Corp., 3174 APPN Implementation Guide, GG24-3702. Provides guid-
        ance on implementing the 3174 APPN functions in various scenarios.

[12]   IBM Corp., AS/400 APPN with PS/2 APPN, 3174 APPN, 5394 and
        Subarea Networking, GG24-3717.   Provides several scenarios of inter-
        action of these nodes including sample definitions and traces.

[13]   Rong-Feng Chang, James P. Gray, and Lap Huynh, Comparison of Con-
        gestion Control Performance of APPN+ and TCP, IBM Corp. Unclassified,
        Technical report 29.1490, December 1992. Describes dynamic problems
        of TCP and shows that APPN/HPR with ARB significantly outperforms TCP
        congestion control. Appendix B contrasts APPN and TCP/IP.

[14]   Israel Cidon and Inder Gopal, "Paris: An Approach to Integrated High-
        Speed Private Networks," *Int. Jour. of Digital and Analog Cabled Sys. 1*,
        pp. 77-85, 1988.

[15]   Israel Cidon, Inder Gopal, and Shay Kutten, "New models and algorithms
        for future networks," *0-89791-277-2/88/0007/0075*, ACM, 1988.

[16]   Douglas E. Comer, *Internetworking with TCP/IP: Principles, Protocols, and
        Architecture*, Prentice Hall, 1991.

[17]   IBM Corp., Client/Server Computing: The New Model for Business, IBM
        Corp., February 1992.

[18]   Rudy K. Cypser, *Communications for Cooperating Systems: OSI, SNA,
        and TCP/IP*, Addison-Wesley Publishing Co., 1991.

[19]   W. Doeringer, D. Dykeman, M. Kaiserswerth, B. Meister, H. Rudin, and R.
        Williamson, "A Survey of Light-Weight Transport Protocols for High-Speed
        Networks," *IEEE Transactions on Communications*, vol. 38, no. 11, pp.
        2025-2039, November 1990. Surveys and classifies high-speed transport
        protocols

[20]   P. E. Green, R. J. Chappuis, J. D. Fisher, P. S. Frosch, and C. E. Wood,
        "A Perspective on Advanced Peer to Peer Networking," *IBM Systems
        Journal*, vol. 26, no. 4, pp. 414-428, 1987.

[21]   IBM Corp., AS/400 Communications: APPN Network User's Guide, Publi-
        cation number SC41-8188. Describes the APPN support provided by the
        AS/400 system. Also describes APPN concepts and provides information
        for configuring an APPN network. APPN advanced considerations and
        configuration examples are included.

[22]   IBM Corp., IBM Local Area Network Technical Reference, SC30-3383.
        Describes token ring LANs, including source routing bridging.

[23]   IBM Corp., Systems Network Architecture Type 2.1 Node Reference,
        SC30-3422-2 (March 1991). Defines the architecture for APPN end node
        and LEN end node at the pre-HPR level of function.

[24]   IBM Corp., Systems Network Architecture APPN Architecture Reference,
        SC30-3422-3 (March 1993). Defines the architecture for APPN network
        node, APPN end node, and LEN end node at the APPN/ISR level of func-
        tion.

[25]   IBM Corp., Networking Services/2 Installation and Network Administrator's
        Guide, SC52-1110. Describes the APPN support provided by Networking
        Services/2 for OS/2 Extended Edition. Also describes APPN concepts and
        provides information for configuring an APPN network.

[26]   IEEE, Project 802—Logical Link Control. This standard defines the 802.2 Logical Link Control protocol.

[27]   Protocol Engines, Inc., XTP Protocol Definition, 1989. This defines the Express Transfer Protocol, one of a new class of optimistic transport layer protocols.

[28]   ISO, Information Processing Systems—Open System Interconnection—Basic Reference Model IEEE 7498, 1984. Defines the ISO 7-layer reference model.

[29]   V. Jacobson, "Congestion Avoidance and Control," *ACM SIGCOMM*, vol. '88, pp. 314-329, August 1988. Describes the "collapse of the internet" before the invention of TCP slow-start.

[30]   Steven T. Joyce and John Q. Walker II, "Advanced Peer-to-Peer Networking (APPN): An Overview," *ConneXions--The Interoperability Report*, vol. 6, no. 10, pp. 2-9, October 1992.

[31]   Steven T. Joyce and John Q. Walker II, "Advanced Peer-to-Peer Networking (APPN): An Overview," *IBM Personal Systems Technical Solutions*, no. G325-5014-00, pp. 67-72, January 1992.

[32]   Matthias Keiserswerth, "A Parallel Implementation of the ISO 8802-2.2 Protocol," *IEEE Tricomm '91*, April 1991.

[33]   David Kushi and Roy C. Dixon, Data Link Switching: Switch-to-Switch Protocol RFC 1434, 1993.

[34]   D. Perkins and R. Hobby, The Point-to-Point Protocol (PPP) Initial Configuration Options RFC 1172, 1990.

[35]   Robert M. Sanders and Alfred C. Weaver, The Xpress Transfer Protocol (XTP) - A Tutorial, Computer Networks Laboratory, Dept. of Computer Science, Univ. of Virginia, TR-89-10, January 1990.

[36]   Robert A. Sultan, Parviz Kermani, George A. Grover, Tsippi P. Barzilai, and Alan E. Baratz, "Implementing System/36 Advanced Peer to Peer Networking," *IBM Systems Journal*, vol. 26, no. 4, pp. 429-452, 1987.

[37]   Kenneth C. Traynham and Robert F. Steen, Data Link Control and Contemporary Data Links, IBM Corp. Unclassified, Technical report 29.0168, June 1977. Analyzes the relationship between bit error rate, packet size, and number of frames outstanding on throughput efficiency. Based on mathematical models, with many graphs.

ANALYSIS OF THE SPANNING TREE AND SOURCE ROUTING LAN INTERCONNECTION SCHEMES

M. Munafò, F. Neri, C. Scarpati Cioffari, and A. Vasco

Dipartimento di Elettronica
Politecnico di Torino
Corso Duca degli Abruzzi 24
10129 Torino – Italy

The paper presents an analysis of the performances of local area networks interconnected according to the Spanning Tree and Source Routing techniques proposed in the framework of the IEEE 802 committee. The study is mainly based on simulation models; some approximated analytical results are also presented.

The two LAN interconnection techniques are analyzed and compared by means of curves plotting performance figures, such as the frame delay or the network throughput, as functions of the total traffic offered to the interconnected system. Indications on how the performances of the interconnected system depend on the placement and configuration of interconnection devices (bridges) are provided.

1   INTRODUCTION

Local area networks (LANs) are nowadays found worldwide and there is a strong demand for higher capacity and geographical coverage. A commonly used solution is the *interconnection* of LANs, both of the same kind and of different kinds, in order to obtain a single networking environment. The interconnection is possible at different levels of the ISO/OSI reference model using different devices. Four kinds of devices can be recognized at a first cut: repeaters, bridges, routers and gateways.

Repeaters allow the interconnection at the physical level of the OSI model, providing a simple regeneration of the received signal. They don't allow the presence of loops in the topology and their number can be restricted because of the signal processing delay they introduce (the IEEE 802.3 standard [1] doesn't allow more than two repeaters between each station pair).

Bridges interconnect two or more LANs operating at the OSI MAC (Medium Access Control) sublevel. They are store and forward devices, that read frames from the LANs they are connected to and possibly forward them, recognizing local (intra-LAN) and global (inter-LAN) data traffic [2, 3]. Bridges allow the existence of loops in the network topology but these loops might have to be broken, possibly switching off some bridges, depending on the adopted bridging technique. Bridges are available that allow an interconnection transparent to the end stations, providing for frame forwarding through an automatic phase of configuration and address learning; these are called *transparent bridges* [4, 5].

Routers allow interconnection at the OSI network level, providing a set of more complex functions than bridges, such as the possibility of alternate routing and traffic balance. In general, the presence of routers in the network is known by the end stations, that have to use a particular frame format to send data out of a single LAN. Brouters [6] are interconnection devices that were recently proposed with the aim of merging the best features of bridges and routers.

Gateways perform the more general kind of network interconnection. They may interconnect heterogeneous environments operating at the higher OSI levels. Gateways are in general special purpose and complex devices.

This paper focuses on communication systems where several LANs providing access to several stations are interconnected by bridges. The case of *two-port bridges*, i.e., bridges interconnecting only two LANs, is given particular attention. Suitable models of these systems are proposed, and some possible approaches to obtain meaningful performance figures are described.

The proposed approach is mainly simulation. This allows an almost arbitrary level of detail in the description of the interconnected system, but provides performance figures whose statistical fluctuations can be reduced only at high computational costs. The performance figures obtained by simulation can be validated in some particular cases (i.e., under suitable hypotheses) with analytical results. In the paper the latter follow two lines.

On one hand, a simple traffic analysis allows the identification of congestion points and of the maximum throughput for an interconnected system; the results can be generalized in the case of regular topologies, leading to some interesting considerations. On the other hand, under suitable hypotheses, queuing networks can be used to model an interconnected system, and to analytically derive performance figures.

The paper will discuss these approaches, and compare the results that can be obtained. The organization is as follows. The two main proposed bridging techniques [namely Spanning Tree (ST) and Source Routing (SR)], are briefly recalled and compared in Section 2. Section 3 describes how a model of the interconnected system can be defined, and describes how a simulator of this model was implemented. Section 4 introduces, mainly for the Spanning Tree case, the bottleneck analysis that permits to compute the maximum throughput for a given interconnected system. In Section 5 the model of Section 3 is further simplified in order to obtain the representation in terms of a queuing network with different classes of customers. Section 6 presents and compares simulation results for some selected scenarios, providing hints for an efficient design of LAN interconnections. Section 7, finally, concludes the paper.

2 TWO LAN INTERCONNECTION SCHEMES

This section reviews the two main proposals of LAN bridging, both born in the framework of the IEEE 802 standardization activity. The Spanning Tree (IEEE 802.1) approach features transparent bridging, thus not making end stations aware of belonging to an interconnected environment, at the cost of constraining the placement of bridges and the achievable connectivity degree. The Source Routing (IEEE 802.5) approach allows much more freedom in the use of interconnection devices, but asks the end stations to be aware of the interconnected environment; as such, it can be said to be something intermediate between bridging and routing.

2.1 Spanning Tree

The IEEE 802.1 Committee has selected transparent bridges supporting the Spanning Tree algorithm as the single standard for the interconnection of 802 Local Area Networks [7].

Transparent bridges perform three basic functions: frame forwarding, learning of station addresses, and resolving existing loops in the topology using the Spanning Tree algorithm.

Bridges have two or more ports attached to different LANs; they receive and transmit frames on the LANs they are connected to. Station addresses are observed and maintained in a forwarding data-base: frames are forwarded or discarded according to the information contained in this data-base.

When a frame is received on a bridge port, its destination address is compared with the information contained in the forwarding data base. If the *destination* address is unknown, the frame is transmitted on every bridge port except the one on which the frame was received. If the address is found in the forwarding data base, the bridge forwards the frame retransmitting it at a specific output port only if the information in the data-base is not calling for a transmission on the same bridge port on which the frame was received.

For the learning activity, if the *source* address of the frame is not found in the forwarding data base, it is added to the database together with the address of the bridge port on which the frame was received, and it will be used for subsequent transmissions in the opposite direction. The bridge maintains a timer for each data base entry, which is reset every time the corresponding source address is observed, so that informations can be removed when a certain period of time since their last updating is elapsed.

The forwarding and learning processes assume that the topology of the network is loop free, so only one single path exists between each station pair. If there are loops in the network topology, the learning process gets confused about the position of the end stations. This means that, although the interconnected system may provide redundancy in the number of available bridges in order to ensure higher overall network availability, transparent bridges must implement the so called Spanning Tree algorithm to transform any arbitrary mesh topology into a single, acyclic spanning tree, possibly disabling some bridges or bridge ports.

To work correctly, the algorithm requires the following.

1. Each bridge must have a unique *bridge identifier*; this is built combining the bridge MAC address with a configurable bridge priority.

2. There must exist a unique well-known group address that is always recognized by all bridges attached to the interconnected network.

3. Each bridge port must be identified uniquely within the bridge with a *port identifier*.

The algorithm is deterministic and the logical topology will always be the same regardless of the order in which bridges are powered up and activated. In the case of failures, the topology is automatically reconfigured.

We now briefly describe the actual calculation of the spanning tree. First, a unique root bridge is selected. The root is the bridge with the highest priority bridge identifier, i.e. the bridge with the identifier with the lowest numeric value. Next, each bridge decides which bridge port lies "in the direction" of the root. This is the bridge port through which the least-cost path to the root is found (according to some cost function: the number of traversed bridges will be used in the paper, but other cost functions can be considered as well) and is called the *root port*. Finally, a unique *designated bridge* is selected for each LAN. This is the bridge that offers the least cost path to the root from the LAN.

Each bridge places the root port and all its ports connected to LANs for which it is the designated bridge into a "forwarding" state. The other bridge ports are put in a "blocking" state. The state of a bridge port is never changed directly from blocking to forwarding. Intermediate states, called "listening" and "learning" are used to ensure that no temporary loops are formed while the topology is stabilizing after some change. In the listening and learning states the Spanning Tree protocol information is received and transmitted, but data frames are not forwarded to or from the bridge port. In the learning state the frames are also submitted to the learning process, thus creating the forwarding data-base. A bridge port remains in the listening state for a period called *forward_delay;* then it passes to the learning state, where it stays for another *forward_delay;* finally it passes to the forwarding state. While the bridge port is in the listening or learning states, it can be placed in the blocking state again if new information is received about the network topology.

The information necessary to operate the Spanning Tree algorithm is exchanged between bridges using suitable *bridge protocol data units* (BPDUs). These messages contain the root bridge identifier, the cost of the path to the root, the age of the carried information and the bridge and port identifier of the sending bridge. BPDUs are transmitted by the root bridge once every *hello_time* (usually 1 to 4 seconds). Every designated bridge transmits a BPDU on each of its designated ports each time a BPDU is received on its root port.

Bridges must be able to respond quickly to changes in the network topology, changing the state of their ports. If a bridge fails, it stops transmitting BPDUs; this situation is recognized as follows. When a bridge doesn't receive any BPDU for *max_age* seconds, it sends a topology change notification towards the root bridge and asks to become the designated bridge for that LAN. After a short period of time, depending only on the diameter of the interconnected system, possible conflicts are resolved, the designated bridges are identified, and the topology stabilizes.

2.2 Source Routing

The Source Routing bridging technique was developed by IBM for the interconnection of IEEE 802.5 LANs (Token Ring) [8], but it can be applied to all IEEE 802 LANs.

In Source Routing [9, 10, 11, 12], the route (sequence of LAN segments and bridges) to be followed by each frame is specified by the source station: every station associates a route with each possible destination station and includes one such route in a suitable routing information field inside the headers of transmitted frames. The routing information for Source Routing consists in a sequence of identifiers of the local network segments that the frame must traverse; this information is read and recognized by the bridges that will provide for the routing.

Bridges following the Source Routing approach don't have to maintain address tables to forward frames; instead, they have to execute a simple string matching during a linear scan of the routing information stored in frame headers.

In addition, some route discovery mechanism is required. A source station can determine the route towards any other station in a number of different ways. The most practical method for route determination is by using a dynamic discovery procedure. A station that has no route to a certain target station sends a special broadcast frame to the target station. This broadcast frame is transmitted by bridges in such a way that all the possible routes between the source and the target station (*all-routes* broadcast) are explored, so that the target station receives as many copies of the frame as there are possible routes. When a bridge retransmits one of these route-searching frames, it records in the Routing Information field of the header the information about the last traversed LAN segment. This dynamic route recording also allows one to recognize and discard copies of the broadcasted frame looping in the interconnected network.

The target station transmits all received copies of the frame back to the source station, routing them along the path recorded in the received header. In this way, the source station receives all the possible routes to the target and stores the best route (according to some criterion: shortest path, largest frame size or any other rule) in a table.

Another dynamic discovery procedure provided by the Source Routing proposal uses a *single-route* broadcast, in which the broadcast frame is sent along the bridged local area network, but never more than once per segment. In this way, the target station receives a single copy of the frame; and transmits the frame back to the source station along all the possible routes, so that the source station receives as many copies of the frame as there are routes. To properly operate using single-route broadcast, bridges have to be configured in a spanning tree in order to avoid frame duplications. This route discovering procedure will not be considered in this paper.

2.3 Comparison of Source Routing and Spanning Tree Bridges

In this section the main differences between the two interconnection methods mentioned above are summarized; more details are available in [13] and [14]. The comparison is detailed for some significant characteristics of the interconnected system.

Transparency – As regards the interconnection transparency for end stations, Spanning Tree is to be preferred to Source Routing, since Source Routing requires the explicit modification of end stations in order to be implemented. In fact end stations must become aware of the interconnection, maintain the routing tables, and transmit frames with a different format (i.e., carrying the routing information).

On the other hand, Source Routing, since it operates at a higher architectural level, actually allows more complex functions, such as multiple paths or maximum frame length, and this can make up for the transparency loss.

End-to-end frame delay – Source Routing can be superior because it always uses the shortest path between stations, while Spanning Tree provides a non-optimal routing. Moreover, link flows can be balanced, thus reducing congestion and delays. We must however consider as a drawback the traffic caused in Source Routing by the dynamic path discovery procedure, which might congest the network and cause higher delays.

High traffic – For low global inter-LAN traffic, both techniques have a linear variation of network utilization when traffic increases. LANs can be more uniformly used with Source Routing than with Spanning Tree, where traffic is constrained along the active topology and LANs close to the tree root are more used. Therefore, an interconnected network using Spanning Tree becomes congested for a lower inter-LAN traffic level with respect to the same network using Source Routing.

Many authors wonder about the effectiveness using of alternate routes between stations, as allowed by Source Routing. Often the actual network topology under Source Routing is similar to a tree-like topology or, anyway, it is very simple, so that Spanning Tree performances are not too dissimilar from the optimum ones. Nevertheless, the possibility of using alternate routes offers a lot in terms of traffic balancement and reliability.

Topology changes – Spanning Tree reacts better than Source Routing to topology changes and network failures. In fact, the Spanning Tree algorithm provides for the periodic exchange of BPDUs in order to test the topology consistency and the correct operation of bridges. In case of error the topology change procedure is started, and a new active topology will be running in a short time, depending only on the topology radius. In Source Routing the network or bridge failures can be noticed only when the end stations, not receiving frame acknowledgements, decide to start a new route discovery phase. The route discovery phase is however very burdensome for the interconnected system, due to the large traffic of broadcasted frames.

Bandwidth – Both techniques assume that bandwidth is a cheap resource in the bridged LAN, which usually is not charged for. Spanning Tree allows transparency to end stations and very simple routing management using non-optimal routes: the exchange of BPDUs leads to a (usually marginal) traffic increase. Source Routing allows optimal routes and improved functions at the cost of increased frame length and of a possibly large amount of route discovering broadcast traffic.

Bridge processing time – Source Routing calls for a modification of the header of forwarded frames, with a consequent recomputation of the frame check sequences; such computation is however commonly done in hardware. Moreover, as already noted, bridges must perform a linear scan of the routing information stored in the variable-length frame headers: this may have some computational cost. Spanning Tree asks for inspecting the forwarding database to properly forward frames, and this operation can be costly in terms of processing time when performed in software, but it also can be performed very efficiently in hardware. Processing times do not seem to be a problem, if bridges are properly designed, with either approach.

Management – The Spanning Tree approach allows lower management costs: the network manager only has to place the transparent bridges, and this can happen also in a system

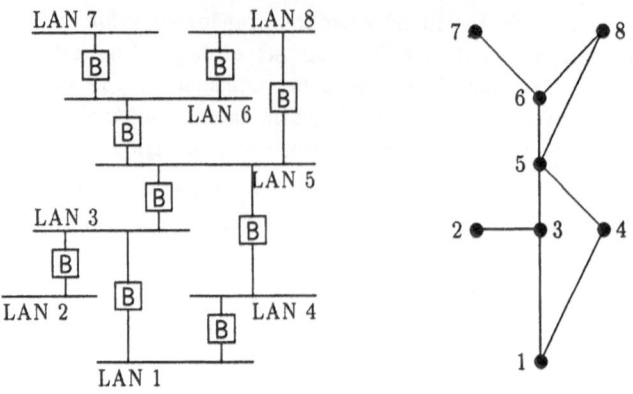

Figure 1: Example of interconnected network and corresponding graph

in operation, without affecting existing stations. The only care is to properly choose bridge priorities (or to properly place the available bridges), since bridge identifiers affect the shape of the spanning tree, and thus the performance of the interconnected system, as we shall see.

The Source Routing approach, on the other hand, requires modifications (possibly both hardware and software) of existing stations. Bridges must be configured, as they need to know the identifier associated with the LANs they are connected to. Moreover, Source Routing may cause overload problems due to route-searching traffic, so that the placement of bridges must be more careful, although better performances with respect to the Spanning Tree can be achieved, due to the possibility of exploiting alternate routes between stations.

Network cost – The implementation of the interconnected system may be very costly with Source Routing, due to the modifications in end stations.

3   MODELLING LAN INTERCONNECTIONS

This section first describes how an interconnected system can be represented by means of an undirected graph whose nodes are LANs and whose edges are associated with bridges. A model of the generic interconnected system is then proposed. This model assumes an idealized behavior of individual LANs, since it does not deal with the details of the access protocol. An implementation of our LAN model by means of a computer program is finally described.

3.1   Graph Representation of the Interconnected System

If we restrict our attention to bridges with two ports, an interconnected system can be represented by an undirected graph, where each LAN is a vertex of the graph and each bridge is an edge. An example is shown in Fig. 1.

If the active topology of the bridged LAN is a spanning tree the graph becomes a tree. Given a number of vertices (i.e., a number of LANs), there is a finite number of topologically different trees with so many vertices: these trees will be used as the basis for a discussion on the effectiveness of the Spanning Tree approach.

Note that the proposed mapping of the bridged LAN onto a graph permits one to apply known results to the analysis of the topology of interconnected systems.

3.2 Our Model for the Interconnected System

A model of an interconnected system can be defined as follows.

Each LAN can be modeled by one single-server queue, assuming an ideal access protocol: the information on the state of the LAN is assumed to be available to an ideal access manager, so that no overhead must be paid to perfectly schedule frame transmissions in a first-come-first-served basis. This might appear as a coarse approximation, but it is close to reality for LANs with limited geographical coverage (small propagation delays) and low to moderate loads. Different queue models can provide more accurate representations of the LAN access protocol, and could be considered in our discussion, but this will not be done, as the focus of the discussion is more on interconnection problems than on access protocols.

All stations and bridges that are connected to a given LAN and have a frame to transmit, queue it in the transmission queue of the LAN. These queues have a finite capacity (this account for a finite number of stations with finite buffering capacity): at most N_q frames can be waiting for transmission on a LAN; exceeding frames are discarded and are considered lost. The LAN queue server transmits each frame on the LAN, keeping the LAN busy for the frame transmission time and then delivering it to each bridge connected to the LAN and to the target station, if it is on that LAN. Propagation times and access overheads are not taken into account, since they are assumed to be negligible when compared with transmission times and bridge delays.

Bridges can be modeled by means of single-server queues too: frames received by all bridge ports are queued to this Forwarding Process queue (one queue per bridge), which also has finite buffering capacity. The Forwarding Process serves the queue at the bridge operation rate, and queues, if required, frames to the queue representing the appropriate output LAN.

Kleinrock's *independence assumption* [15] can be introduced to further simplify the model. In the real world the length of each frame is not changed while it traverses the interconnected system; the independence assumption allows instead a new length to be associated with the frame every time it is transmitted on a LAN. This assumption is known to be fairly accurate under light and moderate loads, and it will show to be so also in our case. The independence assumption will be taken in Section 5 to analytically study LAN interconnections.

3.3 Simulation Programs for the Interconnected System

To study the performances of the two considered LAN interconnection schemes, a simulator was built for both Spanning Tree and Source Routing, using two programs written in the C language under the DEC VMS operating system. The simulator follows a process-oriented, discrete-event simulation paradigm similar to the one provided by the Simula class SIMULATION [16], thus using pseudo-concurrent processes. Efficiency is granted by a particular scheduling technique developed by E. G. Ulrich [17] for the simulation of digital devices.

One process is used to implement the arrival of transmission requests to the LANs. Inter-arrival times are pseudo-randomly extracted from a probability distribution specified by the user at initialization time; the average arrival rate depends on the global load offered to the interconnected system. The arrival process randomly chooses a source/destination station pair, and creates a frame, with a suitable pseudo-random length, to be transmitted between them. Frames are queued in the finite capacity queue associated with the LAN to which the source station is connected, wherefrom they will be extracted in a FCFS order. When the queue is full, the frame is discarded, and the frame loss statistics are collected.

One second class of processes is used to model transmissions on each individual LAN. These processes remove the first transmission request from the LAN queue, wait for the frame transmission time associated with the request, queue a request to all the queues associated with bridges with a port connected to the LAN, and then consider the next request in the LAN queue. Frame transmission requests may be lost when bridge queues are full. If the destination station belongs to the LAN, the delay statistics are collected.

A last type of process is used as a server for bridge queues. One such process is associated with each bridge. It implements the selected interconnection algorithm with a speed also provided by the user at initialization time, possibly queuing transmission requests to the queues associated with the LANs to which the bridge is connected.

An interconnected system with m bridges and n LANs is thus simulated by $m + n + 1$ pseudo-concurrent processes exchanging messages (frames). The differences between the two programs for ST and SR are in the processes associated with bridges, that must observe the protocols specifications. The ST simulator explicitly models the timed exchange of BPDUs, with a full implementation of the Spanning Tree algorithm; the cost function used to build the spanning tree over the topology is, according to the suggestions in the standard, proportional to the inverse of the LAN data rate: if all data rates are the same, this corresponds to a cost function equal to the number of bridges to be traversed.

The SR simulator models the propagation of route-searching frames, that are generated on the source LAN when the source station does not know the route to the destination station. The SR simulator allows two different behaviors with regard to the route-searching frames generation and the criterion followed in the best route selection. These different behaviors can be selected by the user when the simulation experiment is activated. For the generation of route-searching frames, one option is to send all the route discovering traffic (for every station pair) when the simulation is started, before generating the regular data traffic, thus separating in time the two types of traffic. The other option is to mix the two types of frames, by generating a route-searching frame every time the path to a given destination is unknown to the source station. In either case, due to the large time-persistence of the information in routing tables, the route discovering traffic loads the network mainly in the initial phase of the simulation experiment, when stations need to build their routing tables. For the best route selection criterion, it can be either shortest path (in number of traversed bridges) or quickest path (i.e., stations, after sending a route-searching frame, store the route contained in the first returned frame). In any case, the routes chosen by the simulation program depend on the pseudo-random generation of frame transmission requests and reflect the actual dynamic traffic in the interconnected system under simulation.

The information stored in the routing tables (inside end stations for Source Routing, and inside bridges for Spanning Tree) is timed, in the sense that entries are deleted if not refreshed within a given interval of time. The routing tables have a finite capacity for both interconnection schemes.

The simulation program collects several performance figures; among these, we have the mean and variance of end-to-end frame delays, the LAN utilizations (i.e., the fraction of time for which each LAN is busy transmitting frames), and the fraction of frames lost owing to the saturation of the (finite capacity) LAN and bridge queues. An approximate regenerative method [18] was implemented to estimate the accuracy of the measured end-to-end frame delay, in order to automatically obtain a stopping rule for the simulation experiment: the simulation is run until the average of the measured delay falls within a user-supplied confidence interval with a given confidence level. Samples collected in an automatically estimated initial transient are discarded.

4   BOTTLENECKS ANALYSIS

As the Spanning Tree algorithm allows reliable frame transmission in the bridged local area network by routing the data traffic along the active tree-like topology, it is easy to identify in a particular spanning tree which LAN might be a bottleneck for the bridged system when the data traffic increases, thus deriving the maximum achievable throughput, i.e., which is the maximum capacity offered by the interconnected system. This figure provides a simple but meaningful indication of the effectiveness of the interconnection. The same approach can be extended to Source Routing if the routing between any station pair is known, or probabilistically characterized.

This bottleneck analysis is based on the graph representation of the interconnected system seen in Section 3.1. The following (not all strictly necessary) assumptions will be done in the sequel:

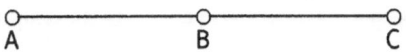

Figure 2: Graph (tree) representing three LANs interconnected by two bridges

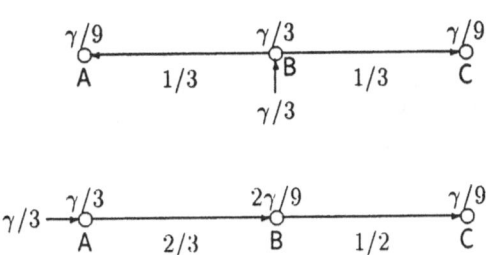

Figure 3: Traffic flows for the 3-nodes graph with equally distributed stations

- Kn stations are equally distributed on the n LANs, K on each LAN;

- all LANs are alike, i.e. they have the same speed C bit/s;

- the global traffic offered to the interconnected system is γ bit/s;

- the offered traffic is equally distributed on the interconnected system, i.e. each station offers the network a traffic equal to γ/Kn bit/s, and destinations are uniformly chosen (this is the simplest scenario, but any traffic pattern can be taken into account).

In the description that follows we refer to Fig. 2, where $n = 3$ LANs are connected by two bridges; the discussion can be easily generalized to any tree. Each vertex receives external traffic equal to γ/n: a fraction $1/n$ of this traffic is directed to stations on the same LAN, while the rest is directed to other vertices, consistently with the topology. See now Fig. 3. In the first part of this figure, the traffic offered to node B must traverse the corresponding LAN, but only $1/n$ of that traffic reaches the other two LANs. In the second part of the figure, an analogous partitioning of the offered traffic is shown for LAN A: a fraction $\gamma/3$ is offered to vertex A, but only $2/3$ of this traffic [generally $(n-1)/n$] is inter-LAN traffic, and is offered to vertex B as well. Part of the traffic coming from A and offered to B is directed to stations on B, and part of it to stations on different vertices (in our case $1/2$ to C). Among these vertices we must not consider vertex A because the traffic generated by vertex A cannot return to that vertex due to the filtering action of bridges.

A similar procedure must be carried on for each vertex, exploiting the symmetries of the graph where it is possible. After this step we can add the different contributions and obtain the total traffic offered to each vertex (see Fig. 4).

For the correct operation of the bridged LAN, it is necessary that no vertex receives a traffic larger than the LAN speed, otherwise the traffic cannot be put through, and the network becomes congested.

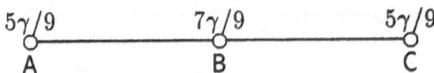

Figure 4: Total traffic offered to each node for the considered 3-nodes graph

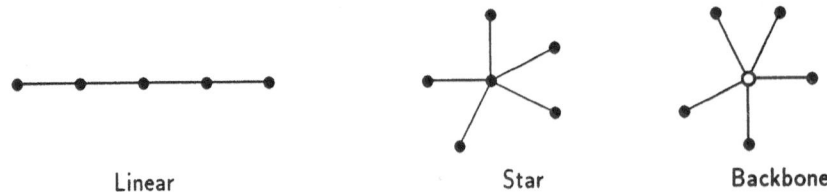

Figure 5: Graphs corresponding to some regular topologies

In our example we have

$$\frac{5\gamma}{9} \leq C \qquad \frac{7\gamma}{9} \leq C \qquad \frac{5\gamma}{9} \leq C$$

We are interested only in the constraint for the bottleneck LAN, in our case LAN B

$$\frac{\gamma}{C} \leq \frac{9}{7}$$

In this case, the global traffic offered to the bridged LAN can't exceed the 9/7 of the individual LANs speed.

Following this technique we have obtained general results for some regular topologies, whose corresponding graphs are shown in Fig. 5. The resulting capacities will be used as checkpoints in Section 6, where the same interconnected systems are studied by simulation. The asymptotic values for large n are also observed and commented.

- **Linear Topology** – LANs are serially connected. This is the topology that gives the maximum distance (in terms of traversed bridges) between any two stations. Permitted load values are

$$\frac{\gamma}{C} \leq \begin{cases} \dfrac{2n^2}{n^2 + 2n - 2} & n \text{ even} \\[2mm] \dfrac{2n^2}{n^2 + 2n - 1} & n \text{ odd} \end{cases} \qquad n \geq 2$$

The maximum traffic tends to 2 for large values of n; this is the maximum throughput gain that can be obtained by any tree-shaped topology with equal capacity LANs and uniform traffic. This result, that has been derived in a very simple way, can also be interpreted as the well known maximum achievable throughput gain in the case of DQDB metropolitan area networks with slot reuse [19]. Note that other authors derive this value for the maximum throughput gain in DQDB-like networks following a totally different approach.

- **Star Topology** – LANs are all connected to a central LAN. This topology gives the minimum distance between two stations. We obtain

$$\frac{\gamma}{C} \leq \frac{n^2}{n^2 - n + 1} \qquad n \geq 3$$

The maximum traffic tends to 1 for large values of n: this is the price that must be paid to obtain the small delays allowed by the minimum number of traversed bridges.

- **Backbone** – It is similar to a star topology, but the speed of the central LAN is k times the speed of the other LANs. We further suppose that there are no stations directly connected to the backbone.

$$\frac{\gamma}{C} \leq \min \left(k\frac{n}{n-1}, \frac{n^2}{2n-1} \right)$$

where n is the number of LANs excluding the backbone. The maximum traffic tends to $\min\left(k, \frac{n}{2}\right)$ for large values of n, showing the large performance gains attainable by this widely used topology.

As already mentioned, the approach proposed in this section can be extended to more general situations: different LAN speeds, uneven distributions of stations, asymmetric traffic patterns can be easily accommodated.

5 ANALYSIS BASED ON THE THEORY OF QUEUING NETWORKS

Under suitable simplifying assumptions, bridged LANs can be analytically studied applying the results of queuing networks theory. The routing of frames must be known, hence the approach more naturally fits systems running the Spanning Tree algorithm.

A suitable model for the analysis of bridged LANs is the BCMP queuing network model [20]. The main assumptions to be introduced to build a BCMP queuing network model of an interconnected system are the following:

- each LAN is an M/M/1 queue with external arrival rate equal to the LAN average arrival rate; call λ_i the external arrival rate to LAN i; in balanced and symmetric conditions $\lambda_i = \lambda = \gamma/n$;

- bridge delays are not considered; the is equivalent to assume these delays to be negligible when compared to LAN traversal delays;

- there are n classes of customers, one per LAN;

- any customer arriving to the queuing network belongs to the class corresponding to the source LAN queue, where he was queued the first time;

- a customer is never subject to a class change;

- any customer may not be queued to each queue more than once;

- frame lengths are exponentially distributed with mean $1/\mu$;

- the queue service times, equal to the average frame length divided by the LAN speed C, are independent of the number of customers in the queue; the average queue service time is $1/\mu C$.

Kleinrock's independence assumption [15] must be introduced: a new service time (i.e., a new frame size) is drawn from the corresponding probability distribution every time a customer (i.e., a frame) traverses a queue (i.e., a LAN).

Four further assumptions are necessary with respect to the model described in Section 3.2: the offered traffic must be Poisson, frame lengths must be exponentially distributed, the exchange of BPDUs for ST and of route-searching frames for SR are not considered, and bridge delays must be neglected. The last assumption can be quite strong (although most commercial bridges today have a frame transfer rate equal to the full speed of the connected LANs), but it is required for BCMP results to be applicable. In fact, a model in which customers are models for transmitted frames, and bridges (in addition to LANs) are modeled with queues, should consider the possibility of customer duplication, i.e. of letting the same customer simultaneously enter two or more different queues. This is due to the fact that bridges must read all the frames on the LANs they are connected to, and more than one bridge can be active in the same LAN. This observation holds true for both Spanning Tree and for Source Routing.

For BCMP queuing networks, we define $p_{i,r;j,s}$ as the probability that a class r customer leaving queue i can pass to queue j as a class s customer. In our case, $p_{i,r;j,s}$ represents the probability that a message generated by LAN r and just transmitted on LAN i is queued for transmission on LAN j as a message generated by LAN s (obviously r must be equal to s). Owing to the assumptions above, the following relations hold

$$p_{i,r;j,s} = 0 \qquad \forall r \neq s$$

$$p_{i,r;i,r} = 0$$

Let $\lambda_{i,r}$ be the arrival rate of class r customers to queue i. For external arrivals, $\lambda_{0,r}$ is the external arrival rate for class r customers and $p_{0,r;i,r}$ is the external arrival probability of class r customers to queue i.

For external traffic uniformly distributed among n LANs, calling γ the global external arrival rate, the following relations hold true

$$\lambda_{0,r} = \frac{\gamma}{n} \qquad \forall r$$

$$p_{0,r;i,r} = \begin{cases} 1 & \text{if } i = r \\ 0 & \text{otherwise} \end{cases}$$

The latter equality simply states that customers belong to the class corresponding to (i.e., having the same index of) the queue where they arrive from the external world.

The arrival rates for each queue and each class can be obtained from the system of linear equations

$$\lambda_{i,r} = \lambda_{0,r} p_{0,r;i,r} + \sum_{j=1}^{n} \lambda_{j,r} p_{j,r;i,r} = \sum_{j=0}^{n} \lambda_{j,r} p_{j,r;i,r}$$

and the global arrival rates for each queue, independently of the customer class, are

$$\lambda_i = \sum_{r=1}^{n} \lambda_{i,r}$$

The BCMP theorem [20] states that a product-form solution exists, and queues can be studied separately, as isolated M/M/1 queues with arrival rates $\lambda_{i,r}$. The queuing network allows steady state solutions only if

$$\frac{\lambda_i}{\mu C_i} < 1 \qquad \forall i$$

where C_i is the speed of LAN i.

After having obtained the solutions for the queues in isolation, the value for the mean customer waiting time (mean frame delay) and global queuing network utilization can be computed. The global queuing network average utilization * is

$$U = \sum_{i=1}^{n} \frac{\lambda_i}{\mu C_i}$$

while the mean customer waiting time can be obtained using Little's theorem. In fact, the average number of customers is

$$E[N] = \sum_{i=1}^{n} E[N_i] = \sum_{i=1}^{n} \frac{\lambda_i}{\mu C_i - \lambda_i}$$

---

*Note that this figure is not normalized: it can range from 0 up to the number of LANs, n.

and the average waiting time

$$E[T] = \frac{E[N]}{\gamma}$$

6 NUMERICAL RESULTS

Several simulation experiments were run using the simulator described in Section 3.3. For the Spanning Tree interconnection scheme the results were checked against the analytical techniques proposed in Sections 4 and 5. For the solution of the queuing networks proposed in Section 5, the package described in [21, 22] was used.

Before proceeding with the description of the simulation experiments, we summarize the reference scenario, i.e., the parameters used for all simulations, unless stated otherwise. They are the following:

- frames have exponentially distributed sizes, with an average of 1250 bytes (i.e., $1/\mu = 1250 \times 8$ bits), with the addition of the (variable) size of the Routing Information field in the Source Routing case;

- LAN speeds are equal to 10 Mbit/s (i.e., $C_i = C = 10$ Mbit/s);

- the bridge working times, i.e. the times required by bridges to route or filter one frame, are exponentially distributed, with an average equal to 0.1 ms;

- the buffering capacity of each bridge is 100 frames;

- the memory of bridges or stations is large enough to maintain all stations addresses; this accounts for an amount of route-searching frames for SR decreasing with simulation time, since the information in routing tables decays very slowly with time;

- the information in the routing tables is discarded if not refreshed within 300 s (according to the IEEE 802.1 standard);

- the offered traffic is equally distributed among LANs;

- destinations are uniformly chosen;

- stations are equally distributed among LANs;

- the confidence interval for each simulation is equal to 2%, with a confidence level of 98%;

- the capacity of every LAN queue is equal to $N_q = 200$ frames.

For Spanning Tree, BPDUs are issued every 2 s (again according to the standard), and the cost function to build the spanning tree is, for equal data rate LANs, equivalent to the number of bridges to be traversed (unit cost per bridge).

For Source Routing, route-searching frames are generated whenever a station does not know the route to a given destination; route-searching traffic is thus interleaved with regular traffic. The route selection criterion is shortest route.

The simulation program was first validated against the results provided by the queuing networks analysis for two simple topologies comprising 3 and 4 LANs, as described in Section 6.1. All other sets of simulation experiments are instead based on topologies with 6 LANs. All the possible different 6-LANs interconnections allowed by the Spanning Tree algorithm are compared in Section 6.2. Some topologies with 6 LANs and different levels of connectivity are compared in Section 6.3 for the Source Routing interconnection scheme. Finally, in Section 6.4, Spanning Tree and Source Routing are compared when operating in similar environments.

Results are presented in terms of graphs plotting the mean and the variance of end-to-end delays, the overall network utilization (sum of individual LAN utilizations), and the fraction of frames lost owing to (LAN and bridge) queue saturations, as functions of the total load offered to the interconnected system. The total load is normalized with respect to the LANs speed: a load equal to 1 fully saturates one LAN.

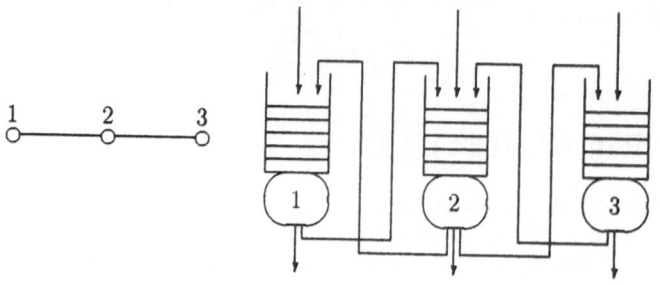

Figure 6: Three LANs topology and corresponding queuing network

6.1 Queuing Networks Results

The agreement between the queuing network model and the simulation model for Spanning Tree has been verified comparing the results for the two simple topologies L3D2T1 and L4D2T1, shown in the first part of Figs. 6 and 7. In the second part of these figures, the resulting queuing network models are shown. Remember that in the queuing network customer classes select the routing of customers among queues. With respect to the simulation model, the queuing network model neglects the delays introduced by bridges, and the finiteness of LAN and bridge queues; moreover, independence must be assumed. For the results presented in this section, the average time required by a bridge to process one frame is 0.1 ms, while LAN queues can store at most 50 frames and bridge queues con store at most 5 frames.

For the considered topologies we compare the end-to-end delay (mean waiting time in the queuing network) and the overall network utilization within the range of validity of the queuing network model: the simulations were run with a normalized offered traffic ranging from 0.1 to 1.2.

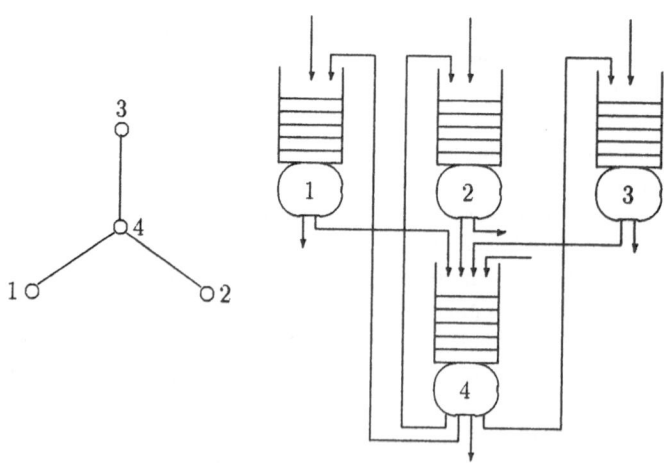

Figure 7: Four LANs topology and corresponding queuing network

Table 1: Numerical results (simulation and queuing network) for the 3-LAN network of Fig. 6

| Offered traffic | Delay (ms) | | | | | | Utilization | | | | | |
| | Figure 6 | | | Figure 7 | | | Figure 6 | | | Figure 7 | | |
	Simul.	Q.N.	Diff.	Simul.	Q.N.	Diff.	Simul.	Q.N.	Diff.	Simul.	Q.N.	Diff.
0.1	2.037	2.018	−0.9	2.202	2.258	2.5	0.1878	0.1886	0.4	0.2071	0.2124	2.6
0.2	2.21	2.1575	−2.4	2.41	2.409	−0.0	0.376	0.3756	−0.1	0.4192	0.425	1.4
0.3	2.399	2.3483	−2.1	2.568	2.585	0.7	0.5684	0.5668	−0.3	0.6232	0.6374	2.3
0.4	2.624	2.5573	−2.5	2.813	2.804	−0.3	0.7543	0.7554	0.1	0.8365	0.85	1.6
0.5	2.867	2.8028	−2.2	3.12	3.048	−1.1	0.9476	0.9424	−0.5	1.0743	1.0624	−1.1
0.6	3.194	3.1258	−2.1	3.352	3.365	0.4	1.1335	1.1334	−0.0	1.2548	1.275	1.6
0.7	3.524	3.5261	0.1	3.782	3.776	−0.2	1.317	1.322	0.4	1.4745	1.4873	0.9
0.8	4.136	4.0488	−2.1	4.42	4.340	−1.8	1.518	1.509	−0.6	1.6997	1.7	0.0
0.9	4.745	4.8173	1.5	5.284	5.188	−1.8	1.6961	1.7	0.2	1.9266	1.9124	−0.7
1.0	5.937	6.0062	1.1	6.73	6.666	−1.0	1.889	1.8888	−0.0	2.1319	2.125	−0.3
1.1	8.287	8.2181	−3.7	9.995	10.177	1.8	2.0856	2.0758	−0.5	2.3406	2.3374	−0.1
1.2	13.277	15.049	13.3	19.144	35.263	84.2	2.267	2.2668	−0.0	2.5405	2.55	0.4

The average end-to-end delay for the queuing network was obtained using Little's theorem as described in Section 5. The average number of customers in each queue is calculated by the package used for the solution, described in [21] and [22].

The results are presented in Table 1, together with the relative differences.

The network utilization grows linearly with the offered traffic, as we can expect if the latter remains below the saturation point of the LANs. A good agreement between the two sets of utilization results can be observed, despite the different assumptions of the two models. The differences between numerical results stay in general within the confidence interval of the simulator.

A good agreement between simulation and analysis can be observed in general also for the average end-to-end delay. In this case, however, some differences arise at high loads, when the central LAN approaches saturation, and the finiteness of queues comes into play (losses can be seen to become non-null for an offered traffic equal to 1.1 and 1.2).

6.2 Spanning Tree vs. Topologies

Our next set of experiments is aimed to analyze the behavior of bridged LANs using Spanning Tree; in particular the different topologies resulting from the possible bridge labeling (i.e. from different assignments of bridge identifiers and priorities) are compared.

Considering all the topologically different trees with n vertices, we can compare the performances of differently bridged LANs. Note that the bridged LAN topology can be adjusted by the network manager in order to obtain particular features, by placing the available bridges in different positions, or by assigning different bridge priorities.

In the simulation experiments we compare the six topologies shown in Fig. 8. These are all the possible trees with six vertices and they also correspond to all the possible active topologies in which a bridged LAN with six segments can be configured by Spanning Tree. The main characteristic of these topologies is the maximum distance, in terms of traversed bridges, between any two LANs: this figure ranges from 2 to 5 in the considered cases.

Fig. 9 shows the average of the end-to-end delay, the variance of the end-to-end delay, the overall network utilization, and the fraction of lost frames, as functions of the normalized total offered load. The results of Section 4 can be used to compute the maximum load for the star (L6D2T1) and the linear (L6D5T1) topologies: in the first case we have $\gamma/C < 1.16$, in the second case $\gamma/C < 1.56$. These load values are shown as vertical lines in the graphs of Fig. 9.

The curves show remarkable performance degradation (note the logarithmic vertical scale) as the offered load approaches the maximum load tolerated by the corresponding interconnected system. The mean and variance of the end-to-end delay tend to a vertical asymptote, the utilization ceases to increase linearly with the offered traffic, tending to an horizontal asymptote (corresponding to the maximum throughput) and the frame loss probability becomes non null and starts growing almost linearly with the offered traffic.

The curves for the mean and the variance of the end-to-end delay do not keep growing

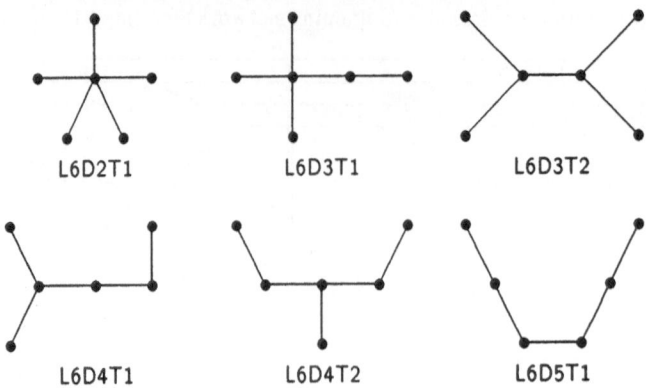

Figure 8: All possible 6-nodes trees; labels have the number of nodes in the first digit, the maximum distance (in traversed bridges) in the second digit, and a progressive number (for trees with the same first two indicators) in the last digit.

towards the vertical asymptote due to the assumption of modeling LANs with finite capacity queues, and of losing frames in overload conditions. When all LAN queues are full, the delay cannot increase further, and delay curves flatten towards a constant value.

The utilization curves show that, for the given offered traffic, the topologies with larger maximum distances between end stations load the interconnected network more, since more LANs must be traversed in the average, and the same frame must be transmitted more times. On the other hand, the shorter the maximum distance, the sooner LANs close to the root of the spanning tree become congested, since they have to carry a comparatively higher traffic.

The frame loss probability curves give a clear indication of the load values at which the networks begin to be congested. The larger the maximum distance of the interconnected system, the higher the load values where congestion begins. Note how the results of the bottleneck analysis of Section 4 match simulation results. Saturation points can be observed also on utilization curves as the values of offered traffic where the curves cease growing linearly.

The comparison of the delay curves of the six considered spanning trees unveils different behaviors for low and high traffic values. As a matter of fact, the curves can be seen to intersect for a value of the offered traffic between 0.75 and 1. For low and moderate traffic values, the interconnected networks with smaller maximum distance between end stations behave slightly better, since less LANs must be traversed by frames in the average. While the offered traffic grows, these topologies tend to become congested before the topologies with larger maximum distance, and delays grow faster. The behavior around the congestion point subdivides the studied topologies into three groups: the two star-shaped topologies (L6D2T1 and L6D3T1) reach congestion first, the linear L6D5T1 topology reaches congestion last, while the other topologies exhibit a similar, intermediate behavior. The curves for the variance of the delay show a behavior similar to the curves for the average delay.

From the results we thus observe that delays and the global network utilization increase with the maximum distance, but LANs are more uniformly used. In fact, with low distance between LANs, there are LANs in which a greater part of the global traffic flows and these become the bottleneck of the network. Considering the extreme cases, the star topology L6D2T1 gives us better delay performances than the linear topology L6D5T1, but it becomes congested for lower traffic. In Spanning Tree network design we thus have to trade off between performance and network congestion.

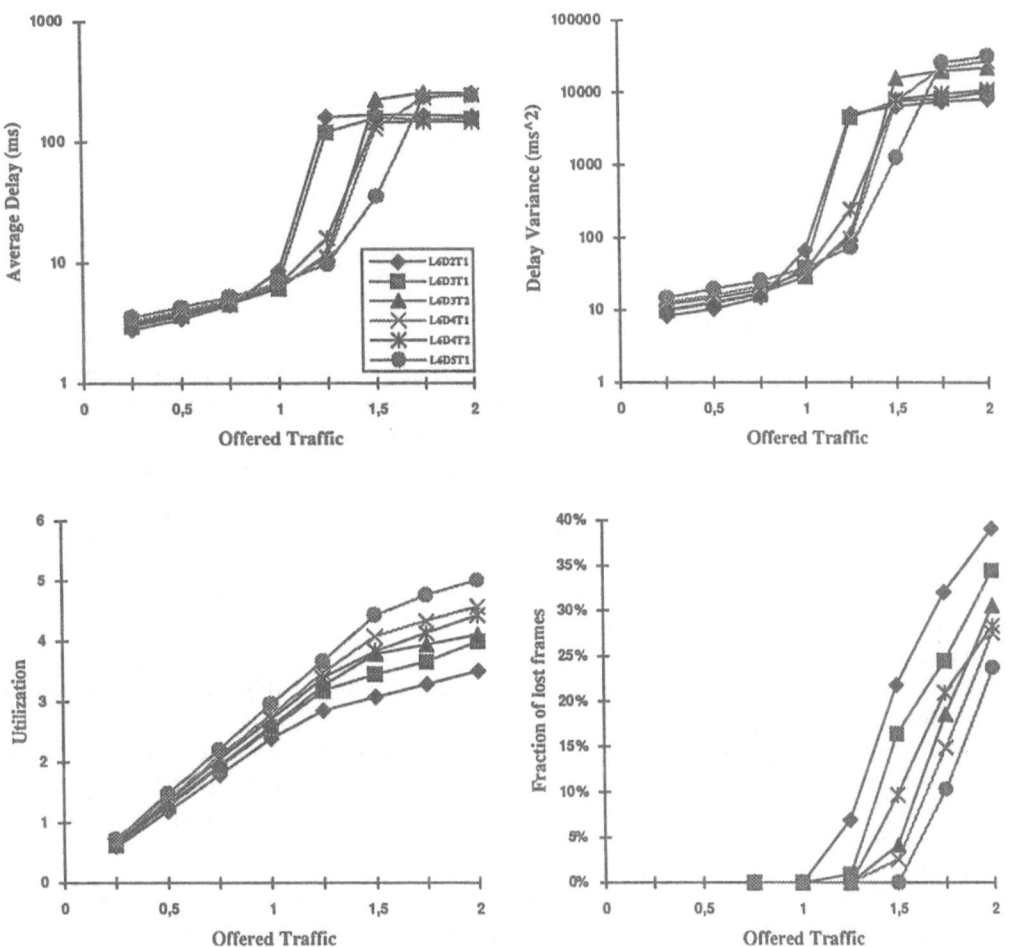

Figure 9: Performance figures as a function of the offered traffic for Spanning Tree with the six topologies of Fig. 8

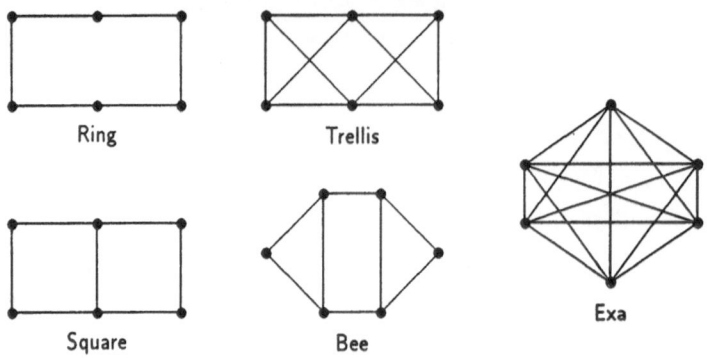

Figure 10: Topologies for the simulation of Source Routing

6.3 Source Routing vs. Connectivity

The set of experiments presented in this section is aimed to analyze the performance of the Source Routing scheme for topologies with different connectivity levels. We have analyzed the five topologies formed by six LANs shown in Fig. 10. These topologies are different from one another in the number of internal connections: Ring has the minimal number of connections that ensure a loop, Exa is completely connected; the number of bridges (i.e., number of edges in the graph) now ranges from 6 to 15; it was always equal to 5 in the previous simulation experiments.

For the results presented in this section, the information stored in routing tables by end-stations is deleted after 10 s.

From the results shown in Fig. 11 we can see, as expected, that the bridged network performances tend to increase with the connectivity level, but probably not as rapidly as one could imagine.

The completely connected topology Exa obviously outperforms the other networks: it becomes congested only for an offered traffic equal to about 3 times the capacity of one LAN, and shows a quite low utilization of individual LANs. Note that this topology asks frames to traverse at most one bridge.

For the other topologies saturation occurs for values of the offered traffic around 2 (which are not too far from what we have observed for the spanning tree with the highest congestion point), and the utilization is generally better for larger connectivity levels. Delay curves also tend to show better performances for higher connectivities, but in a less evident manner: this is due to the fact that, for more connected topologies, a greater number of possible data paths is available, and a larger amount of broadcast route-searching traffic is generated. The high volume of broadcast traffic present in the network may congest the bridged system and reduce the advantages of highly connected topologies.

6.4 Spanning Tree vs. Source Routing

A performance comparison between Spanning Tree and Source Routing can be carried on only in particular cases. Actually, Source Routing uses all available paths, while Spanning Tree is configured in a tree-like active topology that uses only few links.

The comparison between the two interconnection approaches is thus attempted by considering the two extreme cases (smallest – L6D2T1 – and largest – L6D5T1 – maximum distance) for Spanning Tree, and the Source Routing topology with the lowest connectivity level (Ring).

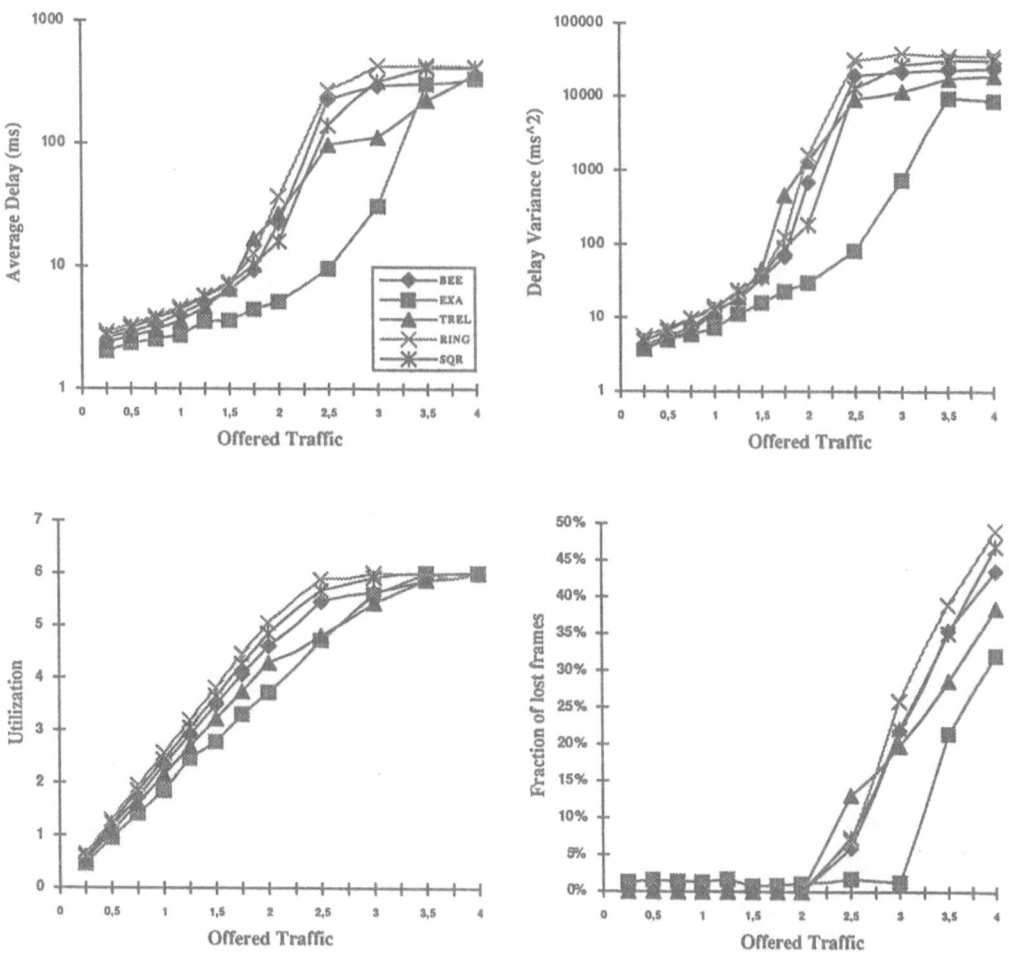

Figure 11: Performance figures as a function of the offered traffic for Source Routing with the five topologies of Fig. 10

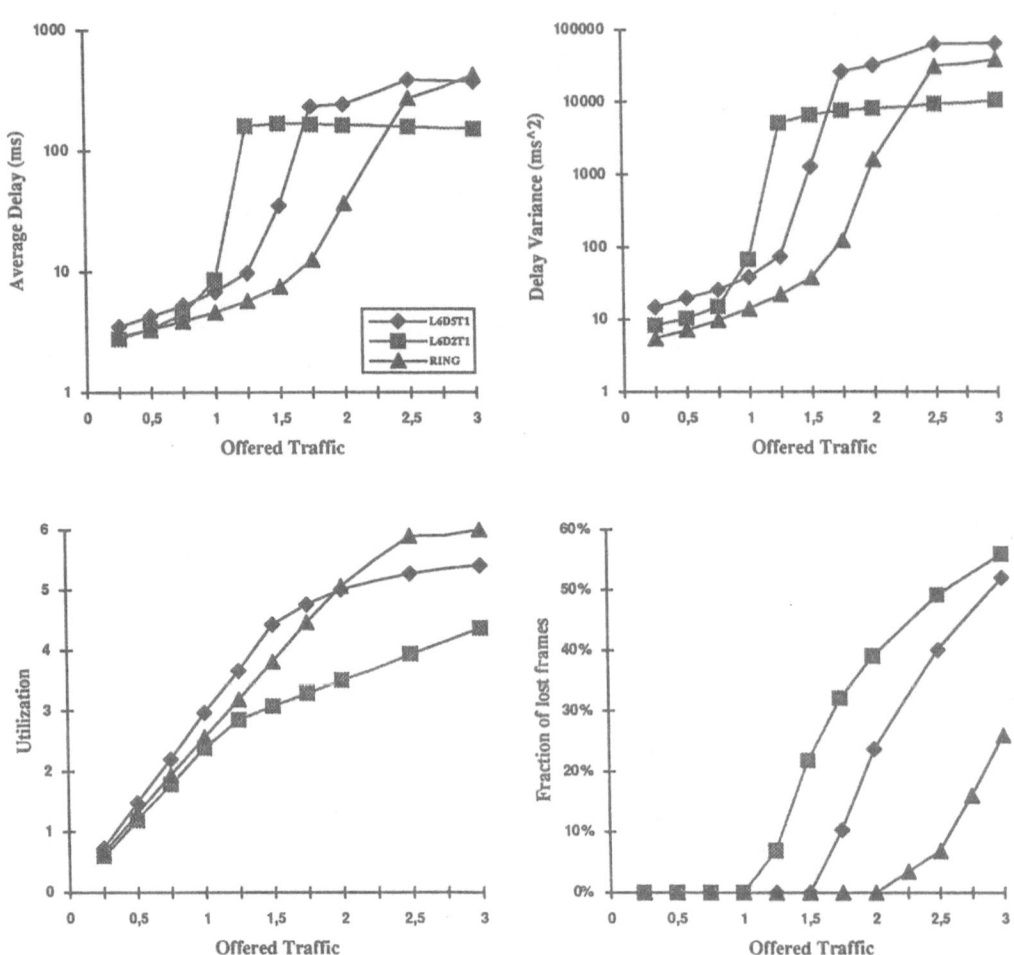

Figure 12: Performance figures as a function of the offered traffic for the L6D2T1 and L6D5T1 trees of Fig. 8, and for the Ring topology of Fig. 10

In the curves of Fig. 12, Ł6D5T1 has a behavior generally more similar to Ring, although Ł6D2T1 behaves very similarly to Ring for low traffic values.

As a general comment, Source Routing behaves better with respect to the considered performance figures, thus making good use of the additional complexity required with respect to Spanning Tree, as pointed out in Section 2.3.

7 CONCLUSIONS

The paper has presented an analysis and comparison of the Spanning Tree and the Source Routing LAN interconnection schemes, based both on simulative and analytical approaches.

Results show that Spanning Tree, even in the case of a fixed and simple cost function, can have a behavior dependent on bridge identifiers: different spanning trees can be formed, with different maximum distances between end stations. The larger this distance, the higher the mean and the variance of frame delays, but the maximum throughput of the interconnected system also becomes higher because of a slower congestion of the LANs closer to the root of the spanning tree. This result translates into interesting hints to network managers and designers.

Source Routing shows increasing performances for increasing connectivity levels (and hence for increasing network costs), although the gain seems to be not as large as expected. As already pointed out (see [3]), the amount of broadcasted route-searching traffic may cause severe congestion problems with Source Routing. However, for the considered topologies, a reasonable persistence of the information stored in routing tables (i.e., reasonable values for refresh timers, and for the size of routing tables) leads to a tolerable amount of route-searching traffic also for highly connected topologies.

Numerical results also show that BCMP models based on Kleinrock's independence assumption provide a quite accurate representation of the behavior of the interconnected system in the case of realistic system parameters.

The proposed analysis can be extended to consider other aspects of LAN interconnection, such as the behavior under asymmetric traffic patterns, the effect of finite capacity routing tables in stations and bridges, or the fault tolerance of the system.

References

[1] IEEE Project 802 – Local and Metropolitan Area Network Standards, *IEEE Standard 802.3: Carrier Sense Multiple Access with Collision Detection*, New York: IEEE, 1985.

[2] B. Hawe, A. Kirby and B. Steward, "Transparent Interconnection of Local Networks with Bridges," *Advances in Local Area Networks*, K. Kummerle, J.O. Limb and F. Tobagi (Eds.), New York: IEEE Press, pp. 482–495, 1987.

[3] Special issue on internetworking, *IEEE Network*, Vol. 2, n. 1, Jan. 1988.

[4] R. Perlman, "An Algorithm for Distributed Computation of a Spanning Tree in an Extended LAN," *Proc. 9th Data Communication Symposium*, Vancouver, Canada, Sept. 1985.

[5] F. Backes, "Transparent Bridges for Interconnection of IEEE 802 LANs," *IEEE Network*, Vol. 2, n. 1, pp. 5–9, Jan. 1988.

[6] Tsung-Yuan Tai and M. Gerla, "LAN Interconnection: A Transparent, Shortest Path Approach," *Proc. IEEE ICC '91*, Denver, CO, June 1991, pp. 1666–1670.

[7] IEEE Project 802 – Local and Metropolitan Area Network Standards, *IEEE Standard 802.1 (D): MAC Bridges*, Sep. 1988.

[8] IEEE Project 802 – Local and Metropolitan Area Network Standards, *IEEE Standard 802.5: Token Ring Access Method*, New York: IEEE, 1985.

[9] D.A. Pitt and J.L. Winkler, "Table-Free Bridging," *IEEE JSAC,* Vol. SAC-5, n. 9, pp. 1454-1462, Dec. 1987.

[10] D.A. Pitt, K.K. Sy and R.A. Donnan, "Source Routing for Bridged Local Area Networks," *Advances in Local Area Networks,* K. Kummerle, J.O. Limb and F. Tobagi (Eds.), New York: IEEE Press, pp. 517–530, 1987.

[11] R.C. Dixon and D.A. Pitt, "Addressing, Bridging, and Source Routing," *IEEE Network,* Vol. 2, n. 1, pp. 25–32, Jan. 1988.

[12] IEEE Project 802 – Local and Metropolitan Area Network Standards, *Draft Addendum to ANSI/IEEE Standard 802.5-1988: Token Ring MAC & PHY Specification Enhancement for Multiple-Ring Networks,* Nov. 1988.

[13] M. Soha and R. Perlman, "Comparison of Two LAN Bridge Approches," *IEEE Network,* Vol. 2, n. 1, pp. 37–43, Jan. 1988.

[14] L. Zhang, "Comparison of Two Bridge Routing Approches," *IEEE Network,* Vol. 2, n. 1, pp. 44–48, Jan. 1988.

[15] L. Kleinrock, *Communication Nets,* Dover, New York, 1964.

[16] G.M. Birtwistle, O.-J. Dahl, B. Myhrhaug and K. Nygaard, *Simula Begin,* Studentlitteratur, Lund, Sweden, 1973.

[17] E.G. Ulrich, "Event Manipulation for Discrete Simulations Requiring Large Numbers of Events," *Communications of the ACM,* Vol. 21, n. 9, pp. 777–785, Sep. 1978.

[18] G.S. Fishman, *Principles of Discrete Event Simulation,* J. Wiley & Sons, New York, 1978.

[19] M.W. Garret and San-Qi Li, "A Study of Slot Reuse in Dual Bus Multiple Access Networks," Proceedings *INFOCOM 90,* San Francisco, CA, June 1990.

[20] F. Baskett, K.M. Chandy, R.R. Muntz and F. Palacios, "Open, Closed and Mized Networks of Queues with Different Classes of Customers," *Journal of the ACM,* Vol. 22, n. 2, pp. 248–260, Apr. 1975.

[21] S.C. Bruell, G. Balbo and P.V. Afshari, "Mean Value Analysis of Mixed, Multiple Class BCMP Networks with Load Dependent Service Stations," *Performance Evaluation,* n. 4 (1984), pp. 241–260.

[22] S.C. Bruell, G. Balbo and S. Ghanta, "A Mean Value Analysis Based Package for the Solution of Product-Form Queueing Network Models," *International Conference on Modeling Techniques and Tools for Performance Analysis,* Paris, France, May 1984.

Efficient, Real-Time Address Resolution in Backbone Networks

W.A. Doeringer, H.D. Dykeman, M.L. Peters*, H.J. Sandick*
and K.V. Vu*

IBM Research Lab Zurich
Saeumerstr. 4
CH-8803 Rueschlikon, Switzerland

IBM Network Systems Architecture(*)
200 Silicon Drive
Research Triangle Park, NC 27709

Abstract

This article presents a novel, dynamic address resolution procedure for high-speed, general-topology backbone networks with group management and hardware-supported multicast facilities. By making efficient use of these particular network features, the described procedure induces only minimal communication and processing overhead independent of the size of the backbone network and the number of attached external networks. The procedure further imposes no architectural limit on the set of supported address spaces, and it meets the timing requirements for address resolution of a very wide range of interconnected external networks, including but not limited to, LANs and MANs.

1.0 Introduction

As is particularly exemplified by the tremendous success of multi-protocol routers [28] and the plethora of solutions for LAN interconnection [10, 13, 26], recent years have witnessed a growing trend among providers of network services to use their installed equipment to provide connectivity between like external networks of a variety of architectures [1, 4, 21]. Irrespective of the external networks to be interconnected, the basic structure of a general-topology backbone network is as depicted in figure 1.

The diagram illustrates two different types of external networks with incompatible protocols, whose individual components attach to the backbone network via *access agents* residing in *access nodes*. These nodes are native network or end nodes of the backbone, and the access agents represent users of the native *data transport system* of the backbone network. The access agents are hence identified by suitable *transport addresses* in the backbone's native address space.[1] Observe, however, that external networks use a backbone's

---

[1] Following the definitions in [16], the term *(transport) address* denotes here a unique identifier of (transport) network entities and service access points.

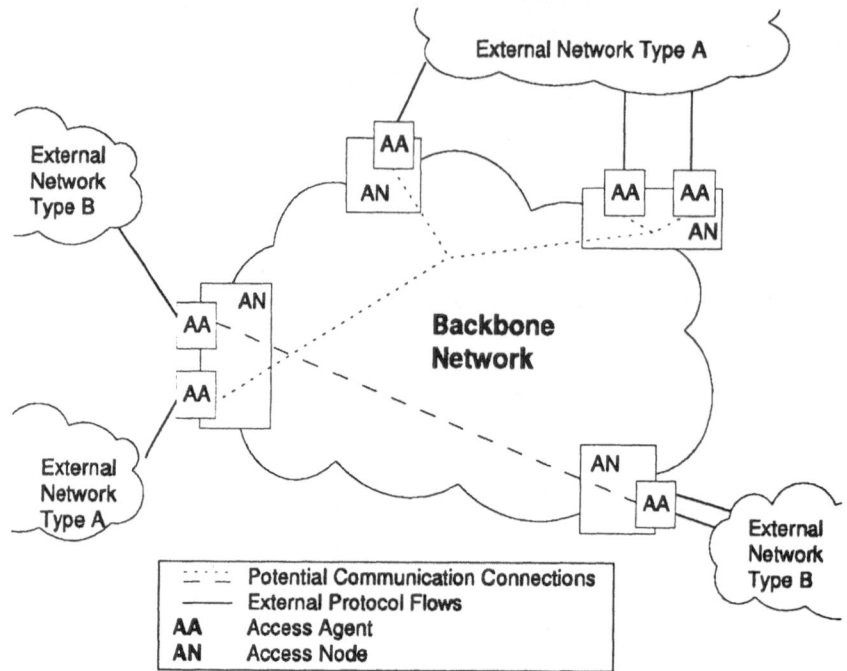

Figure 1. Backbone connectivity between groups of like networks

communication facilities as suitable for their respective purposes. The back-bone network may hence be utilized in parallel as a provider of clear channel services (physical layer), as a pseudo-LAN segment (MAC layer), or as a subnet for a larger internet, to name only the most typical cases, two of which are discussed in detail in the Example section below.

When signalling procedures indicate that a station on one of the intercon-nected external networks is attempting to communicate with a remote partner, the local access agent of the backbone network needs to derive from the sig-nalled address information which peer access agent(s) to contact in order to reach the desired destination resource. However, the backbone transport system in general does not understand nor can it route by the addresses employed by the external networks. Additional procedures are hence required to map dynamically the addresses of external resources to the native back-bone transport address(es) of peer access agent(s) through which communi-cation with the respective resources may be accommodated. Such procedures are referred to as *address resolution*, and are particularly well established in LAN/MAN environments where network-layer entities are allowed free mobility among attached stations, and therefore their addresses need to be mapped dynamically to respective MAC-layer addresses of the stations in which they reside [15, 17, 23]. Let us note here that the particular procedures employed by the LAN/MAN protocols have very strict timing requirements which must be met by the backbone address resolution procedures if the interconnection is to be transparent to the external protocols and the LAN/MAN characteristics are to be preserved. That is, the address resolution across the backbone network has to be performed in *real time*, or in other words, with such low latency as is expected by the LAN/MAN-based protocols.

1.1 The Problem

In a backbone network address resolution in the above sense is the responsi-bility of the access agents which must solve the following *"scoping dilemma"* to accommodate arbitrary network sizes: On one hand, it is not feasible to distribute information regarding the servicing access agents for each resource, due to the explosive growth of mapping tables and the attendant communications overhead. Hence, in cases where the exchanged information is insufficient to determine the exact location of an external resource, the access agents must execute an algorithm to distribute searches to suitable sets of access agents, which may then need to relay such address resolution procedures into the external networks to locate the resource. On the other hand, it is well-known that the all-station broadcasts used in LAN environments do not provide a suitable framework for search distribution in large backbone networks [10, 30]. Hence, the design of a useful address resolution procedure for backbone networks of general topology must optimize the problem of locating resources in real time with acceptable overhead, while providing full connectivity to the external networks without inducing excessive storage complexity.

However, all current approaches to dynamic address resolution for transparent network interconnection suffer from various shortcomings which prevent their use as a general-purpose solution: Beginning with the well-known (intelligent) *split bridges*, one recognizes that the product offerings do not address the technical problems of dynamic distribution of MAC-layer address information between peer bridging devices in a way that could be extended to an arbitrary number of protocols and to arbitrary network sizes. Their mapping tables of MAC-addresses to pertinent peer bridges have to be configured or are acquired through all-station broadcasts. Both approaches do not scale to large number of end stations.

Next we see that even *routing protocols* like IDRP [18] with its very open addressing structure cannot meet the address resolution requirements of a general backbone network: First of all any routing protocol restricts the end-user mobility. A backbone network, however, must not impose any such restrictions above those of the external networks themselves. This issue becomes most pronounced for interconnected MANs and LANs whose most salient feature exactly is the unrestricted user mobility. And secondly, some external networks, such as those using the NetBIOS protocols, do not advertise reachability information at all. Hence, when a routing protocol is to be used to build the address mappings, pertinent information needs to be configured into the access agents. This certainly complicates the system management and severely restricts the user mobility.

With respect to *distributed directories* [31], our analysis showed that they operate in a general environment using hop-by-hop hierarchical search protocols that do not offer the sub-second response times required for high-speed interconnection. Furthermore, since the external network addresses incorporate only very few hierarchical layers (typically only two: network and station identifier), address resolution procedures would in very many cases have to be navigated through the first, second or third level of the gigantic world-wide directory tree. It cannot be reasonably expected that such approaches will yield acceptable performance. This view is shared by recent surveys of X.500 and related directories such as in [29].

When we finally turn to *research prototypes*, we notice that current experimental test-beds for the interconnection of some selected networks [5, 12, 22, 27] either do not address the issue of address resolution, or assume that the user-supplied address of the destination end station includes the address of the servicing access agent. Thus an algorithmic address resolution becomes possible but the transparency of interconnection is lost.

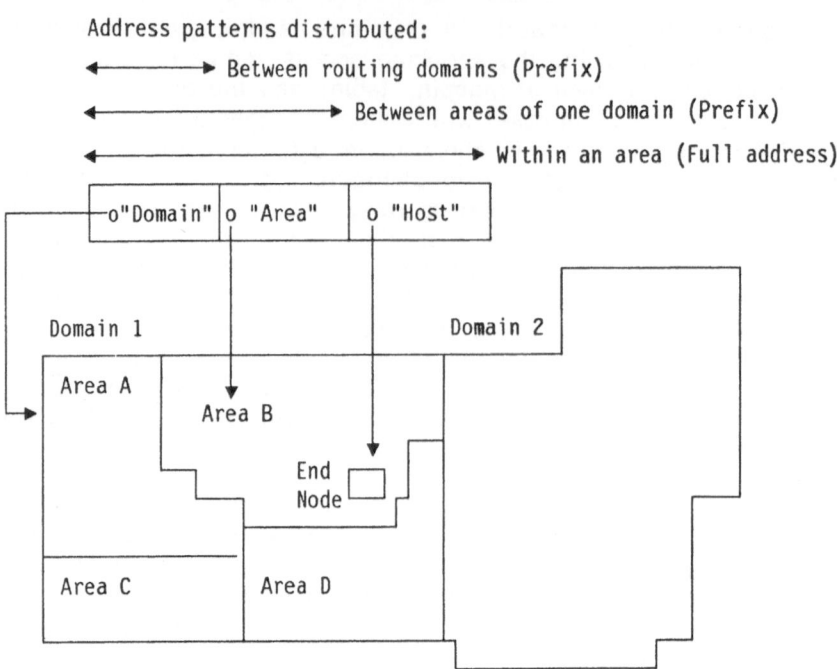

Figure 2. Typical use of OSI network address hierarchy

1.2 The Environment

Having realized that current approaches to large-scale address resolution all fall short of offering a general solution, it proves worthwhile to recall that an analogous scoping problem has to be solved by the routing protocols of all network architectures designed to scale well to large networks of general topology. All such network architectures perform routing based on structured addresses which closely mirror the network topology as exemplified by [9, 14, 16]. The routing protocols of these networks distribute or configure *reachability information* in the form of *address patterns*, which grow increasingly generic (i.e., less specific, and therefore representative of larger sets of end nodes) the further they are distributed. Routing to individual network entities is guided by "best" matches of the specified individual addresses and the advertised address patterns. A typical use of network addresses by OSI routing protocols is depicted in the figure 2. Note that the address patterns of these routing protocols take the form of variable length prefixes whose generality increases with decreasing length. The OSI network nodes base their routing on the "longest matching prefix" rule.

Now that we have seen how external reachability information may be sufficiently compressed, we need to address the problem that a backbone network cannot assign topological semantics to it. That is, any given external address pattern may describe end stations which are distributed across several disjoint external networks attached to different access agents. Hence a pertinent search needs to be multicast to all such access agents, that is, to all agents which share connectivity to a given external address pattern.

1.3 Assumptions about Transport Services

Our design builds on the clear trend in modern network architectures to include dynamic group management and multicast communication within and between such groups into their transport services abstraction [6, 19, 25]. Hence we may assume for the following discussion that the backbone's transport system supports:

- the dynamic formation of groups of transport users based on globally unique *group addresses*

- high-speed multicast communication within and between the above groups based on hardware-supported distribution trees such as described in [2, 3, 7]

1.4 The Solution

Given the above concept of reachability information described by suitable address patterns, and the stated assumptions about the transport services of the backbone network, we define our address resolution procedure as follows:

1.4.1 Step 1 - Acquiring Resource Identifiers

Prior to any address resolution, the access agents need to acquire suitable *resource identifiers* describing the resources in the external networks to which they are locally attached. These identifiers consist of a *protocol type (p)* and an *address pattern (a)*, the latter of which are learned by the access agents through participation in the external routing protocols, or where such participation is not possible, via configuration information supplied by the network administrator.

Observe that such identifiers cover the full range from those that consist of only the protocol indicator, thus indicating potential reachability to all addresses of the specified protocol type, to those indicating reachability to groups of addresses with a common address pattern, and those that contain a full specific, i.e. individual, address. The structured addresses of external network protocols thus allow a sufficient reduction in the amount of information to be supplied to and managed by the access agents.

However, in cases where external network protocols use unstructured addresses,[2] which do not reflect the network topology, reachability information cannot be meaningfully condensed since only complete addresses are useful in locating resources. For such protocols, dynamic routing and address resolution become infeasible for any sizeable network irrespective of the particular procedure. A static decision must therefore be made as to where their frames may be distributed to limit the dissemination of address information and the scope of address resolution procedures. In order to integrate such networks into our general design based on structured addresses, a network administrator must statically identify collections of subnetworks that are to be interconnected, forming a *virtual network*. This virtual network is assigned a *virtual network identifier* that is then used by the access agents in the address resolution procedures disclosed below as the only address pattern describing all resources located on the virtual network. Such a virtual netid is not part of the subnetwork's native non-structured address, however the access agents attached to the virtual network treat all frames as if they had specified the assigned virtual netid as part of the address information. Thus an external network which uses unstructured addresses can only be a part of a single virtual network.

---

[2] For example LAN protocols with their unstructured (IEEE) MAC addresses, or standard NetBIOS protocol implementations.

1.4.2 Step 2 - Deriving Group Identifiers and Joining Multicast Groups

For each acquired resource identifier (p,a), the access agents which attach to external networks of protocol p where resources reside with external addresses matching the address pattern a, use a common algorithm to map the resource identifier to a **globally unique group address G(p,a)** of the backbone's transport system. The access agents then invoke the group management services to join the group identified by this derived group address. This group is flagged to the transport system as one with which a multicast distribution tree should be associated which is dynamically adapted to the group membership . This tree is assumed to be optimized in the sense that it supports multicast between the group members with minimal use of communication and processing resources and negligible queueing delays. Hence, the performance of the communication within and between such groups in a high-speed backbone network is essentially bounded only by the propagation delay of the transmission media.

1.4.3 Step 3 - Multicasting Query Packets

Once the above two steps have been completed, an external address A of protocol p which matches a resource identifier (p,a) known in any of the backbone's access agents, can be resolved to the transport address of an access agent through which the given external address may be reached. The address resolution proceeds as follows:

1. The local access agent , i.e., the one which received from the attached external network a request for communication with the destination address A, maps the resource identifier (p,a) to the corresponding group address G(p,a) using the same mapping function as above. The access agent then transmits a multicast query to the group address specifying that access to the resource (p,A) is required.

   While this transmission proceeds, the reverse path through the backbone network is accumulated in the multicast packets such that every receiving access agent can respond to the query without having to undergo the calculation of a route through the backbone network back to the search originator [2, 7].

   We assume that in the cases where the local access agent is not part of the group identified by the derived group address G(p,a), the multicast message is transmitted using a *remote-access-to-tree routing mode* described in [2, 7]. I.e., the message is transmitted as a unicast packet until it reaches the nearest node on the respective distribution tree, and only then is the message turned into a multicast. Thus again only minimal communication and processing resources are used, and the transmission performance of a unicast packet is preserved.

2. Once the access agents of group G(p,a) receive the multicast query packet, they check their connectivity to the resource (p,A) according to the reliability requested in the query and then report their findings back directly to the search originator using the accumulated reverse path information. Where multiple access agents have connectivity to the destination resource (p,A), some multicast-based coordination procedures may be required among them to avoid the injection of multiple queries into external networks or to provide load-balancing.

1.5 Two Examples

In this section we provide a brief discussion of how the above procedure may be applied to the interconnection of telephone networks, and the use of a backbone network as a subnet for a particular internetworking protocol such as the Internet Protocol [24]. For a detailed description of the benefits of our invention for LAN and MAN interconnection we refer the reader to [11].

1.5.1 Telephone Network Interconnection

When interconnecting telephone networks as in the following example, the access agents act as local telephone exchanges, and the basic structure of the backbone network takes the form shown in figure 3. In this diagram, the access agents attach to four separate telephone subnets covering the indicated areas. Since current telephone networks do not employ dynamic routing protocols which distribute reachability information, the access agents need to be configured with the respective resource identifiers, i.e., (t,724), (t,725), and (t,726), where t stands for the "telephone" protocol, and the numbers represent the address patterns, that is the area codes in our example. This stage corresponds to 1.4.1, "Step 1 - Acquiring Resource Identifiers" above. Once supplied with this information, the access agents join the groups G(t,724) (agent AA1), G(t,725) (agents AA4, AA5, and AA6), and G(t,726) (agents AA2 and AA3) respectively. This completes 1.4.2, "Step 2 - Deriving Group Identifiers and Joining Multicast Groups."

Assume now that access agent AA1 receives a request for a telephone call to destination 725-8911 say. Following the procedure described in 1.4.3, "Step 3 - Multicasting Query Packets," the access agent maps the address pattern, i.e. the area code, 725 to the group address G(t,725) and transmits a multicast query to the group using a remote-access-to-tree routing mode. The multicast is received by the access agents in that group, i.e., AA4, AA5 and AA6, and based on internal information and possibly also private communication among them, one of the agents sends a positive reply back to agent AA1. Access agent AA1 is then in possession of the transport address of a peer access agent through which the telephone call may be routed to its final destination,

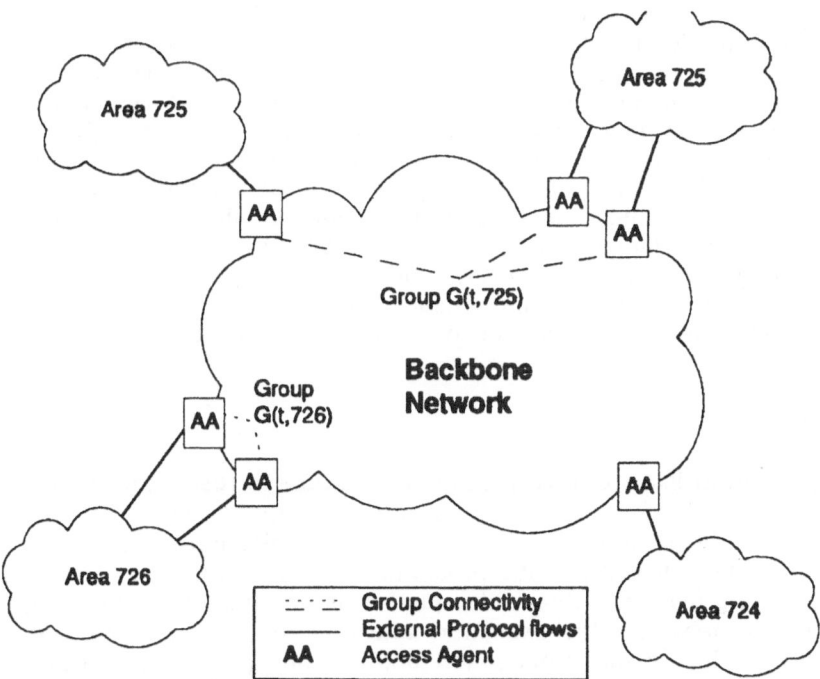

Figure 3. Forming groups based on the area prefix of telephone numbers

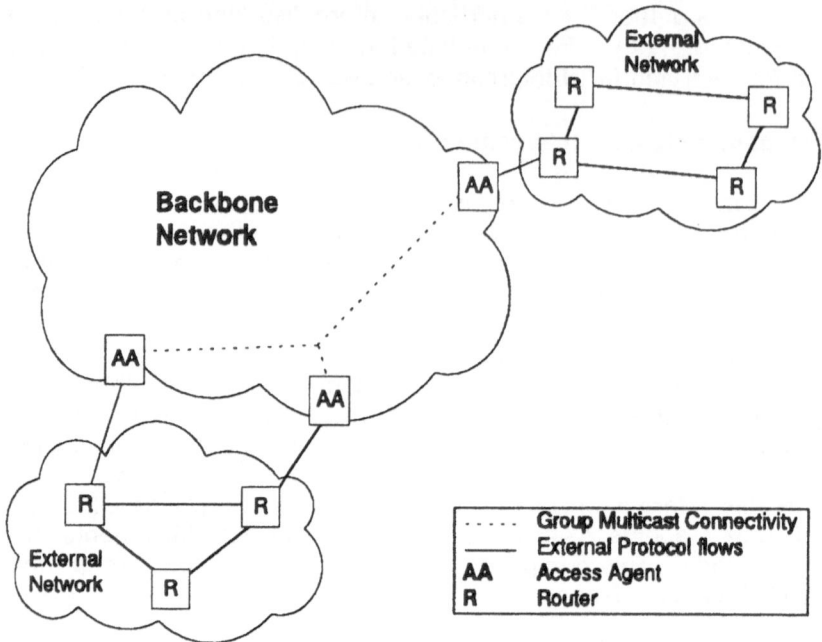

Figure 4. Multicast connectivity over a backbone network

and it may then proceed to establish suitable communication channel(s) across the backbone network to satisfy the external call request.

1.5.2 The Backbone as a Subnet of an Internet

When a backbone is to serve as a subnet for an internetworking protocol [9] its structure takes the shape depicted in figure 4. In this figure, the access agents connect to the external network routing devices, the routers (R), and participate in the full external routing protocol once they have established suitable contact between themselves across the backbone network. However, the specifications of all routing protocols indiscriminately gloss over the problem of identifying adjacent routers, i.e, access agents in our case, attached to a given subnet after system (re-)start. This information is assumed to be "prior knowledge", and will hence in most cases have to be supplied as configuration information. Using the present address resolution algorithm, network layer access agents may simply use a resource identifier of the form (protocol, ANY) to join the group G(protocol, ANY), and use a simple multicast procedure within this group to establish first contact. In this way a substantial part of the typical configuration requirements of network-layer routing devices may be eliminated.

1.6 Summary

In the present article we have presented an address resolution procedure for general-topology high-speed networks with group management and hardware-supported multicast transmission across communication-optimal distribution trees. We demonstrated that the procedure induces very little communication and processing overhead due to its efficient use of these particular features of the backbone network. The execution time was shown to be bounded essentially only by the media propagation delays, thus providing sufficiently low latency even for the transparent interconnection of high-speed LANs and

MANs. As a particularly novel feature, the defined algorithm provides great flexibility to balance the dissemination of reachability information against the scope of multicast search procedures. Hence, it scales well with the size of the backbone network and the number of attached external networks.

In closing we would like to emphasize that the present article represents more than a mere paper study. The issues raised and solutions presented are a result of architectural work for the plaNET network being installed as part of the NREN Aurora test-bed [8, 20]

1.7 References

[1]   ATC, "Special Reports on Multiport Internetworking Systems and Virtual Networks," *LOCALNetter*, vol. 11, no. 2, Architecture Technology Corporation, February 1991.

[2]   B. Awerbuch, Israel Cidon, Inder S. Gopal, Marc Kaplan, and Shay Kutten, "Distributed Control for PARIS," *Proceedings of 9th ACM Symposium on Principles of Distributed Computing*, Quebec, Canada: ACM, August 1990.

[3]   B. Awerbuch, Israel Cidon, and Shay Kutten, "Communication-optimal maintenance of dynamic trees," *Research Report*, IBM Corp., January 1990.

[4]   T. Bradley, C. Brown, and A. Malis, "Multiprotocol Interconnect over Frame Relay," *RFC 1294*, NIC, January 1992.

[5]   R. Callon and H. W. Braun, "Guidelines for the use of Internet-IP addresses in the ISO Connectionless-Mode Network Protocol," *RFC 1069*, NIC, February 1989.

[6]   Israel Cidon and Inder Gopal, "PARIS: An Approach to Integrated High-Speed Private Networks," *Int. Jour. of Digital and Analog Cabled Systems*, no. 1, pp. 77-85, 1988.

[7]   Israel Cidon, Inder Gopal, and Roch Guerin, "Bandwidth Management and Congestion Control in plaNET," *IEEE Com. Magazine*, pp. 54-64, October 1991.

[8]   D.D. Clark, B.S. Davie, D.J. Farber, I.S. Gopal, B.K. Kadaba, W.D. Sincoskie, J.M. Smith, and D.L. Tennenhouse, "An Overview of the Aurora Gigabit Testbed," *Proceedings of the IEEE Infocom '92, Florence, Italy*, pp. 569-581, 1992.

[9]   Comer, *Internetworking with TCP/IP. Principles, Protocols and Architecture*, Prentice-Hall, 1989.

[10] Robert P. Davidson and Nathan J. Muller, *LANs to WANs. Network Management in the 1990s*, Boston, London: Artech House, 1990.

[11] J. Derby, W. Doeringer, J. Drake, D. Dykeman, P. Gopal, S. Kutten, L. Li, M. Peters, H. Sandick, and K. Vu, "Scoping Multicasts in WAN-Interconnected Local Networks," *IBM Technical Disclosure Bulletin*, vol. 34, no. 8, pp. 68-71, IBM Corp., January 1992.

[12] R. Hagens, N. Hall, and M. Rose, "Use of the Internet as a Subnetwork for Experimentation with the OSI Network Layer," *RFC 1070*, NIC, February 1989.

[13] James Herman, "Smart LAN Hubs Take Control," *Data Communications*, pp. 66-79, June 1991.

[14] IBM, "System Network Architecture: Format and Protocol Reference Manual," *Document SC30-3112*, 1980.

[15] IBM, "NetBIOS Application Development Guide," *Document S68X-2270*, IBM Corporation, 1987.

[16] ISO, "Network service definition.," *ISO/8348*, 1988.

[17] ISO, "End system to Intermediate system routing exchange protocol for use in conjunction with the protocol for providing the connection-less mode network service (ISO 8473)," *ISO/DIS 10598*, 1988.

[18] ISO, "Information Processing Systems - Telecommunications and Information Exchange between Systems - Protocol for Exchange of Interdomain Routeing Information among Intermediate Systems to Support Forwarding of ISO 8473 PDUs," *ISO/DIS CD 10747*, 1991.

[19] Special Issue, "Exchange Network-Premises Network Interface," *Journal of Digital and Analog Communication Systems*, April-June 1991.

[20] Special Issue, "Where are we with Gigabits?," in Brian E. Carpenter, Lawrence H. Landweber, and Roman Tirler, editors, *IEEE Network Magazine*, vol. 6, no. 2, March 1992.

[21] N. Lippis and J. Herman, *The Internetwork Decade*, Northeast Consulting Resources and Data Communications, 1991.

[22] L. Morales and P. Hasse, "IP to X.121 Address Mapping for DDN," *RFC 1236*, NIC, June 1991.

[23] Plummer, "An Ethernet Address Resolution Protocol," *RFC 862*, NIC, November 1982.

[24] Jon Postel, "Internet Protocol," *RFC 791*, NIC, September 1981.

[25] RACE Project R1044/2.10, "Broadband User/Network Interface (BUNI)," *Deliverable 44/RIC/UNI/DS/A/005/B1*, RACE Consortium, August 1990.

[26] Roshan L. Sharma, "Interconnecting LANs," *IEEE Spectrum*, pp. 32-38, August 1991.

[27] C. Shue, W. Haggerty, and K. Dobbins, "OSI Connectionless Transport Services on Top of UDP. Version: 1," *RFC 1240*, NIC, June 1991.

[28] Eric Smalley, "Reports cite growth of LAN internet market," *Network World*, p. 19, March 1991.

[29] C. Weider and J. Reynolds, "Executive Introduction to Directory Services Using the X.500 Protocol," *RFC 1308*, NIC, March 1992.

[30] Jeffry S. Yaplee and Robert O. Denney, "Building a Large Network at The Boeing Company," *Proceedings of the 13th Conf. on Local Computer Networks, Minneapolis*, pp. 20-26, October 1988.

[31] Stefano Zatti, "Naming in OSI: Distinguished Names or Object Identifiers?," in Vito A. Monaco and Roberto Negrini, editors, *Proceedings of the CompEuro Conference, Bologna*, IEEE Computer Society Press, May 13-16 1991.

ARCHITECTURE AND PERFORMANCE EVALUATION FOR HIGH SPEED LAN INTERNETWORKING PROCESSORS

Kunihiro TANIGUCHI, Hiroshi SUZUKI, and Takeshi NISHIDA

C&C Systems Research Laboratories
NEC Corporation
MIYAMAE-KU, KAWASAKI
KANAGAWA 216, JAPAN
taniguti, hiroshi, nishida@nwk.CL.nec.co.jp

ABSTRACT

High-performance internetworking processors are inevitable in realizing multimedia communications on gigabit internets. This paper compares various alternative architecture in term of performance and functionality, and proposes an optimal architecture of high-performance internetworking processors, referred to as INP (Internetwork Nodal Processors), which can forward packets at the network layer protocol. INPs are basically composed of two main components: 1) several network I/F modules which perform network layer functions, and 2) a data forwarding module which interconnects the network I/F modules and provides message paths between any pair of them. First, three types of INP architecture are compared in considering allocation methods for various tables used for route decision. Next, showing performance bottleneck caused by a bus based INP architecture, INPs using a switch for data forwarding modules are analyzed. For the switch architecture, three alternative packet buffering schemes are compared. Through these comparisons, this paper proposes that a combination of the table allocation fully distributed on network I/F modules and an input-output buffer switch for a data forwarding module is best-suited for high-performance INPs.

1.   INTRODUCTION

In this decade, computer networks composed of subnetworks, which are called internets, have been growing tremendously in geographical scale, numbers of users and hosts [1], network speed, available services and so on. Especially evolution of link speed is tremendously large. Each LAN has evolved from several Mbps low speed links e.g., Ethernet, token-ring, etc., to high speed links of several hundreds to several Gbps e.g., FDDI, ATM-LAN, etc.,[2][3]. In wide area networks that interconnect separated LANs, have also experienced similar evolution from low speed of T1/T3 and frame relay networks to high speed of ATM and/or SDH networks. Regardless of such a high capacity network infrastructure, however, end users, cannot fully utilize this attractive bandwidth. This mainly comes from the facts that both hosts and internetwork processors executing communication protocols have not made progress so as to handle high speed networks.

Although many researchers have investigated high-speed protocol processors architecture and implementations for protocol processing in hosts[4][5][7], researches on architecture of high-performance internetwork processors are rare. There are various kinds of internetwork processors, e.g., repeaters, bridges, routers and gateways. These are distinguished from a viewpoint that at which protocol layer a processor terminates. To get high speed relaying performance, repeaters and bridges are well-suited internetwork processors, because they do not have to execute higher layer complicated protocols processing. Considering future securable reliable and expandable LAN internetworking, however, the internetwork processor should satisfy the following crucial requirements in addition to high performance:

1. Network segmentation based on administrative domains which make policy based routing available,
2. Reliable and security packet filtering for preventing from unnecessary packet flooding and malicious user packet passing, and
3. Hiding heterogeneous network configurations and technologies, e.g., Ethernet, FDDI, point-to-point, etc., for the purpose of providing user services independent of their physical network infrastructure.

To meet these requirements, internetworking processors capable of network layer processing, i.e., routers, are to be suitable for future LAN interconnections, because of an enhanced addressing and end-to-end service features provided by the network layer. Moreover it is predictable that future internetwork will provide, in addition to none real-time data packet relaying service, reliable real-time data transporting service using new transport and network layers that assure real-time transmissions, e.g., ST-2, XTP, RTP, etc.[6][7][8]. This new internet enables multimedia communication service including audio, video, synchronized data service etc. To realize this, routers should have a new architecture in perspective of performance and functionality. In terms of functionality, the new routers will have TOS (Type of Service) routing and some resource reservation mechanisms to assure quality required by multimedia applications. Regarding performance, the routers should be able to handle several gigabit speed traffic. This new type router is referred to as an INP (Internet Nodal Processor) for discriminating current router architecture. In this paper, only performance aspect of INP is discussed. Our future paper will handle the high functionality for INPs.

INPs are basically modeled as a collection of several network I/F modules and a data forwarding module which interconnects the network I/F modules. Each network I/F modules performs network layer functions which make reference with various tables, e.g., routing, address resolution tables, etc., for route decision. Since location of these tables as well as the table search algorithms is assumed to strongly affect on INP's performance, packet forwarding delay is compared in various table allocation mechanisms.

Furthermore various alternative methods to get faster interconnection between network I/F modules are examined. It is known that memory copy time mainly affects on protocol processing performance [4]. It is shown in this perspective that INPs using shared media, e.g., bus, has serious performance problems. To overcome the problems, we propose switch-based INPs which use variable length packet switches as a data forwarding module. Many researchers have investigated fixed length packet switches, e.g., ATM switch [9], [10]. However, researches on variable length packet switches such as switches for routers are little [13]. Performance for various types of variable length packet using switches is investigated to find most suitable switch architecture for LAN interconnection. Each performance evaluation

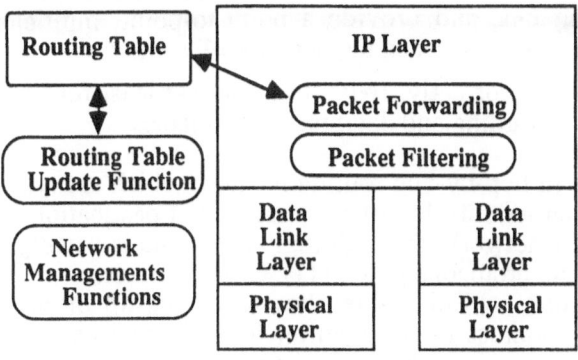

Figure 1   Logical structure of INPs

done in the paper uses internetwork traffic patterns which are experimentally observed in real internets [5].

2.   BASIC STRUCTURE OF INPS

Figure 1. depicts logical structure of an INP is explained using INPs have a protocol stack which is composed of physical, datalink and network layer protocol.
INPs generally have the following four functions:

Packet filtering function to send out only valid and authorized traffic
        beyond it
Packet forwarding function to forward packets to appropriate
        directions,,
Routing table update function to maintain and update routing table, and
Network management function to keep itself and its attached networks
        work correctly.

Fist two functions are involved in network layer protocol. Since the packet forwarding is the most important network service in high speed networks, it is mainly focused in this paper.

Physical structure of INPs is typically configured as shown in Figure 2. To distribute protocol process task, INPs have several networks I/F cards, which are interconnected through a data forwarding module. The network I/F cards have one or some I/F port(s) which is assigned each datalink address and terminate networks at a physical and datalink level. The data forwarding modules are usually

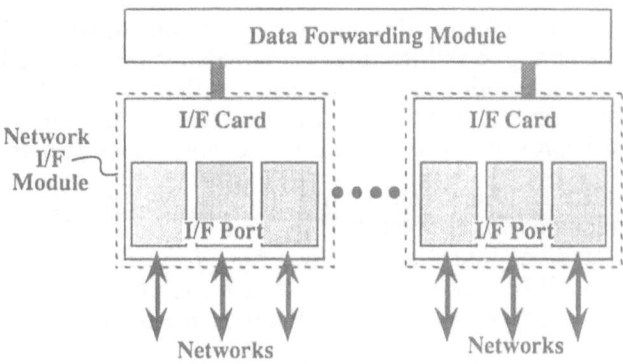

Figure 2   Physical structure of INPs

implemented using bus, and provide a point-to-point, multicast and broadcast communication capability between the network I/F cards.

To route the packet correctly, INPs must have at least four transferring tables, i.e., routing table, forwarding and datalink address tables.

Routing table In hop-by-hop routing methods which are typically used in the current internet, this describes a mapping of destination address to the next hop INP's network layer address. This table is usually managed by dynamic routing protocol e.g. RIP[11], OSPF[12], etc.

Forwarding table This table shows a mapping of next hop INP's network layer address to an identifier for an output network I/F module which includes an I/F card and I/F port connected to the next hop INP.

Datalink address table This describes a mapping of a next hop INP's network layer address to its datalink layer address. This is referred to be as an Address Resolution Protocol Table (ARP Table) in this paper.

These tables are updated in a static and/or dynamic manner when new network nodes are connected or replaced. Dynamic update mechanisms are preferable for large scale internets. Especially routing and ARP tables are preferable to be updated dynamically during network operation, since the network configurations are required to be changed without stopping the whole network.

When an INP receives a packet for forwarding to another INP, it performs a routing function for the packet as the following sequences based on destination network layer address; the INP finds out 1) its next hop network layer address from a routing table, 2) an identifier of an output I/F port in an output I/F card from a forwarding table, and 3) a datalink address of a next hop INP from an ARP table. If the next hop INP is connected to a different I/F card, packets are transferred using a data forwarding module to the target network I/F card and then sent out from the network I/F port with the datalink layer address. If the next hop INP is connected to the same I/F card as the receiver, the packet is sent out from the indicated I/F port directly.

It is easily predictable that future gigabit internets will need high speed INPs such as to accommodate more than ten interfaces with several hundreds Mbps to Gbps rate speed, and forward all packets received within several hundred micro seconds. Although this latency is almost same as current router performance, number of terminated networks and link speed of future INPs are grater than those of current INPs.

3.   PARALLEL PROCESSING

3.1.   Table Allocation

In this section the data forwarding module is assumed to be a shared media, e.g., single bus. Three alternative architectures for table allocation methods are discussed. The shared media assumption does not largely affect on results of performance comparison among these type INPs. It is also assumed that a packet is forwarded across different I/F cards, because this causes the worst delay scenario.

Type A - Centralized Table Architecture. In Type A there is a single central table database which includes all three tables (Fig. 3). This central table DB is located in a physically separated module attached to a data forwarding module and can be referred equally by all I/F cards. In this case, a forwarding process on an input I/F card refers the routing, forwarding and ARP tables in a central table DB.

Type B - Partial Distributed Table Architecture. In Type B, ARP tables are distributed in all I/F cards and when other tables are placed in a central table DB

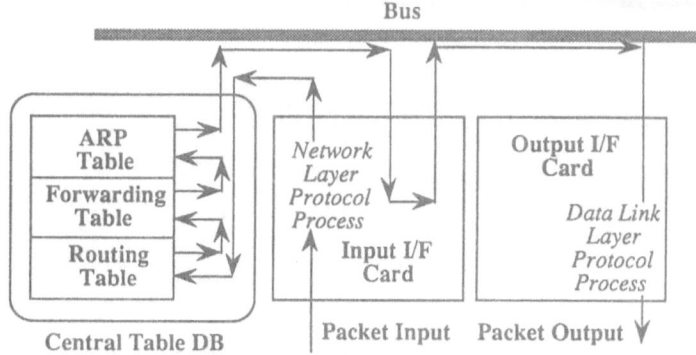

Figure 3   Type A INP structure

(Fig. 4). In this case, a forwarding process on an input I/F card refers the routing and forwarding tables in the central table DB, and sends the received packets to an output I/F card selected. The datalink layer process on the output I/F card determines a target datalink address by referring an ARP table. If it cannot find an entry for the selected next hop INP, it sends a request packet to the next hop for obtaining the datalink address.

As mentioned in the previous section, ARP tables maintain a mapping between a network layer and datalink layer addresses for a next hop which are connected with a specific I/F port within an I/F card. Distributing ARP tables is expected to remove congestion an ARP table search process on a central table DB. In addition, it is expected to eliminate ARP table management traffic on a bus, since ARP tables can be locally updated on each I/F card. Moreover performance improvement can be expected, owing to a parallel processing of network layer and datalink layer process, because they are performed by separate processors.

Type C - Fully Distributed Transferring Table Architecture. In Type C, all tables are distributed and managed by I/F cards locally (Fig. 5). This means that all I/F cards keep an identical routing and forwarding table. In this case, a forwarding process on an input I/F card refers the local routing and forwarding table, and sends the packets to an output I/F card selected. The datalink process on the output I/F card attaches datalink address on the packet by referring the local ARP table.

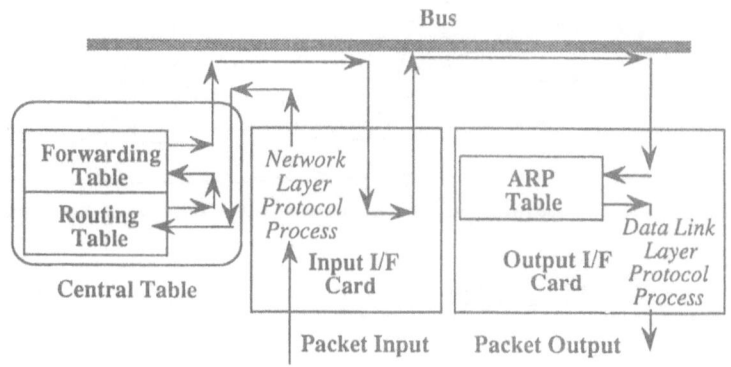

Figure 4   Type B INP structure

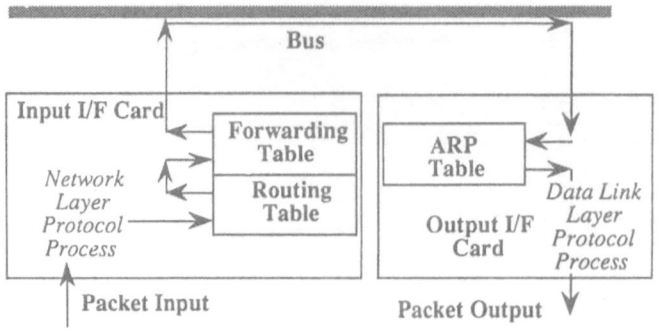

Figure 5   Type C INP structure

3.2   Performance Evaluation

Routing delay for the three INP models can be evaluated using a simple queuing network model. Routing delay is defined as the time from when a packet is received by an I/F card to when it is sent out from an output I/F port: In this evaluation routing delay does not include datalink layer processing time, because we want to evaluate an INP performance independent of a specific datalink layer protocol.

In the model, three major processing components are focused: packets routing process, table access and search process and data transfer process between I/F cards. Figure 6 shows a queuing network model for Type A. Performance for each type of INPs is numerically evaluated by simulations.

Table 1 shows INP's configuration parameters used in the simulation. This distribution of packet inter-arrival time and packet length is based on the experimental results from packet traffic measurement [5].

Figure 7 shows routing delay for Type A, B and C. In smaller number of I/F cards, although it is difficult to find out differences from this figure, Type A and B provide better performance than Type C. This is because the network layer protocol process and table access process are executed in a parallel manner in those types.

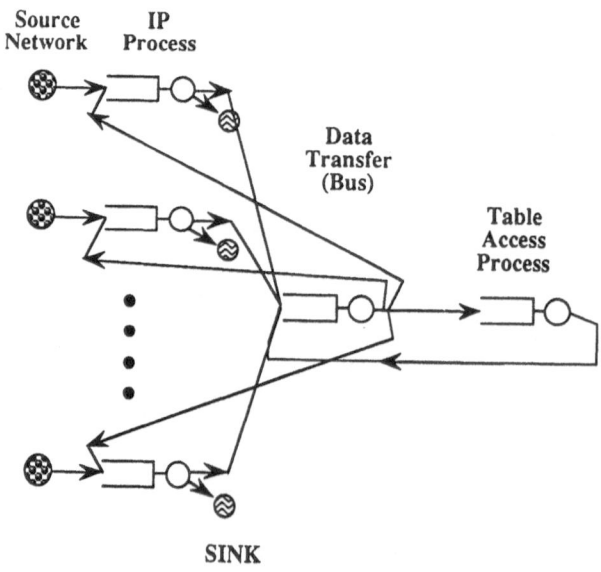

Figure 6   Queuing network of type A INPs

Table 1 Conditions for performance evaluation

Link speed	150 Mbps
Number of the link	2 - 16
Network layer protocol	700 instruction for RISC CPU
Table search	300 instruction for RISC CPU
CPU performance	25 MIPS
Bus transferring speed	600 Mbps
Link load	0.25
Packet inter arrival time density	Shifted Exponential (T(Exp(-x + constant)))
Packet length distribution	Exponential function for short packets + Delta function for long packets Exp(-L) + Delta(L - MPL)

Increasing the number of I/F cards, however, routing delay for Type C becomes smaller than the other two types. This is because, as the number of I/F cards increases, load on data forwarding module becomes large due to increase of traffic caused by table access and packet transfer between I/F cards. In Type C, the load of those processes is distributed on separated I/F cards. Consequently, Type C is best-suited for high performance internets among these three types.

3.3 Modification of Fully Distributed Table Architecture

High performance in Type C architecture is mainly caused by parallel processing, which is realized by distribution of various tables onto each I/F card. Eliminating central processor enhances fault tolerance capability of the systems. This table distribution, however, has a drawback of table integrity problems. As mentioned in Section 2, routing tables are dynamically updated by some routing protocol during network operation. Since each I/F card holds the identical routing table, table update information must be propagated to all table access processes to maintain correct INP operation. Moreover, large amount of memory for routing table may be needed for a large sized internet.

To solve these problems, Type C is modified into Type C' as show in Figure 8. Original routing and forwarding tables are stored in a central site, and each I/F card

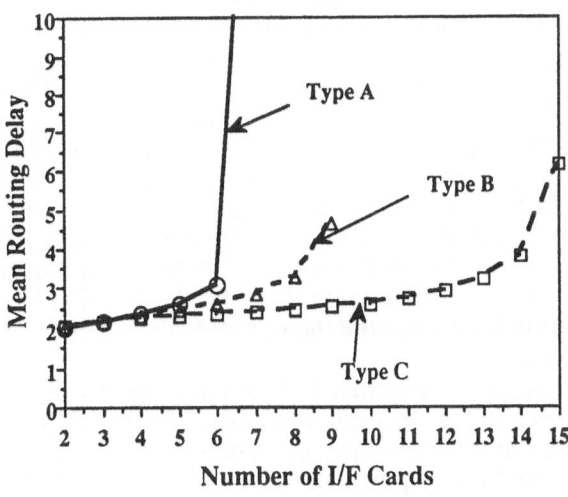

Figure 7   Routing delay comparison (type A, B, C)

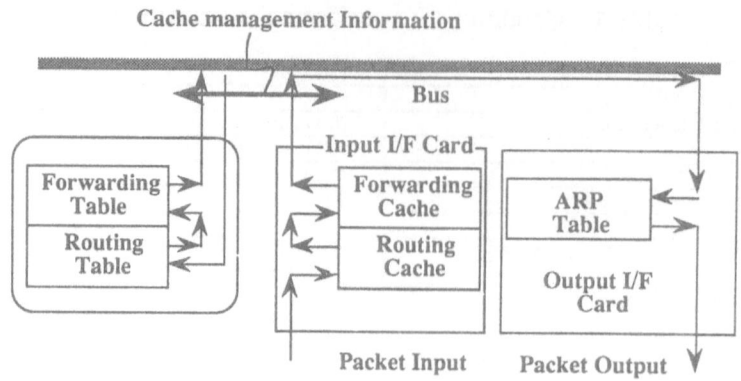

Figure 8   Type C' INP structure

maintains a cache for these tables. When an I/F card receives a packet, it checks its table cache. If it cannot find an entry in the cache, it sends an access request to a central processor. The I/F card stores the entry in the cache after the central processor returns its response. If any entries in an original routing table are updated, the central processor must inform that these entries are invalid. The I/F cards maintaining the corresponding entries delete the entries from the cache. Cache approaches inherently have a cache miss-hit problem. However, it is known that locality in traffic destination is high in real internet environment because of using connection-oriented protocol such as TCP [5]. Thus, in normal operation, the hit rate for routing table cache can be assumed to be large.

In conclusion, the following INP architecture is suited; routing and forwarding tables or caches are distributed among all I/F cards and ARP tables relating the I/F cards are maintained locally in the I/F cards. For high-speed packet processing and table access, powerful CPU should be implemented on each processor. This type of I/F card is referred to as a *router engine* in this paper.

4   HIGH-SPEED ROUTER ENGINE INTERCONNECTION

4.1   Bottlenecks of Bus-based INPs

To get faster INPs, CPU processing power on the router engines and bus bandwidth in INP must be large. Routing delay for INPs with various processing power CPUs and various bandwidth buses are evaluated. Simulation conditions are almost same as those in section 3, except that CPU processing power and bus speed are used as parameters. In this evaluation, 25 MIPS and 100 MIPS CPU and 600 Mbps, 1.2 Gbps and 2.4 Gbps bus are used. Number of router engines is fixed as 16.

Figure 9 shows INP routing delay as a function of link load. At some specific link load points the average routing delay becomes infinity, if CPU and bus speed are fixed. This link load indicates a maximum throughput for INPs. The first performance limit appears at a maximum throughput of 0.25. That is, a 600 Mbps bus is a first bottleneck in this configuration. The second performance limitation at throughput of 0.5 is also caused by bus speed bottleneck (1.2 Gbps bus). The third performance limitation appears at the thought of 0.73, which is caused by 25 MIPS CPU speed bottleneck.

It is shown from these results that INPs connecting 16 networks of 150 Mbps must have more than 2.4 Gbps bus and more than 34 MIPS CPU, in order to achieve maximum throughput of 1.0. Considering current CPU technology, it is expected to be easy to implement more powerful CPU from 34 MIPS. In terms of bus technology, a bus of 2.4 Gbps speed using 256 bits bus width has been standardized,

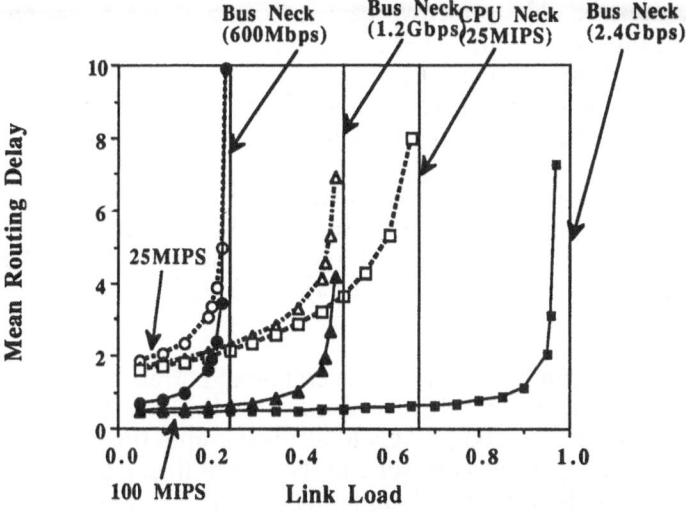

Figure 9   INP performance with various CPU and bus

which is called Futurebus+. Taking the future gigabit LAN into account, however, more high speed bus would be necessary. To increase bus performance, bus width should be increased. This causes physical size and reliability problems. Moreover, higher speed memory and bus driver are necessary. Therefore bus-based architecture is not considered to overcome the bottleneck for realizing gigabit internets. The remainder of the paper proposes switch-based INP architecture which is expected to overcome the bottleneck caused by the data transferring module.

4.2   Variable-length Packet Switch-based INPs

4.2.1   Architecture for Switch-based INPs. Figure 10 shows physical structure for variable length packet switch-based INPs. Significant difference between switch-based and bus-based INPs is features of link I/Fs which interconnect between router engines and a data transfer module. For bus-based INPs, a single but high speed bi-directional link I/F for each router engine is necessary, because a

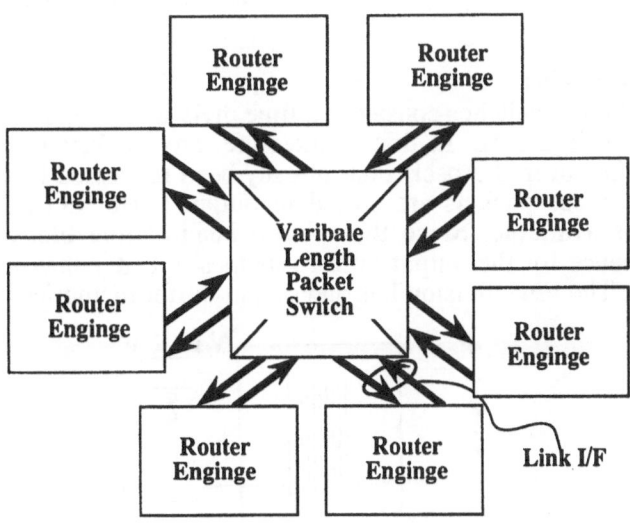

Figure 10   Physical architecture for variable length packet switch-based INP

router engine must be able to receive packet from multiple router engines using a single link I/F. For switch-based INPs, two uni-directional link I/Fs are necessary for receiving and transferring packets. Each link I/F does not need so high speed data transfer, since one link I/F receives packets from a single router engine and another link I/F transmits packets to a single router engine.

4.2.2   Variable-length Packet Switch. There are three alternative types of variable length packet switch architecture; input buffer, output buffer and input-output buffer switches. These types of fixed length packet switches have been extensively investigated in an area of ATM switches[9][10].

Input Buffer Switches

Input buffer switches consist of several input buffers and a cross-bar switch, as shown in Figure 11, where v stands for a link speed. If multiple different input lines want to send packets to a single output line simultaneously, only one of those input lines can send to the output line, and packets in the other input lines must wait in each input buffer. This phenomenon is known as an *output blocking*[9] - [13]. If the first packet in an input buffer cannot be sent and packets are served in a FIFO manner, the subsequent packets must also wait for service in the input buffer, even if their destination output lines are idle. This phenomenon is known as a *head-of-line blocking* [9] - [13]. The head-of-line blocking causes extreme performance degradation.

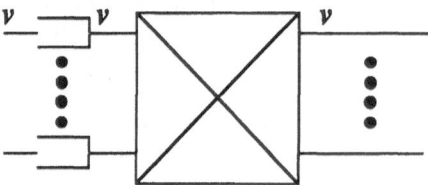

Figure 12   Output buffer switch

To avoid this bottleneck, some improvement schemes have been proposed in ATM world e.g., random-in random-out input with buffers or input buffer scheduling mechanisms [10]. However, it is too complex to implement with such schemes in variable length packet switches, because input buffer memory management for variable length packets becomes complicated. Thus input buffer switches are not suitable for data forwarding module in INPs.

Output Buffer Switches

Output buffer switches consist of a time division bus and output buffers, as shown in Figure 12. In this architecture, any packets can be sent to an output line, even if a packet has already been processed by the output line, because those packets are stored in output buffer until the output line becomes available. Since there is no head-of-line blocking, switching performance for the output buffer schemes is better than the input buffer scheme. The time division bus and output buffer memories within a switch

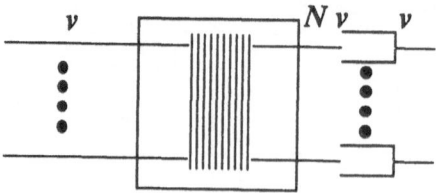

Figure 11   Input buffer switch

must be so high speed, as to handle packets coming from multiple input lines. It is difficult, however, to get memory with both characteristics of high-speed and high-capacity. Thus, the size of this type becomes large, because large amount of buffer is necessary for storing large frames used in LANs, e.g., 1500 byte for Ethernet, 4500 bytes for FDDI, etc.

Input-output Buffer Switches

Input-output buffer switches consist of input buffers, a cross-point switch and output buffers as shown in Fig. 13. Received packets are stored in an input buffer and sent to an output buffer through the cross-point switch. The head-of-line blocking is also occurred in input-output buffer switches, although the blocking rate can be reduced by accelerating input and output buffers speed. The crosspoint switch speed in input-output buffer switches is faster than the one in input buffer switches.

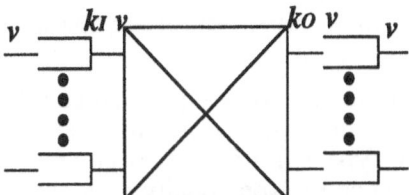

Figure 13   Input-output buffer switch

Memory speed of both buffers is much lower than the one of output buffer switches. Thus it is possible to implement larger capacity memory. This leads to a low packet loss rate and low cost switches.

4.2.3   Models and Evaluation of Variable Length Packet Input-Output Buffer Switches. Since, from the previous discussion, input-output buffer switches are assumed to be most suitable for LAN interconnection, we focus our evaluation on input-output buffer switches. There are two alternatives of variable length packet input-output buffer switches.

Type 1

In Type 1, input buffers speed for sending packet to a cross-point switch, v_{in}, is same as a line speed, and output buffer speed for receiving packets from a cross-point switch, v_{out}, is scaled up by a factor of $k_1 > 1$ in terms of the input buffer speed. That is, $v_{in} = v$ and $v_{out} = k_1 v$, where v stands for line speed. k_1 is a speed-up parameter for Type 1. Since output buffers speed is made scaled-up, the blocking rate can be reduced [13].

Type 2

In Type 2, a cross-point switch speed is scaled up by a factor of k_2. That is, both buffers speeds are scaled-up in terms of a line speed. In Figure 13, $v_{in} = k_2 v$ and $v_{out} = k_2 v$, where $k_2 > 1$ shows a speed-up parameter for Type 2. Packet transferring time is shorter than the one in input buffer switches, because cross-point switch is made speed-up.

These two alternatives are compared from a viewpoint of performance. The speed-up parameter is closely related with memory speed of buffers, and then the cost of INPs. Therefore it is important to determine the speed-up parameter carefully. Mean queue lengths for input buffers are analyzed in changing speed-up

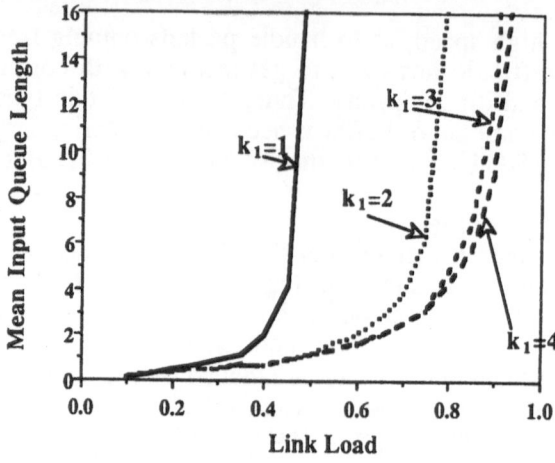

Figure 14   Mean input queue length of type 1 input-output buffer switch

parameters. Necessary minimum speed-up parameter for low blocking rate is estimated from behavior of mean queue length in an input buffer. Since queue length of input buffers in Type 1 has already been analyzed in [13], we evaluate Type 2 switches. To make comparison equally with [13], packet length and inter-arrival times are assumed to be an exponential distribution which is identical with [13].

Figure 14 shows relationship between mean queue length of input buffer and link load in Type 1 switches. These results are obtained from mathematical approximation, and approximately same as simulation results with a condition that number of links is 16 [13]. For the speed-up parameter of 1, the mean queue length increases at the link load of 0.5. For the speed-up parameters of 2 and 3, the performance is drastically improved. On the other hand, so much throughput improvement cannot be seen in the speed-up parameters larger than 4. Thus, the speed-up parameter of 3 is sufficient for Type 1 switches.

Simulation results for input buffer mean queue length for type 2 switches is illustrated in Figure 15. For the speed-up parameter of 1, mean queue length is approximately same as the one in Type 1 switches, whereas, mean queue length is dynamically decreased for the speed-up parameter larger than 2. In this case, the speed-up parameter of 3 is enough for performance improvements, because even if the speed-up parameter is increased more than 3, so much performance improvement cannot be found.

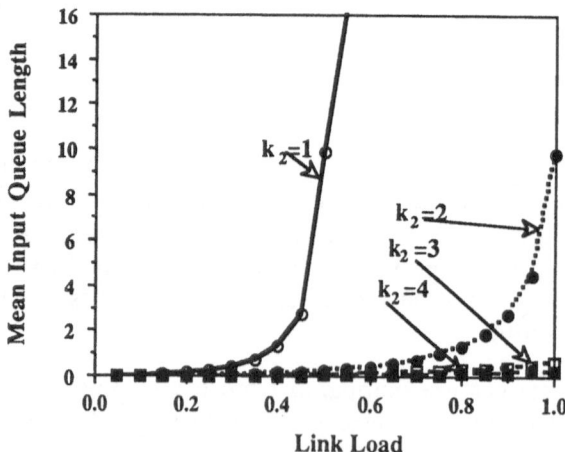

Figure 15   Mean input queue length of type 2 switch

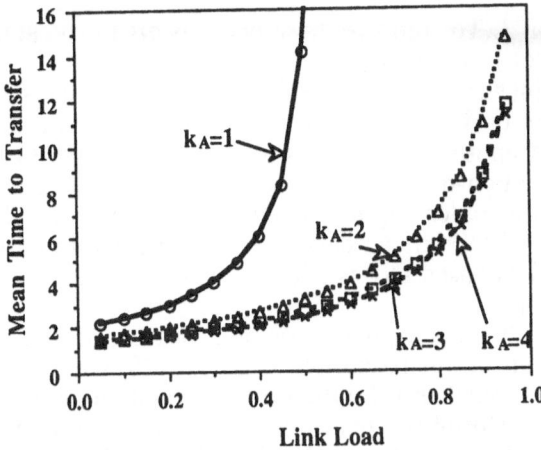

Figure 16   Mean transferring time of type 2 switch

From these comparisons, Type 2 switches are superior to Type 1 switches, in a sense that performance is more highly improved using the same speed-up parameters at the high link usage. Selection of speed-up parameters is crucial, because larger speed-up parameters cause a memory speed problem and then cost problem. In order to examine an appropriate speed-up parameter, packet transferring time is also evaluated. The transferring time means the time from when a packet is put into an input buffer to when the packet is sent out from an output buffer. Figure 16 shows the transferring time as a function of link load. In the speed-up parameter greater than 2, however, delay performance is not affected by on the speed-up parameter strongly, because an output buffer's delay governs the transferring time. From comparison of two types switches, the following conclusions can be lead. The type 1 is superior in the low and middle range of link load. In the high link load, the type 2 is superior. Although the type 2 switches need high speed memory at the input as well as output, they require the small input buffer size, because mean queue length is small even in the high link load region.

4.3   Comparison with Bus-based INPs and Variable Length Packet Switch-based INPs

Routing delay for bus-based INPs and switch-based INPs with input-output buffers is compared through simulations. The simulation conditions are same as

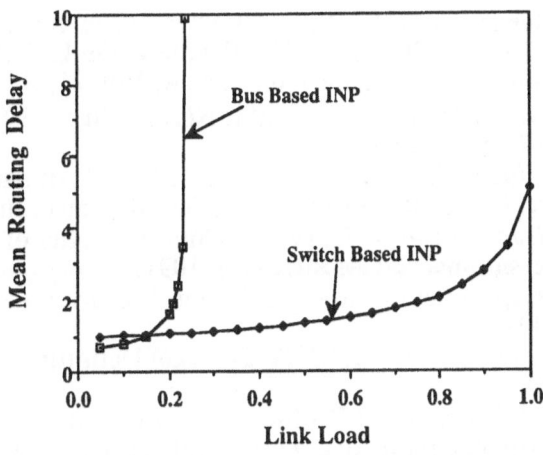

Figure 17   Comparison with bus-based INPs and switch-based INPs

those used in section 3. To compare these two models fairly, simulation conditions and parameters is set to be identical. Hence it is assumed that speed for link I/F between router engines and data forwarding module is equal. In bus-based INPs, a single link I/F treats bi-directional data flow; one for data transfer from a router engine to a bus and another from a bus to a router engine. On the other hand, a single link I/F in switch-based INPs treats only uni-directional data flow; data transfer either from a router engine to switch or from a switch to router engine. Hence, in switch-based INP, router engines must have two link I/Fs. As an example, a link I/F speed for bus-based INP is 600 Mbps, and that a link I/F speed of switch-based INP is 300 Mbps.

Figure 17 shows the routing delay for both bus-based and switch-based INPs. Maximum throughput for bus-based INP is about 0.25, while maximum throughput for switch-based INP reaches 1.0. Obviously, variable length packet switch-based INP shows greater performance than bus-based INP. Consequently, switch-based INPs realize not only low routing delay but also small size and high reliability, because switch modules are small and reliable itself compared to bus.

5   CONCLUSION

We first examined table allocation problems, and proposed to allocate an unique table in a central processor for maintenance and to distribute table caches on each network I/F card. Each network I/F card including full routing functionality and distributed routing table was called as a router engine. This kind of architecture has several advantages, e.g., reliability, expandability, etc., in addition to better performance. Second, various alternate architectures for data forwarding module were discussed. It was shown that bus-based INPs have critical bottleneck for realizing high-performance INPs. Then we evaluated various switches architecture for high speed INP, 1) input buffer, 2) output buffer, 3) input-output buffer switches. It was shown from this discussion that the input-output buffer switch is suited for high speed LAN interconnections. Furthermore, performances for two types of input-output buffer switch were compared in perspective of buffer memory speed and cross-point switch. Finally through the comparison of routing delay for both bus-based and switch-based INPs, it was shown that variable length packet switch-based INP with input-output buffer scheme is best suited for LAN interconnections. In conclusion, INPs with both router engine and variable length packet switches are best-suited for high speed LAN interconnections.

REFERENCE

[1]   *M. Lottor,* "Internet Growth (1981-1991)," rfc-1296, January 1992
[2]   *R.L.Frink and F.Ross,* "FFOL - an FDDI Follow-On LAN," ACM SIGCOMM Computer Communication Review, Vol. 21, No. 2, 1991.
[3]   *C. Patridge (editor),* "Report on Gigabit Networking," IEEE Network Magazine, Vol. 6, No. 2, March, 1992.
[4]   *D. D. Clark, V. Jacobson, J. Romkey and H. Salwen,* "An Analysis of TCP Processing Overhead," IEEE Communication Magazine, June 1989.
[5]   *P. B. Danzig, S. Jamin and D. J. Mitzel,* "Characteristics of Wide-Area TCP/IP Conversations, "ACM SIGCOM, 1991.
[6]   *C. Topolcic,* "Experimental Internet Stream Protocol: Version 2 (ST-II)," RFC-1190, 1990 October.
[7]   *Protocol Engines Incorporated,* "XTP Protocol Definition, Revision 3.5," September 1990.
[8]   *H. Schulzrinne,* "A Transport Protocol for Audio and Video Conferences and other Multiparticipant Real-Time Applications," Internet Engineering Task Force INTERNET-DRAFT, October, 1992.

[9] *M. G. Hluchyj, M. J. Karol,* "Queuing in Space-Division Packet Switching," IEEE Infocom '88, 4A.3, 1988.

[10] *S. Karube, M. Akata, T. Sakamoto, T. Saito, S. Wakasugi, S. Yoshida and H. Shibata,* **"A 250 Mb/s 32 x 32 CMOS Crosspoint LSI for ATM Switching Systems"** ISSCC '90.

[11] *Hedrick, C.L. ,* "Routing Information Protocol," RFC-1058, June, 1988.

[12] *Moy, J.,* " OSPF version 2," RFC-1247, July, 1991.

[13] *J. S..-C. Chen, T. E. Stern,* `"**Optimal Buffer Allocation for Packet Switches with Input and Output Queuing ,**" IEEE Globecom '90, 905.1-905.6, 1990.

IMPLICATIONS OF INTERNET FRAGMENTATION
AND TRANSIT NETWORK AUTHENTICATION

Robert L. Popp[1]

Booth Research Center
Department of Computer Science & Engineering
University of Connecticut
Storrs, CT 06269

ABSTRACT

Presently, the DARPA TCP/IP Internet interconnects various type organizational networks (e.g., government agencies, academic and research environments, and commercial entities) that may have competing and/or conflicting interests. A current problem of interest and research is in providing network access control - administratively heterogeneous networks controlling the flow of information across its boundaries. In particular, enforcing network access control policies via network-level gateway packet authentication has proven to be somewhat difficult. In order for gateways to assume access control responsibilities within an (transit) internet environment, certain orthogonal features of packet-switched networks must be reconciled with network-level access control mechanisms, specifically, packet fragmentation with packet authentication.

In this paper, we begin by describing how the fragmentation process hinders the use of network-level packet authentication as an access control mechanism within a TCP/IP internet. Currently in [18], two protocols are given (i.e., *Delayed Authentication* and *MTU Probing*) that reconcile packet fragmentation and authentication between source and destination access controlled networks. However, fragmentation can also occur within access controlled transit networks. Because of its semantics, ironically, MTU Probing, in fact reconciles packet fragmentation and authentication within transit networks as well. However, it does so with marginal performance. Thus, to this end, we present and analyze two protocols (i.e., one modifies and extends Delayed Authentication, while the other is a novel approach utilizing an alternative fragmentation policy) that reconcile packet fragmentation with packet authentication within the (transit) internet. Results from our current research demonstrate that prototype implementations of our protocols executed within an experimental internet environment offer significant performance improvements over the existing network-level authentication protocol of MTU Probing, and are therefore more than reasonable as alternatives.

1. INTRODUCTION

Current internetworking protocols (e.g., TCP/IP protocol suite [4,16]) can support the interconnection of physically disparate, heterogeneous networks. In an internetwork environment, the details of the underlying networking technologies are made transparent to higher layer protocols, application programmers, and end users. One notable distinction

---

email: rpopp@sparc0.brc.uconn.edu

found between individual network technologies is the maximum size packet allowed to be transported across the network's link. The fixed upper bound on a packet's size for transport is often referred to as the network's *Maximum Transfer Unit* (**MTU**) [4]. In the DARPA TCP/IP Internet, the natural evolution of the many individual networks of which it is comprised led to the various network MTU sizes that can be found (see Table 1) [1,4]. The selection of a network's MTU size is often the result of careful consideration of design alternatives such as hardware constraints, storage requirements, processor performance, throughput, fair and efficient link utilization, store and forward delay, and response time.

Table 1. Various networks and their respective MTU sizes

Network	MTU
FDDI	4500 bytes/packet
ProNET-10	2044 bytes/packet
Wideband Satellite Net	2000 bytes/packet
Ethernet	1500 bytes/packet
Packet Radio Net	254 bytes/packet
Arpanet	126 bytes/packet
Alohanet	80 bytes/packet

In an internetwork environment consisting of many heterogeneous networks with varying MTU sizes, the transport and the effects of oversized packets on small MTU networks are concerns that must be addressed and analyzed. Fragmentation - the process of breaking up an oversized packet (i.e., size(*packet*) > MTU(*network*)) into smaller fragments for transport across a network - is just one mechanism that copes with varying network MTU sizes. Other mechanisms exist. Although it copes with the MTU sizes of various networks, fragmentation may cause problems in certain other internetworking domains. One such domain is network access control via packet authentication.

Currently, the DARPA Internet interconnects various type organizational networks (e.g., government agencies, academic and research environments, and commercial entities) that may have competing and/or conflicting interests. Network access control - administratively heterogeneous networks controlling the flow of information across its boundaries - is a current problem of interest and research. The transparency of the Internet, although advantageous in that it provides for an universal communications environment without restricting individual networks to specific networking technologies and architectures, makes the job of network access control difficult. A discussion about many existing network access control issues is given in [6]. In [7], the *Visa* protocol is described - a network-level cryptographic technique used for network access control and packet authentication. And finally, in [8], an extension of the Visa protocol is described that allows for access control and packet authentication of transit internet traffic.

One issue that recent literature in [6,7,8,9,18] discuss is the need for gateways to bear some of the responsibility for packet authentication, and not simply leave it as an end-to-end host responsibility. In order for gateways to assume this responsibility, certain orthogonal features of packet-switched networks must be reconciled with network-level access control mechanisms. One such reconciliation needed is packet fragmentation with packet authentication.

An authenticating gateway (**AG**) in a packet-switched, access controlled network must have the means for implementing access control policies. Such a policy would require an *AG* to determine if the authenticated parties (i.e., source & destination hosts) are authorized to communicate. One such authentication technique would be the verification of some property of the packet (e.g., a data signature). In order for authentication to be done directly, the *AG* must have access to the entire packet in its original "source host" form [18]. However, given the versatility and disparity of individual networks within an internet, it is often the case that *AGs* receive fragments of a packet, and not the packet itself. Therefore, packets can arrive at *AGs* fragmented. Thus, in order to perform network-level packet authentication (i.e., verify signatures), *AGs* must receive the packets unfragmented or have some mechanism to authenticate them based on their fragments.

In this paper we propose two protocols that allow for network-level authentication of packets in gateways throughout the transit internetwork in the presence of fragmentation. One method is novel in that it utilizes an intra-network fragmentation policy (defined later), curtailing dynamic routing of packet fragments while allowing for immediate packet authentication. The other method combines IP source routing with an extended and modified version of the Delayed Authentication protocol proposed by Tsudik [18], however, state information in *AGs* on a per-packet basis must be maintained. Currently, Tsudik's MTU Probing protocol in fact reconciles packet fragmentation with packet authentication in the transit internetwork. However, its performance is marginal. Thus, in order to demonstrate their behavioral and performance characteristics, our protocols are first rigorously described, analyzed, and compared with each other and with MTU Probing. Afterwards, we provide results from our current research efforts which clearly demonstrate that prototype implementations of our protocols executed within an experimental internet environment offer significant performance improvements over the existing network-level authentication protocol of MTU Probing, and are therefore reasonable alternatives.

The remainder of this paper is organized as follows. In Section 2 we first define and then briefly discuss the characteristics and tradeoffs associated with several fragmentation policies. In Section 3 we discuss both network access control and how it can be employed via the network-level packet authentication mechanism, respectively. In Section 4 we provide a model of an internetwork that we utilize to illustrate the behavior of our protocols and [18]'s MTU Probing protocol. In Sections 5 & 6, we propose two protocols - *Extended Delayed Authentication (EDA)* and *Intra-net Frag Authentication (IFA)* - to implement network-level gateway packet authentication within an access controlled network. In Section 7, we analyze our protocols and compare them with each other and with the MTU Probing protocol. In Section 8, experimental results are described, where an logical internet model was created and prototype implementations of each authentication protocol described in this paper were executed and analyzed. And finally, in Section 9, we conclude the paper.

2. FRAGMENTATION

Before describing the proposed protocols, we first describe two commonly employed fragmentation policies - *inter-network* and *intra-network* fragmentation (see Figure 1). Afterwards, we will discuss some of the tradeoffs associated with having a fragmentation vs. a "no fragmentation" policy. Later we show that tradeoffs associated with the proposed network-level authentication protocols and the existing protocol of MTU Probing simply reflect the advantages and disadvantages corresponding to the fragmentation and "no fragmentation" policies, respectively.

2.1. Inter-Network Fragmentation

Given a fragmentation policy, we define *inter-network* fragmentation (supported by the IP standard [16]) as when the fragments of a packet are treated autonomously as unique, individual packets at the point of fragmentation and are allowed to be routed independently of other packet fragments. Reassembly of the fragments occurs only at the destination host, not at intermediate gateways, unless the intermediate gateway is the destination host of the fragmented packet. One important property of inter-network fragmentation is that of *dynamic routing* of fragments, that is, the fragments of a packet

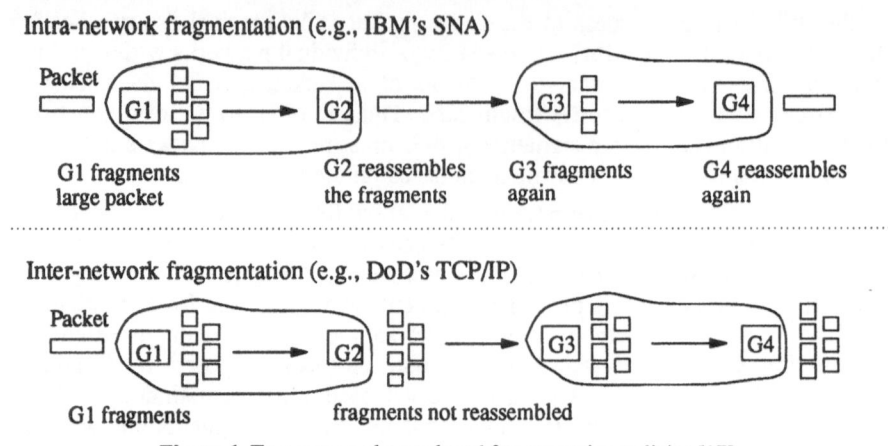

Intra-network fragmentation (e.g., IBM's SNA)

Packet

G1 fragments
large packet

G2 reassembles
the fragments

G3 fragments
again

G4 reassembles
again

Inter-network fragmentation (e.g., DoD's TCP/IP)

Packet

G1 fragments

fragments not reassembled

Figure 1. Two commonly employed fragmentation policies [17].

can follow similar or dissimilar routes from the point of fragmentation to the final destination. Examples of networks utilizing an inter-network fragmentation policy are FRI and ARPAnet [1]. Some of its characteristics are:

- Networks having multiple exit gateways can dynamically route fragments by employing gateway load sharing schemes, and thus, benefiting from fragment routing parallelism.
- Gateways can be streamlined. They do not share the reassembly burden equally with hosts. Since they can be streamlined and utilized more efficiently, gateways can concentrate on their primary task - routing.
- Many smaller fragments may be carried from the point of fragmentation to the ultimate destination, possibly through high MTU networks, thereby decreasing link efficiency.
- Lost fragments are not discovered until the destination host times out. If any of its fragments become lost, the original packet cannot be reassembled, and all relevant buffered fragments are discarded at the destination host after some timeout period.
- Fragments traveling through various routes increases the probability that at least one of the fragments, and, thereby, the entire packet itself, may be lost. The probability that the packet is lost increases, since routing parallelism of fragments also implies the operability and reliability of many intermediate transit networks when routing fragments.

2.2. Intra-Network Fragmentation

We define *intra-network* fragmentation as when all the fragments of a packet are sent along the same link from the point of fragmentation to the next hop, where there they are immediately reassembled. The network provides for the reassembly of packet fragments before the packet enters the next network. Fragmentation is done locally, and is therefore both temporary and transparent to other networks. The current IP standard does not support intra-network fragmentation [16], however, many network environments do support this fragmentation policy by providing an implementation for it between the data-link and network layers of the 7-layer OSI model. Examples of commercial networks utilizing an intra-network fragmentation policy are Xerox XNS and IBM SNA [1]. Some of its characteristics are:

- Since fragmentation is done locally, some overhead incurred from the fragmentation process itself can be minimized (e.g., packet header replication for each fragment generated is not required).

212

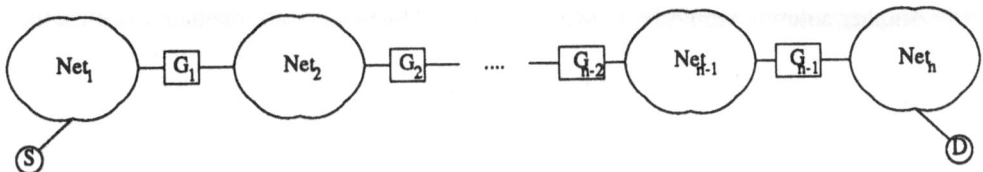

Figure 2. An internetwork of i = 1..n networks Net_i interconnected by j = 1..n-1 gateways G_j

- Increased network reliability (e.g., exit gateways can inform entrance gateways of (un)successful reception of fragments) is possible. In certain cases, network performance can benefit from early detection of lost fragments.
- Overall system throughput and link utilization efficiency can be increased in certain cases. In Figure 2, assume for some packet P,

$$size(P) = MTU(Net_1) = MTU(Net_3) = ... = MTU(Net_n) > MTU(Net_2).$$

Suppose S is the source, D the destination, and S sends P to D. After G_1 fragments P, and sends k fragments of P, say $F_1P, F_2P, ..., F_kP$, across Net_2, the question becomes what to do with all the k fragments arriving at G_2? Certainly, reassembling the k fragments of P back into P and sending it (as opposed to sending $F_1P, F_2P, ..., F_kP$) across the remaining networks will increase throughput and link utilization efficiency.

- Gateway computational cost is proportional to the total number of fragments, not packets, generated. For example, additional costs are incurred as a result of reassembling packets that are not destined for the particular gateway performing the reassembly.
- Packet fragments cannot be dynamically routed. However, although conceptually desirable, pragmatically speaking, most packet fragments follow similar routes anyhow.
- Theoretically possible for packets to be fragmented and reassembled by every gateway (i.e., size*(packet)* = MTU*(source_net)* > MTU*(remaining_nets)*) while traversing the internetwork. Due to duplication of effort, a serious degradation in performance results.

2.3. Arguments for Fragmentation

Certainly one advantage of a fragmentation policy in an internetwork environment is that the underlying detail of a network's MTU size is made transparent to higher layer protocols, applications, and end users. Fragmentation must exist - a necessary evil [11] - if the interconnection of individual networks is desired. Besides this, considering the alternatives, its minimization, and not the alternatives, will give the best results. Even though [11] has suggested that the fragmentation process degrades system performance, fragmentation policy supporters would argue that the "no fragmentation" alternatives are usually not very attractive solutions. In fact, they would argue that, in most cases, a fragmentation policy provides for higher overall system performance over a "no fragmentation" policy. For example, consider the following alternatives to fragmentation.

One solution for implementing a "no fragmentation" policy would be to adopt an *universal* MTU size for the internetwork. If we assume the universal MTU size is the same as the MTU size of the smallest MTU network in the internetwork, this solution has some problems. Obvious drawbacks are inefficient usage of the bandwidth for high speed links, no flexibility in packet sizes, increased traffic due to the smaller sized packets transmitted, and increased gateway store and forward delay due to increased computational resources expended for processing greater numbers of packets. One universal MTU size is simply not very practical for all the disparate needs of interconnected networks.

Another solution would be to avoid small MTU networks all together when routing packets. Again, this solution is problematic. One problem is routing inflexibility. An internet topology is dynamic, routes may change, and such information would need to be propagated back to every affected source host. Another problem is the need for source hosts to have global information of the topology, since the locations of small MTU networks must be known ahead of time by source nodes before sending packets.

Another solution is to use *Probe* packets. Although it has several flavors, the most straightforward approach is as follows: each network that processes a Probe packet stamps its MTU size in the Probe only if its MTU is smaller than the MTU size currently stored in the Probe. Once at the destination host, the actual size of the smallest MTU network in the packet's path has been determined. This size can be relayed back to the source host, who then sends packets no larger than this size to the destination host. Again, similar problems arise here as in previous solutions: increases in traffic, store and forward delay, and system response time.

One last solution would be for the source host to always transmit very small sized packets (e.g., arbitrarily, 576 bytes is reasonable, since this is the IP maximum size packet that a source can send to a destination host without explicit permission [11,16]) in hopes of avoiding fragmentation. Again, this scheme is problematic. Besides the obvious problems, a network must decide what to do if a packet arrives having a larger size than its own MTU size - retransmit or simply drop the packet? Obviously both possibilities create a wealth of other problems to contend with.

2.4. Arguments for "No Fragmentation"

With current trends in high-speed computing and communications, resource utilization must be streamlined and kept at a premium. Fragmentation causes considerable overhead in and utilizes inefficiently valuable network resources. Costs associated with sending packets increase with each additional fragment, since we incur these costs for each fragment generated. The two main components of the fragmentation cost are node (i.e., host & gateway) processing time and performance degradation [1]. Processing time consumption and performance degradation are both proportional to the number of fragments generated, and the minimization of generated fragments can only improve both of these components (e.g., the best case, the minimum number generated would be 0 fragments - no fragmentation). Node processing time component costs include packet header processing, buffer management, and reassembling packet fragments. Performance degradation component costs include delays associated with packet transmission and wasted bandwidth usage due to the replication of packet headers in fragments.

Fragmentation can also affect throughput. In packet-switched networks, sending fewer, larger packets will improve throughput as compared to sending many smaller packets. The reason is that most costs associated with packetized communication is per-packet, and not per-byte [11]. Thus, increasing the number of packets (recall that IP treats a fragment as a separate packet) to transmit only degrades network throughput.

The probability of packet loss increases with fragmentation because the loss of any one fragment of the original packet results in the loss of the entire packet. If any fragment gets lost en route from the point of fragmentation to the destination host, the original packet cannot be reassembled. The destination host simply discards all buffered fragments of the original packet after some timeout period. Thus, in order for it not to be discarded, the packet must rely on the operability and reliability of all the different possible networks that its fragments may traverse.

As was stated previously, fragmentation results in degraded performance. The loss of just one fragment out of say k possible fragments for some packet P may cause an IP or higher-layer protocol to retransmit all k fragments over again. The combination of fragmentation and gateway congestion can have a profound affect on system performance. Consider the following example.

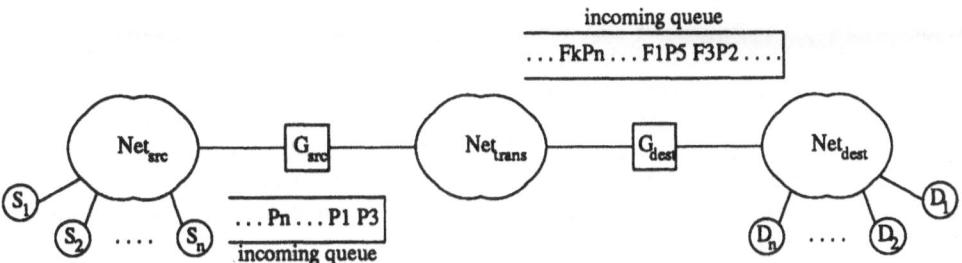

Figure 3. An internet of three networks interconnected by two gateways. Gateway G_{src} enqueues n packets in its incoming queue, fragments each packet since $MTU(Net_{trans}) < MTU(Net_{src}) = size(P_i)$, generates k fragments per packet $F_1P_i, F_2P_i, \ldots, F_kP_i$, and routes the total nk fragments to G_{dest}. Gateway G_{dest} enqueues the fragments and proceeds to route these along with other packets from previous traffic.

Let us first make the following suppositions: we have an internet (Figure 3) consisting of three networks, Net_{src}, Net_{trans}, and Net_{dest}, interconnected by two gateways G_{src} and G_{dest}, respectively. Suppose there are n source hosts on network Net_{src}, say, $S_1, S_2, \ldots, S_n$, and n destination hosts on network Net_{dest}, say $D_1, D_2, \ldots, D_n$. Now, for $i = 1..n$, assume each S_i wants to send a packet P_i to D_i, where the size of each packet is equal to the MTU of Net_{src}, that is, $size(P_i) = MTU(Net_{src})$. Assume G_{dest} has packets yet to be processed in its incoming queue from prior traffic. And finally, assume networks Net_{src}, Net_{trans}, and Net_{dest} have the following MTU size relationship:

$$MTU(Net_{src}) = MTU(Net_{dest}) > MTU(Net_{trans}).$$

Suppose each S_i sends packet P_i to destination host D_i simultaneously. Gateway G_{src} enqueues n packets P_i in its incoming queue, fragments them (since, for $i = 1..n$, $size(P_i) > MTU(Net_{trans})$), and routes these fragments to G_{dest}. Since each original packet P_i is the same size, G_{src} will generate the same number of fragments for each of the n packets received. Let us call these fragments F_jP_i, for $i = 1..n$, $j = 1..k$, where packet P_i has k fragments $F_1P_i, F_2P_i, \ldots, F_kP_i$. Thus, n packets came into G_{src}, and it generated nk fragments and sent these to G_{dest}. Remember, if using TCP/IP, each fragment is really its own autonomous packet at the point of fragmentation until reaching its final destination.

Consider the following, recalling that packets (fragments) can arrive at a node not necessarily in the order transmitted. Gateway G_{dest} enqueues $(k-1)n$ fragments in its incoming queue, say all fragments from the n original packets except the ith one. That is, say all fragments get enqueued except fragments $F_iP_1, F_iP_2, \ldots, F_iP_n$. Now, suppose all buffers at G_{dest} become full, and G_{dest} is forced to start dropping any further incoming traffic due to congestion [4,10,16]. Thus, G_{dest} drops fragments $F_iP_1, F_iP_2, \ldots, F_iP_n$, and routes the remaining fragments in its incoming queue to each destination host respectively. Each destination host commences with the reassembly process. However, since each will never receive P_i's ith fragment F_iP_i, each D_i will time out, discard P_i's buffered fragments, and send an ICMP error message back to its respective source host S_i. Assume each S_i retransmits its packet again. Now another nk fragments have been sent back to the already congested G_{dest}. Thus, paradoxically, dropping fragments makes congestion worse. Dropping fragments increases retransmissions, which leads to more fragments and further congestion. Therefore, as the loss rate for fragments increases due to congestion, overall throughput and system performance dramatically decrease.

3. NETWORK ACCESS CONTROL / PACKET AUTHENTICATION

Network access control across organizational boundaries is a current area of interest and research. With the advent of the TCP/IP protocols came the increasing connectivity of administratively heterogeneous and autonomous networks of various organizations. How such organizations can control the flow of information across its network boundaries is precisely the question network access control mechanisms need to address and resolve. Simply stated, network access control specifies a set of hosts and/or networks between which a gateway will route packets. The gateway verifies the authenticity of the authorization between the source host - destination host connection via the access control mechanism. In [6], a very thorough discussion is given concerning many important network access control issues.

Interestingly enough, early internetworks ignored the issue of network access control completely, mainly because such internetworks were either owned and administratively controlled by a single organization (e.g., single corporation, university, or government agency), or because the network interconnections were between research institutions where little need to limit the flow of information was necessary [7]. However, current internetworks connect many dissimilar type organizational networks that may have competing and/or conflicting interests. Thus, the need for network access control can no longer be ignored or overlooked.

To illustrate an occurrence when the employment of network access control may be appropriate, consider the following hypothetical example. The subject of this example will be the University of Connecticut's Booth Research Center LAN (BRC_LAN) that is connected to the DoD's Arpanet. Given that the Arpanet has a policy that the use of its network is strictly restricted to academic and/or research related traffic, in the past, the user population of BRC_LAN could access the Arpanet without any need for network access control (assuming BRC_LAN was stand-alone and autonomous). No special measures by BRC_LAN network administrators were necessary in order to comply with Arpanet's policy. However, as years past, and greater connectivity was both encouraged and implemented, the user population of BRC_LAN logically grew larger with connections to several industry and organizational *external* networks. Thus, the user population that could access the Arpanet was no longer restricted to local BRC_LAN users, but, instead, the potential user population included those users of the external networks connected to BRC_LAN. Therefore, in order to adequately comply with Arpanet's policy, network administrators of BRC_LAN may need to introduce some form of a network access control policy restricting access past the Arpanet gateway to only the local BRC_LAN user population.

To provide network-level access control between authorized pairs of hosts/networks, Estrin *et al.* presented the *Visa* protocol [7]. Conceptually, a visa is an unforgeable signature for a packet cryptographically computed using a secret key and is placed in the packet's header. Visas are signatures used for authenticating the packet's integrity; they are bound to each packet in order to guarantee the authenticity of the packet's contents. An authenticating gateway (AG) in a visa-controlled network uses such a signature to verify the authorization of the packet's transmission between the source and destination host networks.

Before discussing how visas are used to control traffic in visa-controlled networks (i.e., source, destination & transit networks traversed by the packet), we first describe the method used to compute such visas - via the Cipher Block Chaining (CBC) mode of the Data Encryption Standard (DES) algorithm (see Figure 4). The Federal Information Processing Standards (FIPS 46) of the National Bureau of Standards (NBS) specifies DES, which is a cryptographic algorithm utilized for the cryptographic protection of sensitive computer data. FIPS 81 defines four modes of operation for the DES algorithm, of which CBC is one, which specifies how computer data will be encrypted and decrypted. FIPS 81 also describes an DES authentication technique known as the Message Authentication Code (MAC), which is the visa, computed as a cryptographic function of the message and stored with the message when transmitted. We computed the MAC (visa) for the packet to be transmitted using DES's CBC mode of operation (note: we employed 32-bit visas in the prototype implementations of our protocols, so the 32 most significant bits of the last output block of DES's CBC mode are used as the visa).

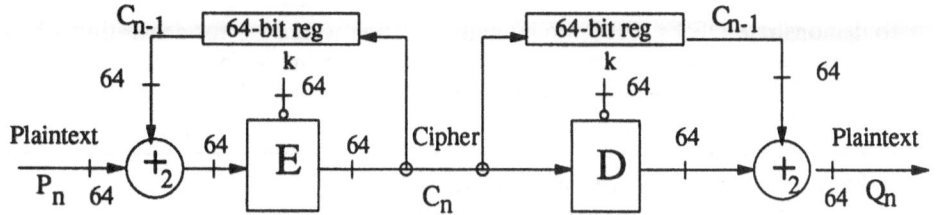

Figure 4. DES algorithm using CBC (Cipher Block Chaining) [5].

In Figure 4, DES's CBC mode works as follows. CBC is a block cipher system in which the output of one encipherment step is used in the input of the next encipherment step, so that each cipher block is dependent, not just on the plaintext block from which it came, but also on all previous plaintext blocks of the previous encipherment steps. The encipherment portion of CBC has a feedback path from the ciphertext output back into the plaintext input. Except for the first 64-bit block, every succeeding block in the plaintext is added (modulo-2) to the ciphertext of the previous block, thus making the *nth* ciphertext block C_n a function of all previous plaintext blocks (inclusive) $P_1, P_2, \ldots, P_n$. Decipherment is just the reverse of this process. After decipherment of the ciphertext, the previous ciphertext is added to it (modulo-2), which corresponds exactly to the operation carried out during the encipherment process. The only detail left to discuss is how DES's CBC operation deals with the initial and final blocks of the plaintext to be encrypted. For the initial plaintext block P_1, the previous ciphertext block is not available (since C_0 is nonexistent). Thus, to resolve this discrepancy, a 64-bit *Initializing Vector (IV)* is stored as C_0. For the final plaintext block P_n, since messages of arbitrary length may not fall neatly on 64-bit boundaries, P_n may not be 64 bits in length. To resolve this, simply pad P_n if it is indeed less than 64 bits in length (e.g., insert $\frac{64 - |P_n|}{8} - 1$ (8-bit) spaces at the end of P_n, and, at P_n's last byte position, store the number of padding spaces inserted, indicating to the receiver how many spaces need be removed to obtain the original message).

Algebraically, the operation of CBC encipherment can be expressed for the *nth-block* plaintext P_n, where each block $P_1, P_2, \ldots, P_n$ is 64-bits long, as,

$$
C_n = \begin{cases} E_k(P_1 +_2 IV) & \text{if } n = 1 \\[2ex] E_k(P_n +_2 C_{n-1}) & \text{if } n > 1 \end{cases}
$$

The operation of CBC decipherment can be expressed for an *nth-block* ciphertext C_n as,

$$
Q_n = \begin{cases} D_k(C_1) +_2 IV & \text{if } n = 1 \\[2ex] D_k(C_n) +_2 C_{n-1} & \text{if } n > 1 \end{cases}
$$

To prove that the result of decipherment, Q_n, is equal to the original plaintext, P_n, note,

$$
\begin{aligned}
Q_n &= D_k(C_n) +_2 C_{n-1} \\
&= D_k(E_k(P_n +_2 C_{n-1})) +_2 C_{n-1} \\
&= P_n +_2 C_{n-1} +_2 C_{n-1} \\
&= P_n +_2 0 \\
&= P_n.
\end{aligned}
$$

To demonstrate DES's (CBC mode) authentication technique for generating a MAC (visa) for a particular message, consider the following example (note DES's input text specified in ASCII for convenience, however DES I/O values in table given in hex).

DES cryptographic key: 0123456789ABCDEF
DES (ASCII) input text: "Deposit $1,000,000.00 in my bank account"
Initializing Vector (IV): 0000000000000000

Time	Plaintext	DES input	DES output
1	4465706F73697420	4465706F73697420	B0573586AD4E9DCB
2	24312C3030302C30	946619B69D7EB1FB	B0AF2D59E85400BB
3	30302E303020696E	809F0369D87469D5	0F71E93E4CE755BA
4	206D792062616E6B	2F1C901E2E863BD1	BC7EAEB85DCCBB06
5	206163636F756E74	9C1FCDDB32B9D572	10CD42F50B754875

The 32-bit MAC (visa) is: 10CD42F5.

Now, returning back to the issue of visas, [8,9] gives a discussion on the two basic approaches to controlling traffic in transit networks. One approach is based on extensions of traditional access control mechanisms (e.g., Visas), and the other is based on policy routing techniques. In this paper, we assume the former mechanism is utilized.

The extension of the Visa protocols in [8,9] for controlling traffic in transit networks is fairly straightforward. The process of establishing authorization in visa-controlled transit networks is essentially the same as it is for the source - destination hosts pair in the Visa protocol. The source host obtains a visa key from each the source, destination, and any visa-controlled transit network in the packet's path from source to destination host. If every transit network is visa-controlled, the source host would have to obtain a visa key and place in the packet's header a signature for each such transit network. Each *AG* en route from the source to the destination host verifies its respective network's signature stored in the packet, that is, each *AG* checks the authorization of the packet to enter or exit the attached visa-controlled network (and not whether the packet is authorized to travel the entire path from source to destination host).

The *AG* must have access to the packet in its original form in order to verify its signature. However, packets can arrive at *AGs* fragmented. Thus, to authenticate packets (i.e., verify signatures), *AGs* must receive packets unfragmented or have some mechanism to authenticate them based on their fragments.

In a visa-controlled network using traditional Visa protocols, [18] provides a solution to the network-level authentication - fragmentation problem (i.e., between source and destination networks only) via *MTU Probing* and *Delayed Authentication*. However, fragmentation can also occur in visa-controlled transit networks as well. Ironically, because of its semantics, the MTU Probing protocol does in fact reconcile packet fragmentation with packet authentication in transit networks. However, its performance is marginal. Below we propose two protocols to perform the reconciliation as well. One solution is a modification and extension of the Delayed Authentication protocol, while the other is a novel approach utilizing an intra-network fragmentation policy. As will be clearly seen later, results from prototype implementations of our protocols demonstrate significant performance improvements over a prototype implementation of [18]'s MTU Probing.

4. INTERNETWORK MODEL

When illustrating later the behavior of our proposed protocols and the MTU Probing protocol, it will be convenient to use the example *internetwork model* specified in Figure 5. In general, we can define an internetwork $I = (V,E)$ as an undirected graph, where the nodes (hosts and gateways) in the internetwork are denoted by the vertex set V, and the networks are denoted by the edge set E, respectively, i.e.,

$$V = \{S,D\} \cup \{G_i: i = 1..n\}$$

$$E = \{(u,v): u,v \in V \ \& \ \text{there exits a network connecting nodes u and v}\}.$$

For academic purposes we consider $n = 6$ gateways in the internetwork I, as illustrated in Figure 5, with

$$V = \{S,D,G_1,G_2,G_3,G_4,G_5,G_6\},$$

$$E = \{(S,G_1),(G_1,G_2),(G_1,G_3),(G_2,G_3),(G_2,G_4),(G_2,G_5),(G_3,G_5),(G_4,G_6),(G_5,G_6),(G_6,D)\},$$

To complete our definition, we define a weight function $\mathbf{w}: \mathbf{E} \to \mathbf{Z^+}$, where, for all $(u,v) \in E$, $w(u,v) = $ MTU of network (u,v) (i.e., using our previous notation, $w(u,v) = MTU(u,v)$). So, for example, in Figure 5 network (S,G_1) represents a network for which the source host S is connected, and the MTU of network (S,G_1) is given by $w(S,G_1) = 1000$ bytes.

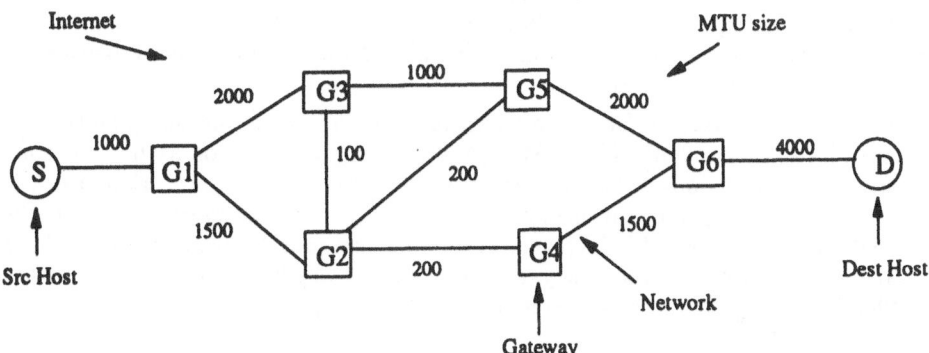

Figure 5. An internetwork model I = (V,E), where the vertices $v \in V$ in I represent the nodes (hosts and gateways) in the modeled internetwork, and the edges (u,v) $\in$ E in I represent the networks. The weight of each edge w(u,v) represents the modeled network's MTU size.

5. EXTENDED DELAYED AUTHENTICATION (EDA)

Before getting to the specifics of the *Extended Delayed Authentication* protocol, we first give a set of assumptions this protocol makes.

- Network access control mechanisms similar to the Visa protocols in [7,9] are utilized. Recall that each *AG* in the packet's path from source to destination uses a secret key shared with the source host to authenticate the packet.

- Packets transmitted to the destination host are plaintext, with the data signatures of each the source, destination, and visa-controlled transit networks placed in the header.

- Data signatures stored in the packet's header are the result of applying a one-way characteristic function F_{char} to ciphertext.

- The characteristic function used to compute the packet signature is **decomposable,** that is, if given packet P, composed of r fragments, F_1P, F_2P,..., F_rP, with packet message P_M, having message fragments F_1P_M, F_2P_M,..., F_rP_M, then

$$F_{char}(P_M) = F_{char}(F_1P_M) \text{ op } F_{char}(F_2P_M) \text{ op } \cdots \text{ op } F_{char}(F_rP_M), \qquad (1)$$

where **op** is some binary operator. Moreover, this property must be true without maintaining copies of fragments or introducing excessive computational delays in *AGs*.

- "Pseudo" dynamic routing of fragments (i.e., fragments of the same packet can follow dissimilar routes in any *nonvisa-controlled* transit network) is allowable. However, all fragments of a packet must pass through the **same** set of authenticating gateways for each of the source, destination, and any visa-controlled transit network traversed by a fragment. For example, given packet P, consisting of fragments $F_1P, F_2P,...,F_rP$, with fragment routes $RF_1P, RF_2P,...,RF_rP$, respectively, where, for $i = 1..r$,

$$RF_iP = \{AG_j : j \in \{src, trans_1,...,trans_n, dest\} \ \& \ AG_j \text{ processed } F_iP\},$$

and assuming $1..n$ visa-controlled transit networks traversed in $P's$ path from source to destination host, the following restriction must hold:

$$\left(\{AG_{src}, AG_{dest}\} \cup \bigcup_{j=1}^{n} \{AG_{trans_j}\} \right) \subseteq \bigcap_{i=1}^{r} RF_iP. \qquad (2)$$

Such a restriction is easily enforceable via the IP *loose source route* OPTION [16], which specifies sequences of IP addresses IP_i, for $i \in [AG_{src}, AG_{trans_1},..., AG_{trans_n}, AG_{dest}]$ for which the packet must follow, but allows for dynamic routing between successive IP addresses in the list.

EDA works as follows. The source host initially engages in the set up authorization and authentication procedure with an **ACS** (Access Control Server) on each of the source, destination, and visa-controlled transit networks. In order to enforce restriction (2), it is during this set up procedure that the source host ascertains the IP addresses of each visa-controlled transit network AG involved. Assuming communication is authorized, each respective ACS distributes a secret key (e.g., assume two-way keys) to the source host. The source host now possesses a set of secret keys, say

$$K = \{k_{src}, k_{trans_1}, k_{trans_2},..., k_{trans_n}, k_{dest}\}.$$

With each secret key $k_j \in K$, the source host encrypts packet $P's$ message P_M to be transmitted (i.e., $E_{k_j}(P_M)$); then, for each encryption $E_{k_j}(P_M)$, the source host computes a characteristic function value (F_{char}) to produce a set of packet signatures (visas), say

$$S = \bigcup_{k_j \in K} \{F_{char}(E_{k_j}(P_M))\}.$$

The source host sends packet $P's$ message P_M in plaintext format, with the set of signatures and the list of IP addresses of all relevant visa-controlled transit networks placed in $P's$ header, say

$$P = <S, IP_{AG_{trans_1}}, IP_{AG_{trans_2}},..., IP_{AG_{trans_n}}, P_M>.$$

The main idea of the protocol is to maintain state information at each AG_j, for $j \in \{src, trans_1, trans_2,..., trans_n, dest\}$, on a per-packet basis (e.g., each AG_j maintains a local Packet State Table PST_j), and once the last **logical** fragment of a fragmented packet arrives at AG_j, it can then verify the authenticity of the entire packet. Note that we make a distinction between the last *logical* fragment and the last *physical* fragment of a packet.

```
EDA()
  loop
    RECEIVE(F_iP)                                        /* receive ith fragment of packet P */
    if F_iP = P then                                     /* F_i is entire packet - no fragmentation */
      if F_char(E_{k_j}(F_iP_M)) = F_char(E_{k_j}(P_M)) then   /* ck packet signature in header */
        SEND(F_iP)                                       /* route to next hop */
      else DISCARD(F_iP)                                 /* packet erroneous and/or tampered with */
    else                                                 /* F_iP ≠ P, that is, F_iP is a fragment */
      if F_iP ∉ PST_j then                               /* when AG_j receives 1st fragment of P */
        INIT(id(F_iP),PST_j); START(id(F_iP),Timer_{id(F_iP)})
      if EXPIRED(id(F_iP),Timer_{id(F_iP)}) then
        FREE(id(F_iP),PST_j); KILL(id(F_iP),Timer_{id(F_iP)})
      else
        if LAST_FRAGMENT(F_iP) then                      /* assume P has r fragments */
          if F_char(E_{k_j}(F_1P_M)) op ... op F_char(E_{k_j}(F_rP_M)) = F_char(E_{k_j}(P_M)) then
            SEND(F_iP)
          else DISCARD(F_iP)
          FREE(id(F_iP),PST_j); KILL(id(F_iP),Timer_{id(F_iP)})
        else    /* F_iP is not last fragment of P; below, LOS - Length, Offset, & Signature */
          INSERT(id(F_iP),LOS(F_iP),PST_j)
          if F_iP = F_1P then     /* packet signature stored in 1st physical fragment */
            INSERT(id(F_iP),F_char(E_{k_j}(P_M)),PST_j)
  end loop
```

The former refers to the last fragment processed by a node, whereas the latter refers to the end fragment of a fragmented packet. For example, given r fragments of P processed by some arbitrary node in the following sequence: $F_2P,F_1P,F_rP,...,F_7P$, the last physical fragment is of course F_rP, whereas the last logical fragment is F_7P. When we use the phrase "last fragment" we are referring to the logical, and not physical, last fragment.

Recall that we do not require any AG_j to locally store copies of the message fragments $F_1P_M,F_2P_M,...,F_rP_M$, instead, AG_j simply processes and forwards all fragments without delay. However, since authenticating individual message fragments is meaningless, AG_j computes and stores partial signatures, that is, the characteristic function F_{char} of each encrypted message fragment $F_{char}(E_{k_j}(F_iP_M))$, for $i = 1..r$ in its local PST_j. Once it has processed the last fragment of the packet, AG_j can then determine the authenticity of the entire packet. When AG_j receives the last fragment, if the composite F_{char} for all the encrypted message fragments (i.e., partial signatures) matches the F_{char} of the packet's original encrypted message, i.e.,

$$F_{char}(E_{k_j}(F_1P_M)) \text{ op } ... \text{ op } F_{char}(E_{k_j}(F_rP_M)) = F_{char}(E_{k_j}(P_M)), \qquad (3)$$

then AG_j forwards the last fragment, otherwise, it discards it (recall that the packet's original signature set S, that is, $F_{char}(E_{k_j}(P_M))$, for $k_j \in K$, is stored in the header of the original packet P, or, in cases that fragmentation occurs, in the first physical fragment F_1P).

The last step is critical since if a match is found, then AG_j allows either the next authenticating gateway, say AG_{j+1}, in the fragment's path to verify the authenticity of packet P, or it allows the destination host D to reassemble packet P (assume that AG_{j+1} and/or destination host D are waiting for the last fragment). If a match is not found, then AG_j simply discards P's fragment. Two situations result from discarding the fragment: (1) the destination host D is unable to reassemble packet P since it will time out, and (2) every succeeding authenticating gateway $AG_{j+1},...,AG_{dest}$ en route to D cannot verify the authenticity of P. Having IP allow D to time out and discard P's fragments is exactly what we want to occur, since we do not wish for D to obtain a copy of a possibly tampered and/or erroneous packet. However, in regards to packet P, what do $AG_{j+1},...,AG_{dest}$ do with

$P's$ entry in their local *Packet_State_Table*? They have no way of knowing that the discarded fragment is not forthcoming. To resolve this, we employ the same mechanism as is used at D. We start a timer in each $AG_{j+1},...,AG_{dest}$ as soon as it receives the first fragment of packet P, and if the timer ever expires, then they simply drop $P's$ entry in the local *Packet_State_Table* and kill the timer.

To see visually how the EDA protocol behaves, consider the following example illustrated in Figure 6. Assume source S wants to send a packet P, say of 1000 bytes, to destination D, and further, assume each gateway G_j for $j = 1..6$, in the internetwork I is an authenticating gateway AG_j (i.e., each network $(u,v) \in E$ is visa-controlled).

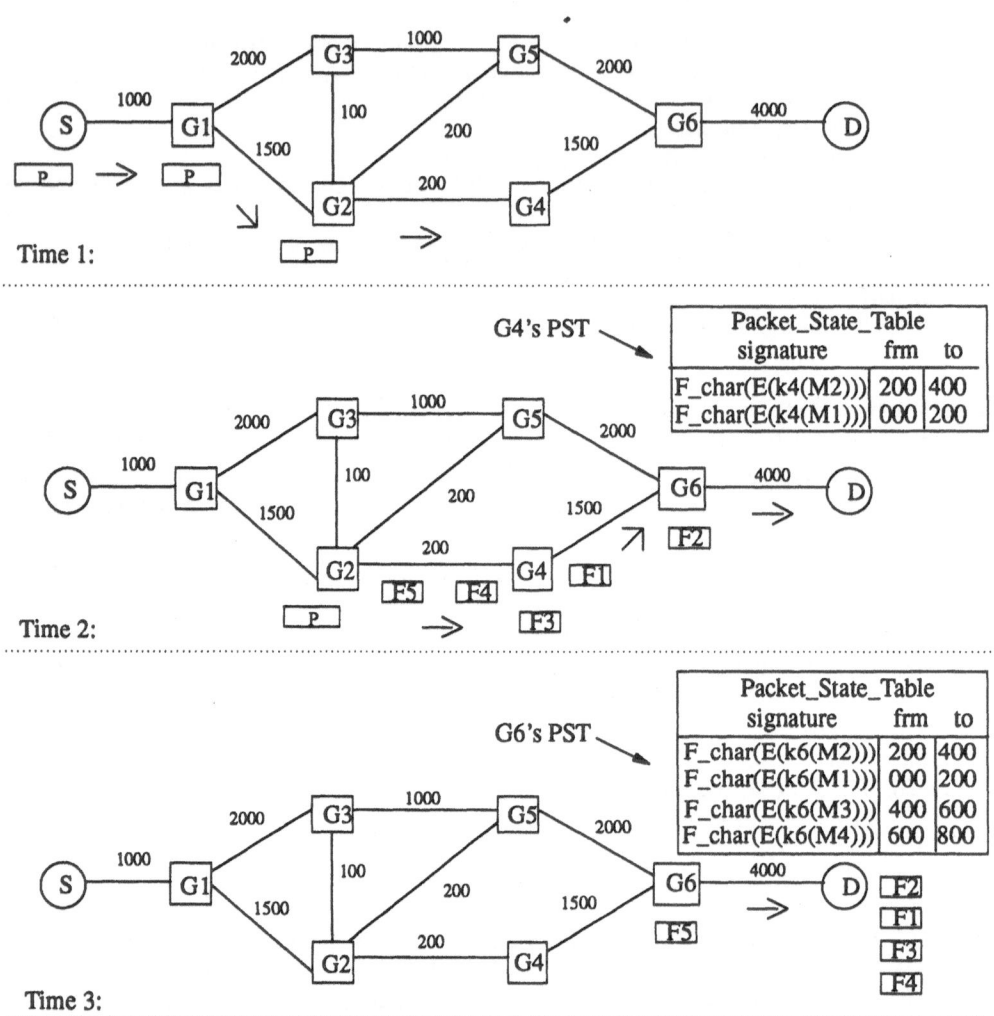

Figure 6. An illustration of the EDA protocol at 3 different snapshots in the Time sequence.

At Time t1, S sends P to G_1, it authenticates P (correctly) and routes it to G_2, where it also authenticates P. Now assume G_2 decides to route P to G_4. In this case, it must fragment P into 5 fragments, say $F_1,...,F_5$ (ignore the issue of header lengths in each fragment). Note that no entry for P in $G_1's, G_2's$ local *Packet_State_Table*, PST_1, PST_2, respectively, was necessary, since each received P in its entirety, and not fragmented.

Now, at Time t2, G_2 sends fragments $F_1, \ldots, F_5$ to G_4. Since G_4 receives P fragmented, it first creates an entry for P in PST_4 upon receiving P's initial fragment F_2. Then, for each additional fragment received, it computes and stores in P's corresponding entry in PST_4 the offset and partial signature of the fragment received. (the length of each fragment is easily computed via its offset fields). In this case, G_4 has processed F_1, F_2, routed them to G_6, and is currently processing F_3. Note how the fragments are not delayed or stored at each gateway, that is, once the gateway has computed the fragment's offsets and partial signature, it routes it to the next hop.

At Time t3, first note since G_4 has sent to G_6 all of P's fragments, $F_1, \ldots, F_5$, it is implied by the illustration that G_4 was able to verify the authenticity of P based on its fragments, in other words, that equation (3) was true (in this case, taking the number of fragments $r = 5$). Now at G_6, two things to note. First, it has processed and sent to D all of P's fragments except the "logically" last one (as it turns out it is also P's physically last fragment as well). The second thing to note is that the crucial point in the EDA protocol is now, when G_6 processes the last fragment. If equation (3) holds in this case as in G_4's case, then G_6 forwards the last fragment F_5 to D, allowing it the opportunity to reassemble P. Otherwise, G_6 discards F_5, thereby preventing D from reassembling the original packet P, since D will eventually time out and discard P's received fragments $F_1, \ldots, F_4$ since it never received F_5.

6. INTRA-NET FRAG AUTHENTICATION (IFA)

An alternative protocol for network-level packet authentication in the presence of fragmentation is *Intra-net Frag Authentication*. IFA makes no assumptions about the authentication protocol (although assume Extended Visas are used), and no state information at *AGs* is necessary; however, IFA incurs all costs associated with intra-network fragmentation. Because of its relatively good performance behavior in various internetwork environments, in addition to being an fairly easily implemented protocol, we consider IFA to be an excellent protocol for network-level packet authentication. The basic idea of IFA is to have all visa-controlled networks enforce the following restriction, where, for $j \in \{src, trans_1, \ldots, trans_n, dest\}$,

$$AG_j \textbf{ must} \text{ employ an intra-network fragmentation policy.} \tag{4}$$

Initially, as was the case in the previous protocol, the source host engages in the set up authorization and authentication procedure to obtain the set of secret keys from ACSs in each the source, destination, and visa-controlled transit networks. Since reassembly is done locally at each AG_j, they now have the ability to verify the authenticity of packets on the spot, even with fragmentation. Once the last fragment of a packet arrives at AG_j, it completes the reassembly process, verifies the authenticity of the packet, and either forwards to the next hop or discards the packet, whichever is applicable. In the former case, succeeding *AGs* en route to the destination host may not even receive fragments, but instead, the packet itself (e.g., if no fragmentation occurs between neighboring authenticating gateways AG_j and AG_{j+1}). If the packet arrives unfragmented at AG_j, not only can its authenticity be verified on the spot, but all costs associated with intra-network fragmentation are not incurred. In the worst case, however, every network in packet P's path from source to destination host is visa-controlled, and, for $j \in \{trans_1, \ldots, trans_n, dest\}$,

$$size(P) = MTU(Net_{src}) > MTU(Net_j).$$

In this case, we incur intra-network fragmentation costs at each AG_j. Note that if the current AG_j discards an erroneous and/or tampered packet, then no succeeding $AG_{j+1}, \ldots, AG_{dest}$ will have knowledge of this. This is desirable since valuable resources of succeeding *AGs* are not wasted. We introduce two *Options* to the basic IFA protocol:

```
IFA()
  loop
    RECEIVE(F_iP)                                    /* receive ith fragment of packet P */
    if F_iP = P then                                 /* F_iP is entire packet - no fragmentation */
       if F_char(E_{k_j}(F_iP_M)) = F_char(E_{k_j}(P_M)) then   /* ck packet signature in header */
         SEND(F_iP)                                  /* route to next hop */
       else DISCARD(F_iP)                            /* packet erroneous and/or tampered with */
    else                                             /* F_iP ≠ P - reassemble packet */
       if not(LAST_FRAGMENT(F_iP)) then  /* determines if F_iP is last logical fragment */
         AFFIX(id(F_iP),F_iP_M,Buffer_{id(F_iP)})
       else                                          /* packet reassembled */
         if F_char(E_{k_j}(Buffer_{id(F_iP)}))= F_char(E_{k_j}(P_M)) then  /* ck packet signature in header */
           SEND(Buffer_{id(F_iP)})                   /* route to next hop */
         else DISCARD(Buffer_{id(F_iP)})             /* packet erroneous and/or tampered with */
  end loop
```

† *(Option 1):* If pseudo dynamic routing of fragments is desired, then we simply employ the IP loose source route OPTION as in the EDA protocol. Thus, dynamic routing of fragments is possible in nonvisa-controlled transit networks, however, by restriction (4) and the semantics of intra-network fragmentation, all fragments must pass through the same set of *AGs* in each the source, destination and visa-controlled transit networks. Notice that because of the semantics of intra-network fragmentation, restriction (2) is also applicable.

† *(Option 2):* If "full transit" authentication is desired, then we simply employ restriction (4) (i.e., in "full transit" authentication, a packet is verified by an *AG* in each transit network hop en route to the destination host). Note the IP loose source route OPTION is not required since it is implied by the semantics of "full transit" authentication that the entire path from source to destination host is well known and already established.

To see visually how the IFA protocol behaves, consider the following example illustrated in Figure 7. Assume similar parameters as was specified for EDA.

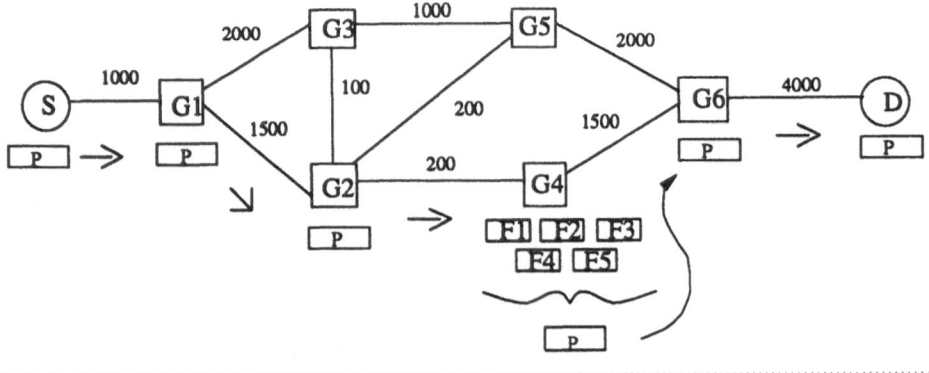

Figure 7. An illustration of the IFA protocol (in a global view).

Similarly to the EDA protocol illustrated in Figure 6, G_1, G_2 authenticate packet P (correctly). Once at G_2, and following a similar route as in the EDA case, G_2 fragments P into fragments $F_1, \ldots, F_5$ and routes these to G_4. As G_4 receives $P's$ fragments, it proceeds to incrementally reassemble them until it receives $P's$ last fragment. Once receiving the last fragment, G_4 completes the reassembly process and proceeds to verify the signature stored in the header. If correct, it then routes P in its entirety to G_6 and eventually onto destination host D. Otherwise, if incorrect, G_4 discards P, thereby preventing any further AG_js en route to D (G_6 in this case) and D from ever processing P.

To see visually how [18]'s MTU Probing protocol essentially behaves, consider the example illustrated in Figure 8. Again, assume the same parameters as before.

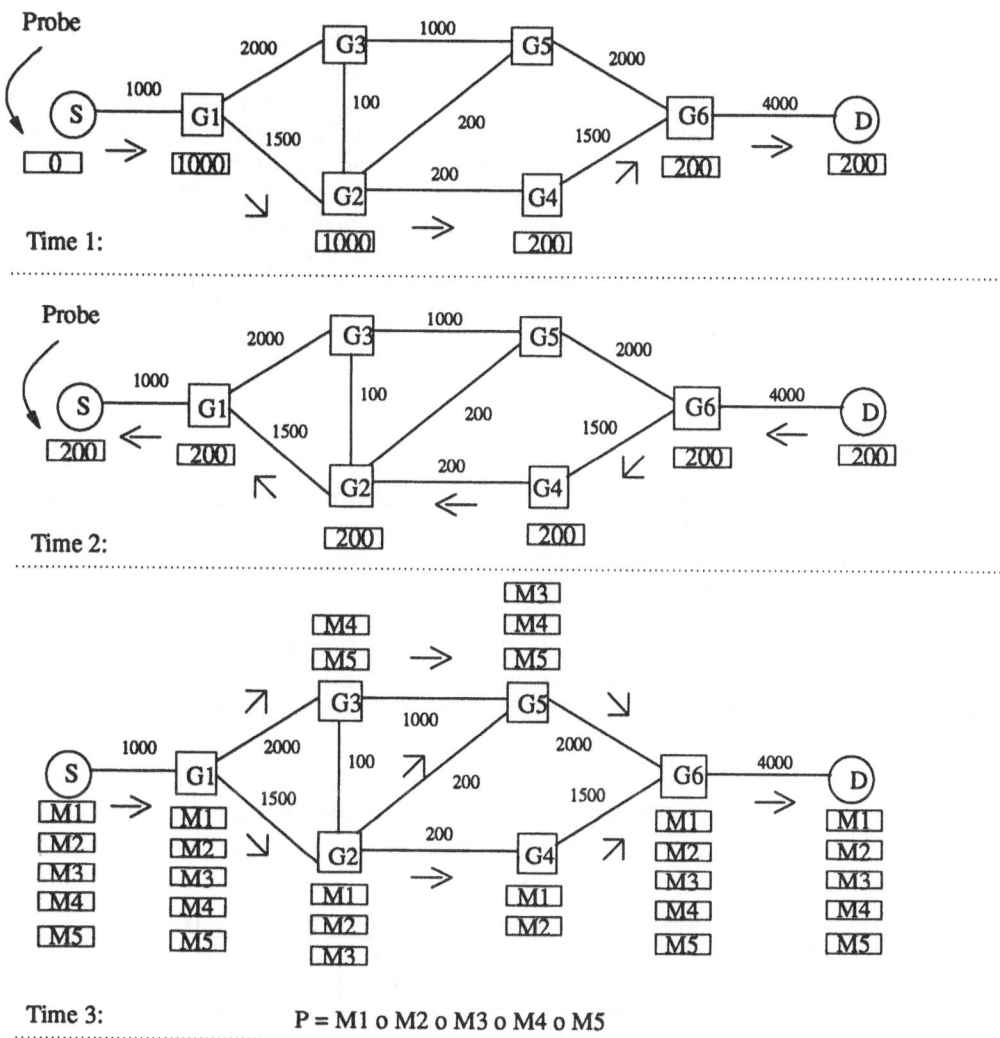

Figure 8. An illustration of the MTU Probing protocol at 3 different snapshots in the Time sequence.

Tsudik's [18] MTU Probing protocol essentially behaves in the following manner. At Time t1, a Probe packet is sent from the source S to the destination D, where each AG_j that processes the Probe will stamp its forwarding network's MTU size in the Probe if this size is smaller than the value currently stored in the Probe. When the Probe reaches D, the smallest network MTU size will have been discovered and stored in the Probe (in our case, network (G_2, G_4) has the smallest MTU given by $w(G_2, G_4) = 200$ bytes).

At Time t2, the value in the Probe must be relayed back to the source S. Thus, destination host D sends the Probe value back to S.

At Time t3, S will send to D the original 1000 byte packet message P_M in 5 smaller packets, $M_1, \ldots, M_5$, of size 200 bytes each. In general, given a packet message P_M, a packet header P_H, and the Minimum MTU (Min_{MTU}) size discovered in the path from source to destination host, MTU Probing will send the original message P_M in

$$\left\lceil \frac{|P_M| + |P_H| * \left\lceil \dfrac{|P_M|}{Min_{MTU}} \right\rceil}{Min_{MTU}} \right\rceil \tag{5}$$

packets (note that $|P_i|$, i = M,H, denotes "the cardinality of"). Some other interesting features of MTU Probing should be noted. First, dynamic routing can be fully exploited, as the illustration tries to show, although, it may be a risky venture (note that network (G_2, G_3) must be avoided, since its MTU size is 100 bytes). To be totally safe, $M_1, \ldots, M_5$ should follow the same path as the Probe. In addition, note that packets $M_1, \ldots, M_5$ will each have the full complement of signatures S (sorry for the possible ambiguity) stored in the header. And lastly, at the network-level, indeed, fragmentation has been avoided, which is consistent with the semantics of MTU Probing. However, at a higher level (e.g., application-level), the original message P_M must be delivered to the destination D, thus, somewhere above the network-level the partial messages $M_1, \ldots, M_5$ must be reassembled back into and delivered as P_M (e.g., somewhat like an application-level fragmentation).

7. ANALYSIS

The analysis of the two protocols will not consider costs incurred by each due to the use of the Visa protocols for packet authentication. For example, the initial set up cost (e.g., source host obtaining secret keys and computing signatures for each AG along the packet's route to verify) is an inherent cost of the Visa protocol. In addition, placing data signatures in the packet's header will require adding a new option to the IP OPTION field to support such a scheme, which in effect will require the modification of existing IP code in both hosts and gateways. A good discussion about these and other costs is given in [6,8,9]. Our analysis only considers costs relevant to each protocol as described above.

Table 2. Comparison of four network-level authentication schemes

Characteristic	MTU Probing	EDA	IFA (Opt 1)	IFA (Opt 2)
instant authentication	yes	no	yes	yes
supports fragmentation	no	yes	yes	yes
fragment dynamic routing	yes	yes	yes	no
gateway modifications	no	yes	yes	yes
host modifications	yes	no	no	no
per-packet overhead	no	yes	yes	no
more packets generated	yes	no	no	no
increased packet size	no	yes	yes	no
control packets needed	yes	no	no	no

Table 2 summarizes the main characteristics of four packet authentication schemes described in this paper (see [18] for a full specification of the MTU Probing protocol). Although Table 2 is general, in the analysis that follows, note that many of the differences between MTU Probing, EDA, and IFA simply reflect the differences in fragmentation

226

policies, specifically, using "no fragmentation", inter-network fragmentation, and intra-network fragmentation, respectively.

• Increased Packet Size:

EDA incurs a per-packet visa cost for each of the source, destination, and visa-controlled transit networks by requiring a signature from each such network to be stored in the packet's header. IFA and MTU Probing do not (recall that no assumptions need be made about the authentication protocol, although, via the Visa protocol scheme, both also incur a per-packet visa cost). In addition, both EDA and IFA (Option 1) require having the IP addresses of each transit network stored in the header. IFA (Option 2) and MTU Probing do not require storing these addresses in the packet's header.

• Fragmentation Costs:

EDA experiences only inter-network fragmentation costs. IFA (Option 2) experiences only intra-network fragmentation costs. However, interestingly enough, IFA (Option 1) may experience both inter- and intra- network fragmentation costs - a *hybrid* fragmentation cost. MTU Probing does not incur any fragmentation cost, at least not at the network level.

• Dynamic Routing:

Each of EDA, IFA (Options 1 & 2), and MTU Probing allow for dynamic routing of packets, however, only EDA and IFA (Option 1) allow for "pseudo" dynamic routing of fragments. Because IFA (Option 2) strictly requires the use of an intra-network fragmentation scheme at each hop, it sacrifices dynamic routing.

• Node (Host and Gateway) Overhead:

Overhead is mainly due to maintaining additional information and requiring extra processing at a node. Both EDA and IFA protocols require no additional information be maintained or extra processing be done at a host node, MTU Probing does (i.e., supporting the Probe packet). However, although quite different, EDA and IFA require significant overhead at each authenticating gateway, MTU Probing does not. IFA requires the support of intra-network fragmentation from every visa-controlled network. EDA requires both state information and a local timer be maintained at each AG on a per-packet basis, plus the additional processing required to update the timer, to compute partial signatures for message fragments, and to determine whether or not the current fragment is the last fragment of the packet (because of possible duplicate and overlapping fragments as a result of retransmission, the cost of determining whether or not a fragment is a packet's "last fragment" is equivalent in cost to partially reassembling the packet itself less the cost of buffering the fragments).

• Other Relevant Costs:

Note that, since fragmentation is supported, neither EDA nor IFA create an increase in the number of packets generated. However, since MTU Probing bypasses fragmentation, it must generate, on a per-packet basis, the Probe packet, plus the number of packets as specified in equation (5). In addition, in the EDA protocol, there exists a potentially significant degradation in resource utilization. For example, consider the case that a packet fragment is tampered with in the very first transit network traversed, say $trans_1$, out of n transit networks. Then each of the $n-1$ succeeding AGs $AG_{trans_2},...,AG_{trans_n}$, plus the destination host, will process all fragments but the tampered one, since it never arrives due to being discarded by AG_{trans_1}. Then not until each of these AGs and the destination host times out will they each discard the resources associated with the packet. Obviously a tremendous effort was wasted by each of the $n-1$ AGs unnecessarily. IFA does not suffer from this, since discovery of a tampered fragment results in its (and the packets) immediate disposal.

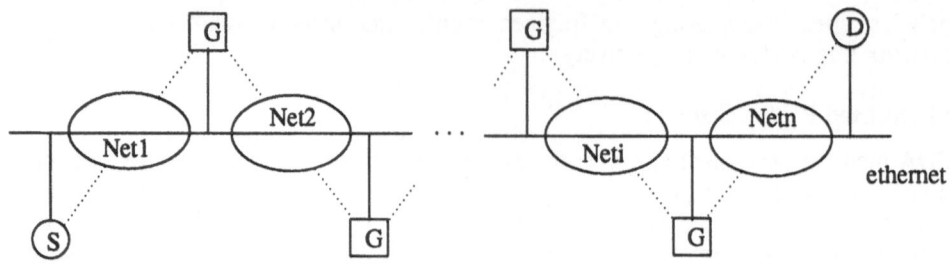

Figure 9. Model of an logical internet. Logically distinct networks on a single physical ethernet bus.

8. EXPERIMENTAL RESULTS

The intent of our experiments was to evaluate and compare the overall performance of the network-level authentication protocols discussed in this paper: [18]'s MTU Probing, EDA, and IFA. We created a logical internet environment (Figure 9), which is nothing more than logically distinct networks built on top of a single physical (dedicated) ethernet bus, and ran prototype implementations of the protocols in different settings. The logical internet provided a nice platform to measure and compare relative costs such as fragmentation, reassembly, Probe packets, increased packet size due to additional header information, and extra node processing with respect to the different protocols.

The configurations of our experiments were as follows: each node in the logical internet was a SUN - 3/80 workstation connected onto a single, dedicated ethernet bus; the size of the message transmitted by the source node was 1000 bytes; the fragmentation algorithm was taken directly from [16]; the reassembly algorithm was taken from [3]; IP

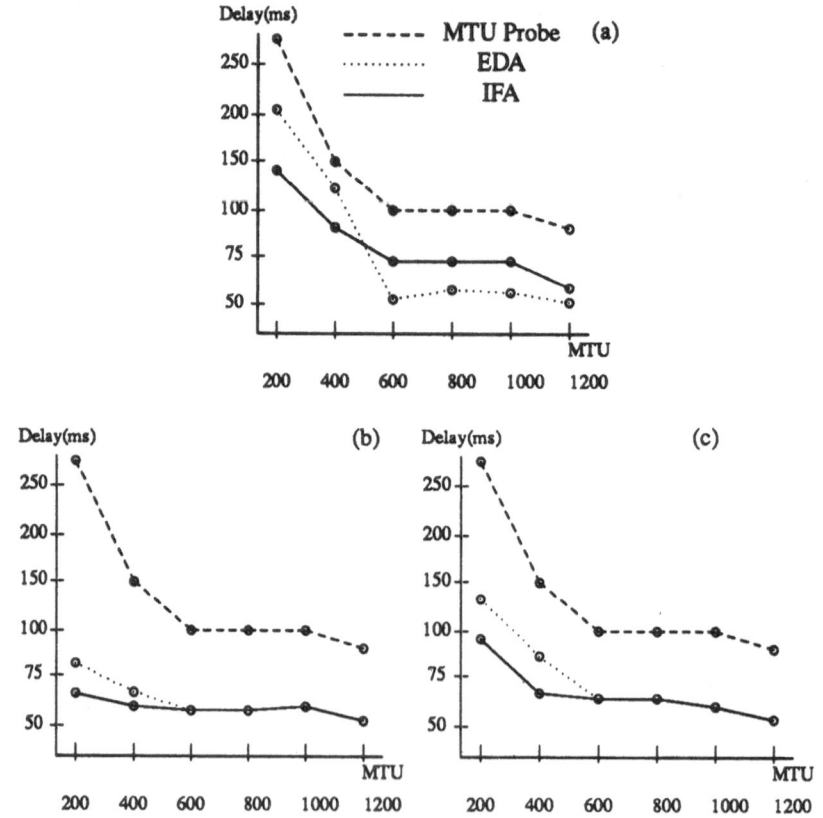

Figure 10. Graphical representation of the performance of three network-level authentication protocols.

routing was fixed at each node (note that the effects of dynamic routing were not tested in our experiments); network access control between logical networks was implemented via the Extended Visa protocol mechanism [9]; the authentication function was a software implementation of the DES based MAC (Message Authentication Code) [5,14,15]. And the metric our experiments captured was the source - destination host message transmission latency (delay) via $n = 8$ transit networks.

In our first experiment (Figure 10 (a)) the configuration of the internet was such that each of the logical networks had the same universal MTU size (i.e., universal MTU sizes denoted by the horizontal axis). In our second experiment (Figure 10 (b)) only one of the logical networks in the internet had an MTU size as denoted by the values on the horizontal axis, while all other networks had an equivalent MTU size of 1200 bytes. In our third experiment (Figure 10 (c)) half the networks in the internet had an MTU size as denoted by the values on the horizontal axis, while all other networks had an MTU size of 1200 bytes.

In each case, the performance of each protocol was quite sensitive to the amount of fragmentation occurring in the internetwork. In a *heavyweight* fragmentation environment (i.e., when a large packet is transported across many small MTU networks; $200 \leq MTU \leq 500$ in Figures 10 (a-c)) IFA performed the best. In this type environment, EDA suffered from the large amount of *AG* processing required for each packet fragment (per-packet state information and timers) per network. Also, irrespective of the processing requirements at *AGs* for IFA (e.g., reassembly, buffering) and EDA, what also helped IFA outperform EDA was from the fact that IFA packets had a smaller header size than EDA packets. In our experiments we chose to implement IFA (Option 2) since we wanted to measure the full effects of fragmentation and reassembly on a per-network basis. As a result, IFA packet headers only had the set of signatures S, whereas EDA packet headers had these in addition to the IP addresses of each of the 6 transit networks traversed by the packet (fragments). Larger packet headers have a greater impact on smaller packets than on larger ones. For small MTU networks, large packet headers result in less data per packet (fragment), which results in more packets (fragments) per message. For MTU Probing, processing cost and transmission delays associated with both the Probe packet and the cumulative sum total of packets per message were major contributors to its poor performance.

However, interestingly enough, when the severity of the "heavyness" decreases in such environments (i.e., when the number of fragments generated by the many small MTU networks diminishes; MTU > 500 in Figure 10 (a)) EDA had very good, in fact, better, performance than IFA. In this case, the reassembly cost at each network hop, although the amount of fragment reassembly was marginal, dominated IFA's cost as opposed to the much less amount of state information and processing required by EDA. In this case, we believe the buffering and data copying costs of IFA were more significant than the computation of the partial signatures and the maintenance of timers, especially since only very few fragments were generated.

In a *lightweight* fragmentation environment (i.e., when the number of small MTU networks generating the fragments diminishes; Figures 10 (b-c)), in this case, IFA still demonstrated excellent performance. However, notice that both IFA and EDA converge to the same point as less fragmentation occurs in the internetwork (i.e., when MTU = 1200 for Figure 10 (a) and when MTU $\geq$ 600 for Figures 10 (b-c)). Again, from the previous paragraph, we believe that the additional costs per-fragment associated with IFA and EDA were insignificant and relatively equal since the level of fragmentation was small. However, even though MTU Probing improved, it still is significantly worse than the others. Clearly, the performance degradation here reflects the previously mentioned transmission delays and processing costs associated with the Probe (data) packet(s). To send the Probe, await a response, and then send the original message in ≥ 1 data packet(s) is simply very expensive.

The IFA and MTU Probing protocols were insensitive to the smallest MTU network in the internet, no matter its location. In such cases, both protocols always resulted in the same performance. However, EDA was very sensitive to the location of the smallest MTU network. For example, Figure 10 (b) reflects EDA's performance when the one network chosen to have the various MTU sizes was the last transit network. However, when we chose the first transit network instead, EDA's performance degraded by a factor of 2. This clearly demonstrates the effect of processing many smaller packet fragments as opposed to just the packet itself. As was stated prior, the cost of packetized communication is per packet, not per byte. Processing a greater number of smaller fragments as opposed to one large packet certainly may have a significant affect on performance.

9. CONCLUSION

The implications of the fragmentation process on network access control via gateway packet authentication are significant within an TCP/IP internet. In this paper, we described various fragmentation policies and their affects on the performance of several network-level packet authentication protocols. In particular, a thorough comparison was given between fragmentation and "no fragmentation" policies, and, given fragmentation, between intra- and inter- network fragmentation. Two protocols were presented that reconcile packet authentication with packet fragmentation throughout the transit internet. One protocol allows for curtailed dynamic routing of fragments, at the expense of keeping state information per-packet in transit network authenticating gateways. The other protocol allows for immediate authentication at the expense of employing an intra-network fragmentation policy. Both protocols were fully described and analyzed, and, via prototype implementations, tested and compared. Each protocol demonstrated varying degrees of complexity, overhead, and performance degradation, however, within an experimental environment, our protocols performed significantly better than Tsudik's [18] existing protocol of MTU Probing, and are thus reasonable alternatives.

REFERENCES

[1]   M.A. Bonuccelli, "Minimum fragmentation internetwork routing," in *Proc. IEEE INFO-COM '91*, April 1991.

[2]   L. Breslau, and D. Estrin, "Design of inter-administrative domain routing protocols," *ACM SIGCOMM '90*, 1990.

[3]   D. Clark, *IP Datagram Reassembly Algorithms*, NIC, RFC 815, July 1982.

[4]   D. Comer, *Internetworking with TCP/IP Volume I*. Prentice Hall, Englewood, NJ, 1991.

[5]   D.W. Davies and W.L. Price, *Security for Computer Networks*. Wiley & Sons, NY, 1989.

[6]   D. Estrin, "Inter-organizational networks: Implications of access control requirements for interconnection protocols," in *Proc. ACM SIGCOMM '86*, Augest 1986.

[7]   D. Estrin, *et al.*, "Visa protocols for controlling interorganizational datagram flow," *IEEE JSAC*, May 1989.

[8]   D.Estrin and G. Tsudik, *End-to-end argument for network-layer access controls*, TR-90-13, USC, July, 1990.

[9]   D. Estrin and G. Tsudik, "Secure control of transit internetwork traffic," *Com Nets ISDN Sys* 22, 1991.

[10] S. Jamaloddin, "Congestion-free communication in high-speed packet networks," *IEEE TOC*, Dec. 1991.

[11] C. Kent and J. Mogul, "Fragmentation considered harmful," in *Proc. ACM SIGCOMM '87*, Augest 1987.

[12] L. Linn, "Practical authentication for distributed computing," *Proc. 1990 IEEE Symp Sec/Priv*, May 1990.

[13] J. Mogul, *et al.*, *IP MTU Discovery Options*, RFC 1063, July 1988.

[14] National Bureau of Standards, *Data Encryption Standard*, Federal Information Processing Standards Publication 46, January 1977.

[15] National Bureau of Standards, *DES Modes of Operation*, Federal Information Processing Standards Publication 81, December 1980.

[16] J. Postel, *Internet Protocol*, RFC 791, Sept. 1981.

[17] A. Tanennbaum, *Computer Networks*. Prentice Hall, Englewood Cliffs, N.J., 1989.

[18] G. Tsudik, "Datagram authentication in internet gateways: implications of fragmentation and dynamic routing," *JSAC*, May 1989.

LOCAL AREA NETWORK TRAFFIC LOCALITY:
CHARACTERIZATION AND APPLICATION

Neeraj Gulati[1]
Carey L. Williamson
Richard B. Bunt

Department of Computational Science
University of Saskatchewan
Saskatoon, SK, CANADA S7N 0W0

ABSTRACT

Network traffic locality is a special case of the locality phenomenon commonly seen in computer systems. The source and destination addresses seen in network packet traffic are observed to be non-uniformly distributed, both in time and space. Several recent research efforts have identified the presence of network traffic locality, and suggested ways to exploit the locality property in the design and operation of computer networks. While the intuitive notion of network traffic locality is fairly well understood, there is still a distinct lack of formal definitions or measures of network locality in the literature.

In this paper, we propose four new measures to characterize network traffic locality: persistence, concentration, address re-use, and reference density. These measures are motivated, in part, by locality characteristics identified in other domains (e.g., memory reference studies of program behaviour, and file referencing characteristics), but reflect properties important to network applications. Definitions are given for each measure, and measurement data collected from a local area network environment are used to quantify each measure. Finally, to demonstrate a potential application for the network traffic locality property, we present a brief analysis of how this property might be used to advantage in the design of a network interconnection device.

1. INTRODUCTION

Network traffic locality is a special case of the locality phenomenon commonly seen in computer systems. The source and destination addresses seen in network packet traffic are observed to be non-uniformly distributed, both in time and space. Successive packets are rarely independent of each other, and aggregate packet traffic is frequently dominated by a few transmitting hosts, even over significant intervals of time.

---

[1]Now at Bell-Northern Research (BNR), Ottawa, Canada.

Local Area Network Interconnection, Edited by R.O. Onvural
and A.A. Nilsson, Plenum Press, New York, 1993

In packet-switched networks, the locality property is a natural artifact of the protocols and applications used for data transmission. Bulk data transfer activities, such as file transfer, generate bursts of packets between a particular pair of hosts. Interactive terminal traffic also generates highly correlated bidirectional traffic streams. Models such as the *packet train model* [9] or the *packet tandem model* [10] have been proposed to explain such behaviour. Such models have proven to be useful not only in the improvement of techniques for modelling workloads, but also in their application to network design.

Several recent research efforts have identified the presence of network traffic locality [8, 9, 11], and suggested ways to exploit the locality property in the design and operation of computer networks [2, 6, 14]. Exploiting locality, usually by caching recently referenced items, has been crucial in the success of virtual memory systems, multiprocessor systems, and distributed file systems, where the access time of the cache is an order of magnitude faster than access to main storage. It remains to be seen how important the notion of locality is to the design of future networks. Several recently proposed network architectures [13, 15] rely explicitly on network locality for efficient operation. Exploiting locality may become even more important as networks become larger and faster [8].

While the intuitive notion of network traffic locality is fairly well understood, there is a distinct lack of formal definitions or measures of network locality in the literature. The purpose of this paper, then, is to precisely characterize the salient aspects of network traffic locality in a local area network (LAN) environment (i.e., in the packet traffic seen by a network interconnection device between LANs). In particular, we propose four new measures of network traffic locality: persistence, concentration, address re-use, and reference density. These measures are based, in part, on locality characteristics identified in other domains, such as program behaviour [4], and file referencing behaviour [12]. Definitions are provided for each measure, and measurement data collected from a local area network environment is used to quantify each measure.

The remainder of this paper is organized as follows. Section 2 motivates and defines our four locality measures. Section 3 presents our measurement methodology, and demonstrates our locality measures on packet trace data collected in a local area network environment. This demonstrates the extent to which these properties exist in a typical (albeit small) network environment. Section 4 discusses a possible application of these locality properties to the design of network interconnection devices, by presenting a brief analysis of the implications of the observed locality properties on the performance of a network interconnection device. Finally, Section 5 presents our conclusions, and suggested directions for future research.

2. LOCALITY CHARACTERISTICS

Studies of memory and file referencing behaviour have found that such references typically display non-random characteristics [4, 12]. References are distributed non-uniformly in both time and space with small subsets of information favoured over relatively long time intervals. The term "locality of reference" has been used to describe this phenomenon.

The General Principle of Locality [5] has been described as:

1. During any interval of time the resource demands of the system are distributed non-uniformly over the total set of available resources.

2. The demand pattern for a given resource tends to change slowly with time.

3. The correlation between the immediate past and immediate future demands tends
to be high; the correlation between disjoint resource demand patterns tends to
zero as the elapsed time between them tends to infinity.

In the networking context, the "resource demand pattern" is the sequence of source
and destination addresses seen in the network packets of an aggregate network traffic
packet trace.

Two aspects of locality have been identified, referred to as *temporal locality* and
spatial locality. Temporal locality means that the information currently in use is likely
to be used again in the near future. Spatial locality implies a high probability of references
to neighbouring addresses. While the notion of neighbouring addresses makes
sense in memory referencing and file referencing behaviour, it may not make sense in
the context of computer networks.

The terms *persistence* and *concentration* have also been used to characterize locality
behaviour [3]. Persistence refers to the tendency to repeat a particular reference (or
group of references) in succession (an example of temporal locality). Concentration, on
the other hand, refers to the tendency for the references to be limited (or concentrated)
to a small subset of the total reference space.

Two similar characteristics are evident in most analyses of network traffic locality.
First, successive packets on the network tend to be highly related. That is, a packet
travelling from host A to host B on the network is nearly always followed by another
packet from A to B, or a packet from B to A [9]. This is primarily a short-term characteristic,
demonstrating temporal locality and persistence. Second, a small number
of hosts tend to account for a large number of the packets transmitted on the network.
This is primarily a long-term characteristic, demonstrating concentration.

Two locality measures are proposed for each of these characteristics. For short-term
characteristics, the measures *persistence* and *address re-use* are proposed. For
long-term characteristics, the measures *concentration* and *reference density* are proposed.
These measures are defined in more detail in the following subsections.

2.1. Persistence

Persistence refers to the tendency for an address, once referenced, to be referenced
again and again, consecutively, on the network. A burst of consecutive packets from
the same sender would be said to have high source address persistence. A burst of
consecutive packets heading to the same destination would be said to have high destination
address persistence. Workload generated by bulk data transfers, for example,
would be expected to yield reference streams with high source and destination address
persistence.

The presence of persistence in network traffic would make very simple caching
strategies effective when handling packets. For example, some implementations of
TCP use a single element cache to store the protocol control block for the most recently
active transport-level connection. With a highly persistent reference stream, only the
first reference causes a cache miss, and all repeated references are resolved in the cache.

Thus our first step towards characterizing network traffic locality is with a measure
of persistence. The aim is to find the probability of seeing w consecutive packets from
the same source, or to the same destination. That is, starting from a given packet at a
random point in the reference stream, what is the probability that the next w packets
are from the same source as the current packet (source address persistence) or to the
same destination as the current packet (destination address persistence).

The persistence measure P_w is computed for various values of w, where w is called the *persistence window size*. The calculation is as follows. Let N_w denote the number of times that the addresses of the next w packets in the trace all match the address of the current packet. Let T_w denote the total number of places in the trace where this persistence could possibly occur (i.e., $T_w = n - w + 1$, where n is the number of packets in the trace). P_w is then simply the ratio

$$P_w = \frac{N_w}{T_w}$$

The persistence measure P_w is computed separately for source addresses and destination addresses.

2.2. Address Re-use

One drawback of the persistence measure defined in Section 2.1 is the requirement for related packets to or from the same address to appear consecutively on the network. In a packet-switched network that is statistically multiplexed among many hosts, the presence of persistence may be completely masked in the aggregate traffic. Thus a new measure, address re-use, is proposed as a generalized form of persistence. Rather than requiring related packets to appear consecutively on the network, they need only appear "close" together.

Address re-use is defined as the tendency for the address used in one network packet to reappear as the destination address of a future packet. Specifically, address re-use determines the probability that the address (source or destination) used in the current packet appears as the destination address of at least one of the next w packets, for small values of w.

Address re-use provides a more general measure of temporal locality than the persistence measure provides. To study this aspect of locality, the complete reference trace is considered w packets at a time. The window size w determines how far ahead in the reference trace you look.

Destination address re-use, D_w, is computed as follows:

$$D_w = \frac{N_w}{T_w}$$

where N_w is the total number of times that a destination address is re-used within the next w packets, and T_w is the total number of places where destination address re-use could possibly occur in the trace. A similar method is used to determine the source address re-use S_w.

2.3. Concentration

Concentration is defined as the tendency for the host addresses seen in a particular period of the trace to be limited to a small subset of the total host addresses seen in the entire trace. During any finite interval of w packets, activity is concentrated on a small number of hosts called the *working set* of active hosts.

In general, concentration is important for applications such as cache management. It is important to know how much of the total address space that is referenced during the entire duration of a trace is referenced during a specific interval. If only a small portion of the total address space is referenced, only a small portion of it needs to be kept in the cache. Our proposed concentration measure addresses this issue. Again, the measure is defined separately for source addresses and destination addresses.

The working set model, used extensively in the characterization of program behaviour [4], seems to provide a good framework for expressing the notion of concentration in a reference stream. If w is the working set window size, then the working set size W_w is the number of different hosts that send (or receive) at least one of the next w packets. So when sets of w packets are considered at a time, W_w shows the number of different addresses that appear in such sets. The smaller the working set size, the higher is the degree of concentration in the trace.

One of the drawbacks of the traditional working set model as a measure of concentration is that the W_w values are not directly comparable for different values of w. For example, if $W_{100} = 20$ and $W_{200} = 28$ for the same trace, it does not necessarily mean that higher concentration is being observed for the window of size 100. In fact, one could argue that the window of size 200 shows higher concentration since there are only 28 different addresses observed (14% of a possible 200), while for a window size of 100, 20 different addresses appear (20% of the possible 100). So the concentration for different window sizes cannot be compared simply by comparing the corresponding values of the working set size. Instead, concentration should be expressed as a function of both the working set size and the window size.

Defining concentration as the ratio of the working set window size w and the corresponding working set size W_w is one alternative, but is also unsuitable. Consider a window size of 10 and a window size of 100. If there is only one address appearing in all the packets in the window, then the concentration measure for the $w = 100$ window would be 10 times higher than that for the $w = 10$ window, even though both examples exhibit the "same" concentration behaviour, in the sense of one active host dominating all activity. Furthermore, the concentration values fall non-linearly as the working set size increases, making comparisons difficult (e.g., the concentration is 100 when $W_{100} = 1$, 50 when $W_{100} = 2$, 33 when $W_{100} = 3$, etc.) In other words, this measure of concentration is still not comparable across working set window sizes.

To overcome these deficiencies, a measure for concentration that is an extension to the working set model is proposed. The new measure is normalized to values between 0 and 1. If the addresses of all w packets in the working set window are the same (i.e., $W_w = 1$), then the concentration is said to be maximum ($C_w = 1$). If all the w packets have different addresses (i.e., $W_w = w$), then the concentration is said to be minimum ($C_w = 0$). In between these two extremes, the concentration C_w is defined by the following linear relationship:

$$C_w = \frac{w - W_w}{w - 1}, w > 1$$

This concentration measure captures the intuitive notion of concentration. The smaller the number of different destination addresses in a given subtrace, the higher is its concentration. Also, the longer a subtrace is for a given working set size, the higher the concentration. The new measure is also reasonably comparable across different window sizes.

Another advantage of the new concentration measure is its direct relationship to cache size. Given a trace of length w and concentration C_w, a cache size of approximately $(1 - C_w)w$ entries is required to keep the active working set in the cache. The longer the trace, or the lower the concentration, the larger the cache size required. The concentration measure is applied separately to both source and destination addresses.

2.4. Reference Density

Reference density is defined as the tendency for a small number of hosts to account for a large proportion of the total network traffic. Given h, the number of hosts active during a trace, and n, the number of packets in the trace, it is rare that all the hosts send and receive an equal number (n/h) of these packets. Over a period of observation, there are usually certain hosts that send or receive more than their share of the packets.

Reference density is a measure of this long-term concentration property. The reference density, R_n, is defined as the proportion of references accounted for by the n most active hosts. Reference density is computed from the traces by sorting the hosts in order of activity, and computing their cumulative contribution to the aggregate traffic. Again, this measure is applied separately to both source and destination addresses.

2.5. Host-pair Locality

Up to this point characteristics of source and destination address locality have been considered in isolation. In this section an aspect of locality that involves both the source and the destination address is considered.

Host-pair locality is defined as the frequency of occurrence of at least one of the next w packets that is traveling between the same host pairs as the current packet. This measure is similar to the address re-use measure discussed in Section 2.2. This measure includes unidirectional as well as bidirectional traffic between pairs of hosts.

3. MEASUREMENT RESULTS

3.1. Methodology

Traces of the activity on an Ethernet local area network were captured using an HP 4972A LAN Protocol Analyzer. The LAN Analyzer is a hardware network monitoring device that operates at the data link level. That is, it understands Ethernet frames (packets) and (48-bit) Ethernet addresses. It has no knowledge of higher level applications and protocols (e.g., TCP/IP). The device was operated in promiscuous mode, attached to one of the LAN segments in our research laboratory. The LAN Analyzer has a 1 megabyte buffer memory for incoming data, and a 20 megabyte Winchester disk. After capturing each trace, the data was transferred to a file server for later analysis.

To provide a variety of samples, ten different traces were obtained during the winter of 1992 and formed the basis for the analysis that follows. Details of the ten traces are given in Table 1.

A trace consists of a timestamp, a source address, and a destination address for each packet seen on the network[1]. The size of each trace was limited by the amount of disk space available on the LAN Analyzer when the trace was collected. In general, the disk provides enough space for about 5 minutes of tracing for aggregate network traffic. The exact duration of each trace was a function of the level of network activity during the trace period.

Since our local area network is connected to the campus network and the Internet, the traffic observed in any trace period involved hosts on other parts of the internetwork. From 503 to 665 distinct Ethernet addresses were seen during the periods traced. Of the hosts observed in the traces, 26 were identified as machines in our research laboratory, and 21 others as part of our department's teaching laboratory. The remaining addresses represent unidentified machines that are elsewhere on the internetwork.

---

[1] The analyses considered in this paper are only for data link level (Ethernet) addresses. Similar analyses could be performed on higher-level addresses, such as IP addresses, or TCP port identifiers.

Table 1. Description of the traces used in this study.

Trace	Date	Start Time	Size(pkts)	Duration	Hosts	Load(pkts/sec)
1	Feb 10	14:46:43	50068	267 secs	612	188
2	Feb 11	15:20:47	50076	287 secs	630	175
3	Feb 12	18:24:46	80134	344 secs	515	233
4	Feb 14	21:29:16	80326	620 secs	633	130
5	Feb 16	15:06:09	80116	332 secs	511	241
6	Feb 18	13:10:11	60065	322 secs	665	187
7	Feb 20	00:32:01	60922	495 secs	503	123
8	Mar 11	09:06:28	39403	300 secs	534	131
9	Mar 12	11:16:06	51573	267 secs	641	193
10	Mar 14	03:25:43	50970	284 secs	535	180
Averages for the ten traces:			60365	351 secs	578	178

To obtain a representative sample of the general activity on the network, traces were captured on different days of the week and at times when different types and levels of activity were expected. Each trace contains a time-stamped reference string of 39,000 to 80,000 packets with a duration varying from 267 seconds to 620 seconds.

Ten traces represent a large body of experimental work. To conserve space, only three of the traces are selected for presentation here[2]. The selected traces are Trace 5, Trace 6 and Trace 8. Trace 5 was chosen because it has the highest average load in terms of packets per second (241 packets/sec). Trace 8 was chosen because of its low load (131 packets/sec). Trace 6 was chosen because it has the highest number of active hosts (665), and a load that is close to average (187 packets/sec).

3.2. Persistence

Our first attempt to characterize network traffic locality is with a measure of persistence. The aim is to find the probability of seeing w consecutive packets from the same source, or to the same destination. That is, starting from a given packet at a random point in the reference stream, what is the probability that the next w packets are coming from the same sender (source address persistence) or heading to the same destination (destination address persistence) as the current packet.

The persistence data obtained for the three selected traces is plotted in Figure 1 and Figure 2. Figure 1 shows source address persistence, and Figure 2 shows destination address persistence. Both source and destination addresses exhibit very similar characteristics.

For all traces, persistence values drop sharply with increasing window size. Long sequences of persistent references are rare. The longest persistent sequence seen in any of the traces was $w = 9$.

Overall, the persistence values for each of the traces are low. For Trace 5 in Figure 2, for example, some destination address persistence is observed only 18% of the time, and no persistence is seen 82% of the time (i.e., 82% of the time the next packet is destined to a different address than the current packet).

From the analysis of persistence it is concluded that there aren't many instances when successive packets are observed on the network destined for the same host. There are two possible explanations for this behaviour. First, low persistence might be due to bidirectional traffic being dominant. That is, packets from A to B are followed

---

[2]The complete results for all ten traces are available in [7].

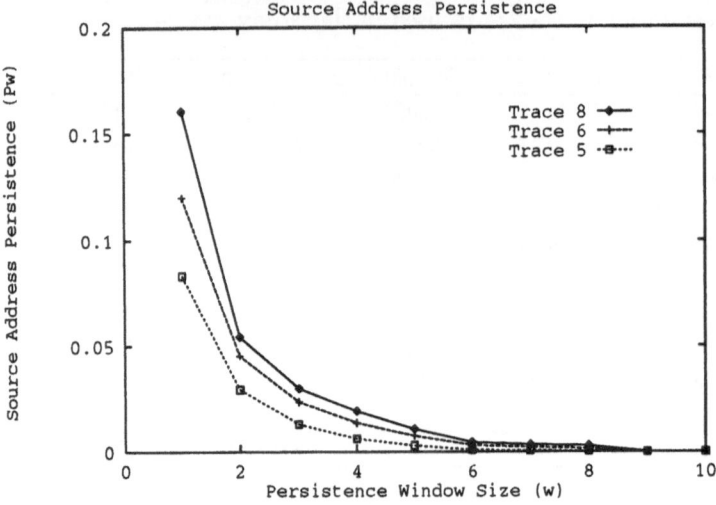

Figure 1. Source address persistence.

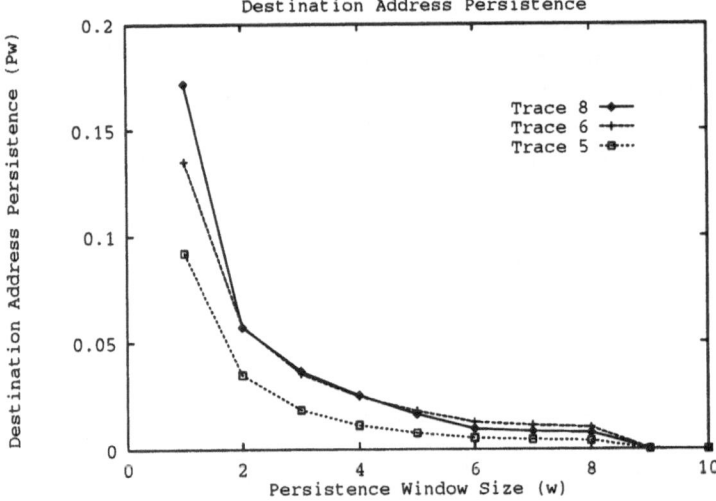

Figure 2. Destination address persistence.

by packets from B to A and vice versa. Second, low persistence may be due to the statistical multiplexing of the network among many hosts. That is, since the traffic observed is the aggregate traffic between various hosts on the network, many individual traffic streams may be interleaved, masking their persistence.

The load on the network appears to have a bearing on the overall persistence observed, providing some basis for the latter hypothesis. Trace 5, which has a much higher load than Trace 6 and Trace 8, shows lower address persistence. The persistent traffic is less frequently masked when the other hosts on the network are not very active (low load). Regardless of the network load, however, our results indicate that persistent traffic is not a dominant characteristic of network traffic locality.

3.3. Address Re-use

Address re-use is defined as the tendency for the address (source or destination) used in a particular network packet to reappear "soon" as the destination address in future packets. These packets need not appear consecutively on the network, but only "close together" in time.

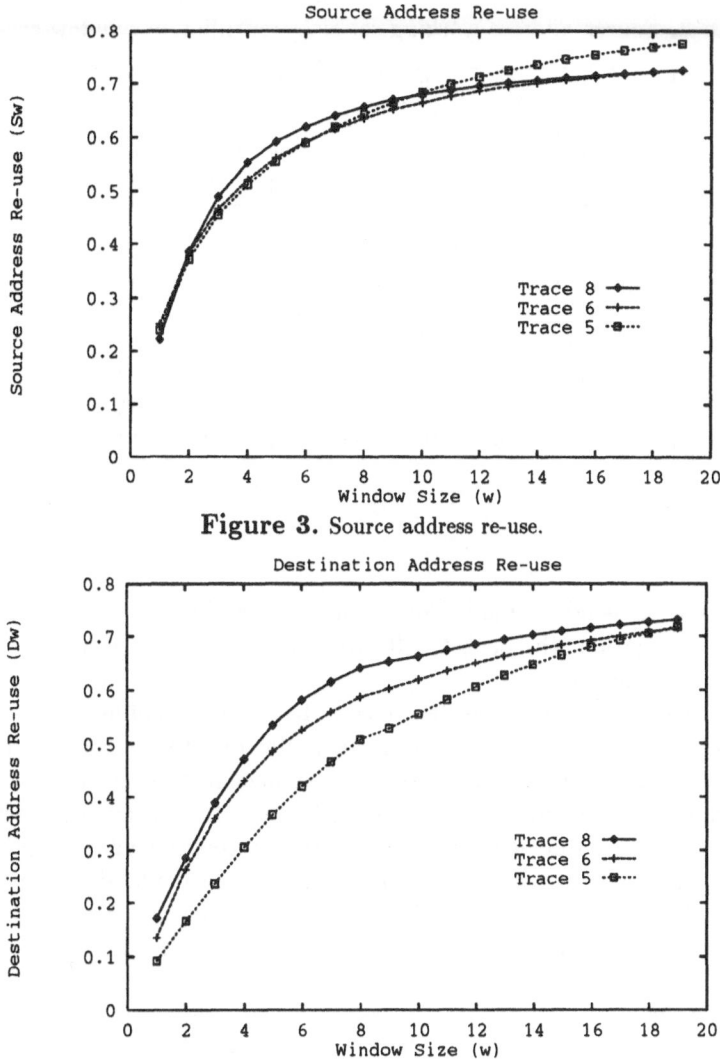

Figure 3. Source address re-use.

Figure 4. Destination address re-use.

Address re-use characteristics for the three selected traces are shown in Figure 3 and Figure 4. Figure 3 shows source address re-use (the probability that the source address in the current packet appears as the destination address of at least one of the next w packets), and Figure 4 shows destination address re-use (the probability that the destination address in the current packet appears again as the destination address in at least one of the next w packets),

The results show significant destination and source address re-use. The larger the window size w, the higher is the probability that an address is used again within that window. The rise in the re-use values is steep for small window sizes, indicating the presence of temporal locality in the traffic. For larger window sizes, the rise is slower: once an address becomes "stale", the probability of it being re-used in the future diminishes, resulting in a smaller increase in the address re-use values. Source address re-use values are higher than destination address re-use values for small window sizes w, as can be seen by comparing Figures 3 and 4.

The load on the network also has an effect on the address re-use results. The trace with highest load (Trace 5) has the lowest destination address re-use values (Figure 4). Trace 5 "catches up" with the other traces at larger window sizes since the active

Table 2. Source address working set sizes for different window sizes w.

w	10	30	60	100	150	200	300	400	500
Trace 5	6	13	18	25	32	38	50	61	71
Trace 6	6	12	20	28	38	46	62	75	88
Trace 8	6	14	22	31	46	58	77	93	106

Table 3. Destination address working set size for different window sizes w.

w	10	30	60	100	150	200	300	400	500
Trace 5	6	12	16	21	24	28	33	37	40
Trace 6	5	11	16	21	25	29	35	39	42
Trace 8	5	10	15	20	24	29	35	39	43

working set of hosts is fairly small. That is, if a sender is going to send another packet to a specific destination, the packet will probably be among the next 20 (aggregate) packets sent on the network.

Overall, the address re-use results suggest that there is significant predictability in the aggregate network traffic. Aside from its application to the modelling of network traffic, many network system design problems can benefit from this type of predictability in the traffic. For example, caching could be used to improve the performance of network interconnection devices. By caching the destination address and the source address of recently received packets, a prediction can be made about the destination addresses of the next few packets. The extent to which this predictability can be exploited was investigated by simulation and is reported in Section 4.

3.4. Concentration

Concentration is defined as the tendency for the host addresses seen in a particular period of the trace to be limited to a small subset of the total hosts. Two possible measures of concentration are the traditional working set size measure and our new concentration measure C_w.

Tables 2 and 3 show the working set size characteristics for the selected set of traces, for source and destination addresses, respectively. The working set size shown is the average working set size (rounded to the nearest integer) over the full duration of the trace, for the indicated value of the working set window size.

These results show that significant concentration exists in the references. For any trace interval of length w, the number of hosts W_w active during that period is significantly less than w, and the increase in W_w with increasing w is clearly sublinear. Our traffic shows even higher concentration than that observed by Jain [8]. For example, for a window size of 100, the working set size was 40 in the traffic studied by Jain. The reason for the difference in results is the lower number of hosts directly attached to the network used in this study.

Our new concentration measure is plotted in Figure 5 and Figure 6 for a variety of window sizes. Only Trace 5 is shown, since the three selected traces show almost identical behaviour with regard to both the source and destination address concentration. The jaggedness of the curves in these figures is due to integer roundoff in the working set size computation.

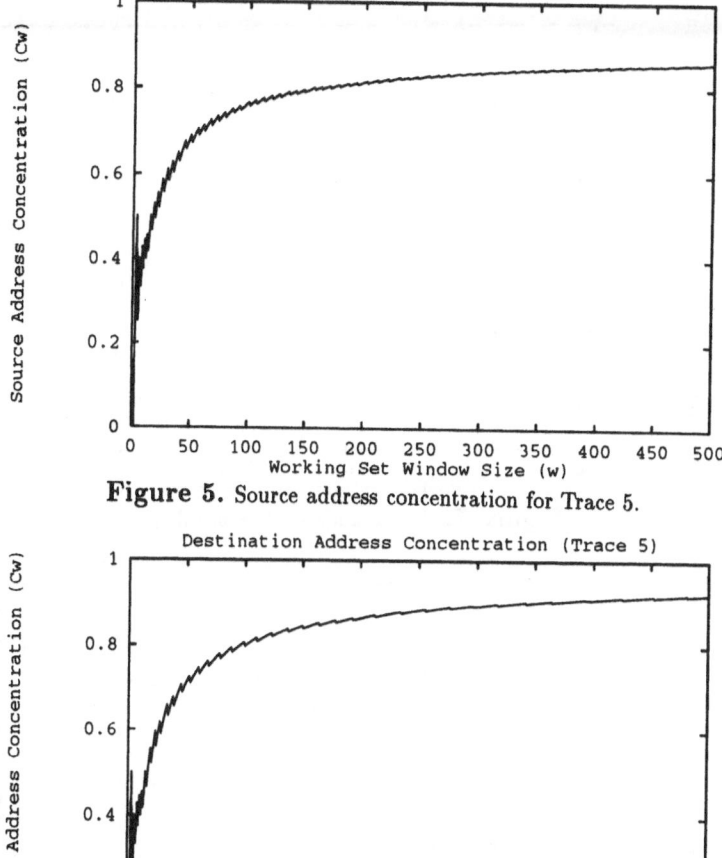

Figure 5. Source address concentration for Trace 5.

Figure 6. Destination address concentration for Trace 5.

Concentration increases with window size initially, but flattens out for larger window sizes. The concentration seen in an interval of 200 packets is almost the same as that seen in an interval of 400 packets. In other words, the rate at which new addresses appear in the trace (causing misses in a cache) is fairly constant throughout the trace. In all the traces studied, both concentration measures approach 1 as $w \to \infty$.

Destination addresses show significantly higher concentration than source addresses. For example, at a working set window size of $w = 500$, destination address concentration is 0.92, while source address concentration is 0.86. Applying the cache size estimate $(1 - C_w)w$ from Section 2.3, this means that a cache for source addresses would need to be roughly twice as large as a cache for destination addresses in order to achieve a similar hit rate. This observation is confirmed by the working set size results reported in Tables 2 and 3.

3.5. Reference Density

Reference density is defined as the tendency for a small number of hosts to account for a large proportion of the total network traffic. Figures 7 and 8 show the graph of

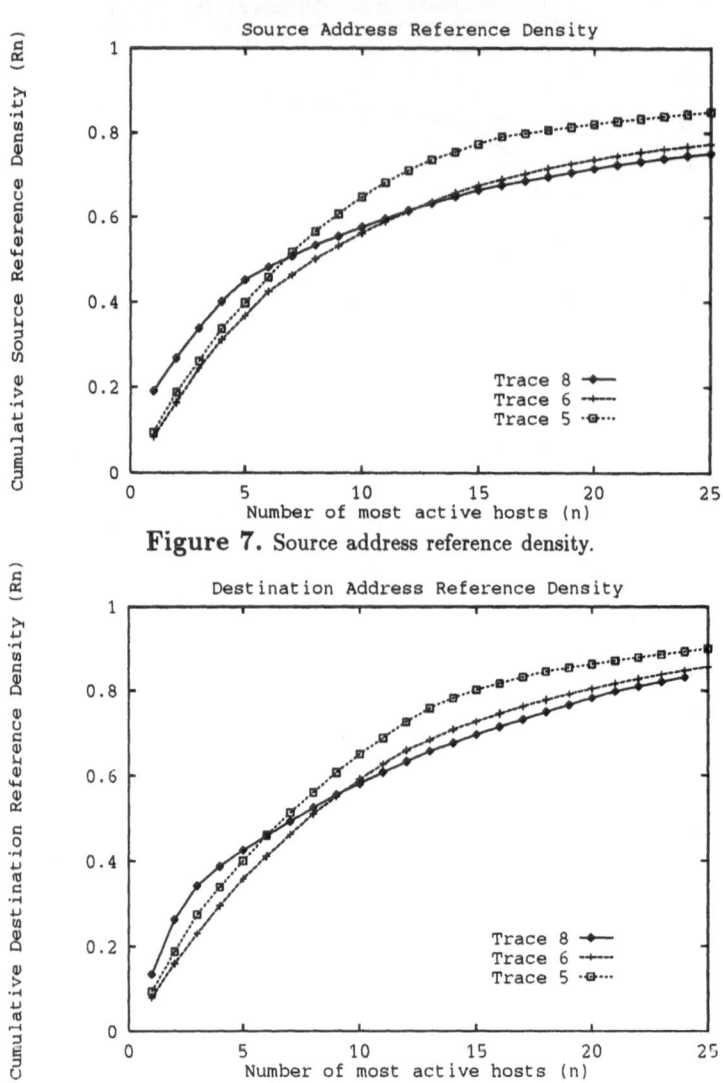

Figure 7. Source address reference density.

Figure 8. Destination address reference density.

cumulative reference density of the 25 most active hosts for the selected set of traces. Source and destination address reference densities are almost identical.

All three traces show similar behaviour for reference density. The curves show a non-linear increase that is rapid for the first few hosts and then levels off. This behaviour is a natural consequence of considering the hosts in order of activity. As more hosts are considered, their contribution to the cumulative reference density diminishes causing the gradient of the curve to decrease.

References are definitely non-uniformly distributed across the address space. In the case of Trace 5 in Figure 8, for example, the 10 most active destinations (2% of the active hosts) receive 65% of the packets, and the 25 most active destinations (5% of the active hosts) receive 90% of the packets. The high reference density values reflect the fact that there are certain hosts on the network that are contributing a large fraction of the traffic. These hosts are the file servers and a few hosts that are used by a large number of users.

The reference density values depend somewhat on the number of active hosts on the network. For example, Trace 6 has more active hosts than either Trace 5 or Trace 8, and has slightly lower reference density.

244

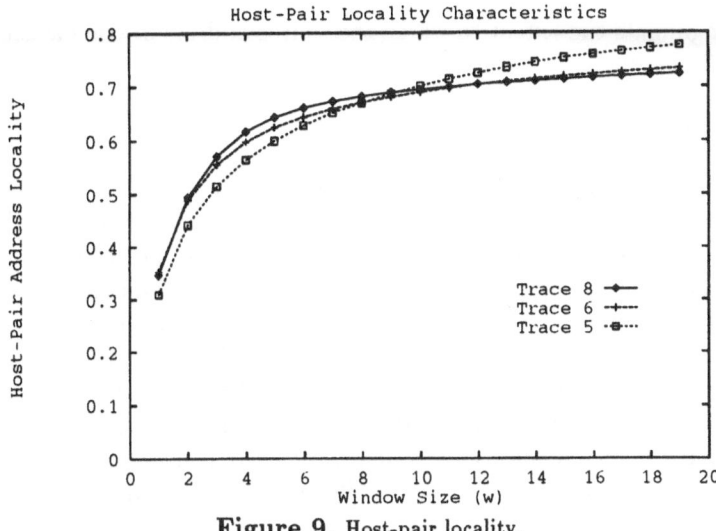

Figure 9. Host-pair locality.

Overall, our results show even higher reference density than previous studies. For example, results obtained by Jain [8] showed 5% of the hosts accounting for 55% of the traffic. Again, our higher reference density results are most likely due to the smaller number of hosts on the network considered in this study.

3.6. Host-Pair Locality

Host-pair locality is defined as the probability that at least one of the next w packets seen on the network is travelling between the same host pairs as the current packet. This measure is similar to the address re-use measures discussed previously, but includes both unidirectional and bidirectional traffic between a pair of hosts.

Results for the host-pair locality measure are shown in Figure 9. Host-pair locality increases quickly for small window sizes, and then levels off, similar to the address re-use characteristics. For small w, values for host-pair locality are even higher than the values for source and destination address re-use.

The measurement results for host-pair locality, combined with the address re-use characteristics, indicate that most of the traffic seen on the network is bidirectional. For example, consider the case where a packet has just been seen travelling from host A to host B. For Trace 5, the destination address re-use for $w = 1$ is $D_1 = 0.09$ (see Figure 4). This means that in 9% of the cases, the next packet on the network is also destined to host B (but not necessarily from host A). At the same time, the source address re-use for $w = 1$ is $S_1 = 0.24$ (see Figure 3), meaning that in 24% of the cases the next packet is destined to A (but not necessarily from B). So in 33% of the cases the next packet is destined for either A or B (but not necessarily from either B or A, respectively). The host-pair locality measure for $w = 1$ is 0.31, indicating that the next packet is between A and B 31% of the time. Thus almost all of the source address re-use is caused by bidirectional traffic. Only 2% of the time $(0.33 - 0.31)$ is address re-use for $w = 1$ caused by a packet to or from some other host on the network.

The calculation above shows that the bidirectional traffic is more than twice the amount of unidirectional traffic (0.24 versus 0.09). This result is not surprising since higher layer protocols, such as TCP, often generate such a pattern (i.e., data packets in one direction, and acknowledgement packets in the other direction, with one acknowledgement packet seen for each data packet). The presence of bidirectional traffic helps to explain the low source and destination address persistence values observed in the traffic and highlights the extent to which the source address of one packet can predict the destination address of the next one.

3.7. Summary

Several observations are evident from our locality measurement results. First, there is relatively little persistence in packet traffic on local area networks. Long sequences of persistent references are rare. Second, the lack of persistence does not mean there is no network traffic locality. Other measures, such as address re-use, do a much better job of quantifying temporal locality. Third, all traces show significant address concentration, address re-use, and reference density. The higher the address concentration, the lower the number of addresses that need to be stored in a cache to predict the destinations of future arrivals. Finally, network traffic locality is both a short-term and a long-term phenomenon. Packets on a local area network are concentrated to small subsets of the address space over significant periods of time.

4. PERFORMANCE OF A LOCALITY-BASED NETWORK INTERCONNECTION DEVICE

There are several possible applications of the phenomenon of network traffic locality [2, 6, 14, 15]. Among these are implications on the design and performance of network interconnection devices. To take specific examples, the presence of temporal locality suggests that caching recently active host addresses could be effective, and the presence of concentration suggests that only small caches would be required. This issue was investigated further.

In this section, we present a brief analysis of a data link level network interconnection device (i.e., a LAN bridge) designed to exploit network traffic locality by caching recently active host addresses. While the performance benefits achievable by exploiting locality at the data link level in a LAN bridge are minimal, the same locality-based principles can be applied at the network layer or higher layers of the OSI model, where significantly more software processing is required for each packet (e.g., routing-table lookups for IP addresses at gateways, ARP caches for IP/Ethernet address translation, or TCP packet processing). A LAN bridge is presented only as a simple example of how network traffic locality could be exploited.

A data link level interconnection device is designed to selectively forward packets (or frames) between two or more networks based on the destination address in each packet. The device contains a *forwarding table* that has a list of data link level host addresses, along with the associated network port on which each host is accessible. For each packet that arrives, the device looks up the destination address in the forwarding table, and then either forwards the packet to the appropriate network, or filters (discards) the packet.

By adding a *forwarding table cache*, it may be possible to reduce the number of forwarding table lookups, thereby reducing the forwarding delay of the device. A trace-driven simulation model was developed to estimate the performance of a data link level interconnection device with and without a forwarding table cache. In this paper, we present only two parts of this study:

(1) what replacement policy to use for the cache, and

(2) what fetch policy to use for the cache

A more complete performance analysis can be found in [7].

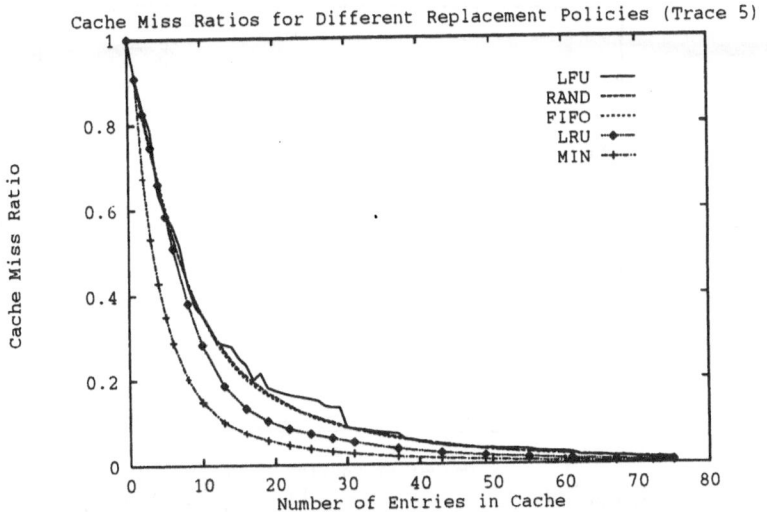

Figure 10. Cache miss ratios for LFU, FIFO, RAND, LRU and MIN (Trace 5).

The cache replacement policy determines which item to remove from a full cache to make room for a newly arriving entry. Five replacement policies were considered: LRU (Least Recently Used), LFU (Least Frequently Used), FIFO (First In First Out), RAND (Random Replacement), and MIN (Optimal Replacement). MIN provides the lower bound for a theoretically optimal cache replacement policy [1], but is not a practically implementable policy since it requires knowledge of future references.

The fetch policy determines what information to put into the cache, and when. Given that the network interconnection device has just seen a packet travelling from host A to host B, is it better to put A (the source address) into the cache, B (the destination address) into the cache, or both A and B into the cache? These fetch policies are called Source Fetch (SF), Destination Fetch (DF), and Source-Destination Fetch (SDF), respectively.

To answer the first question, the simulation program was executed for each of the traffic traces under different replacement policies and cache sizes. In all cases, Destination Fetch (DF) was used for the cache fetch policy.

The simulation results for the five replacement policies are shown in Figure 10. Since all three traces show very similar results, only Trace 5 is shown. The miss ratios drop sharply at first with increasing cache size as active entries are added to the cache. Beyond a certain cache size (here about 15), however, the additional entries retained have a much lower probability of being re-used and the drop in miss ratios becomes more gradual.

As expected, LRU consistently outperforms LFU, FIFO and RAND. LRU works well because of the temporal locality properties (persistence and address re-use) identified in Section 3. LFU performs poorly since the LFU replacement policy does not take into account the fact that a newly fetched entry has a high probability of being used in the next few references. Instead of being retained, this entry is most likely to be the one discarded from the cache when a replacement has to be made, and performance is poor. A FIFO cache does slightly better since it retains recently fetched entries until at least $n - 1$ more addresses are received, where n is the size of the cache. Again, however, the performance of FIFO is relatively poor since replacements are not done on the basis of recent usage. RAND, which indiscriminately picks an entry to be discarded when a replacement has to be done, performs as well as LFU and FIFO. Beyond a certain cache size (here 50), because of the large number of entries in the cache, the difference in performance of all policies becomes marginal.

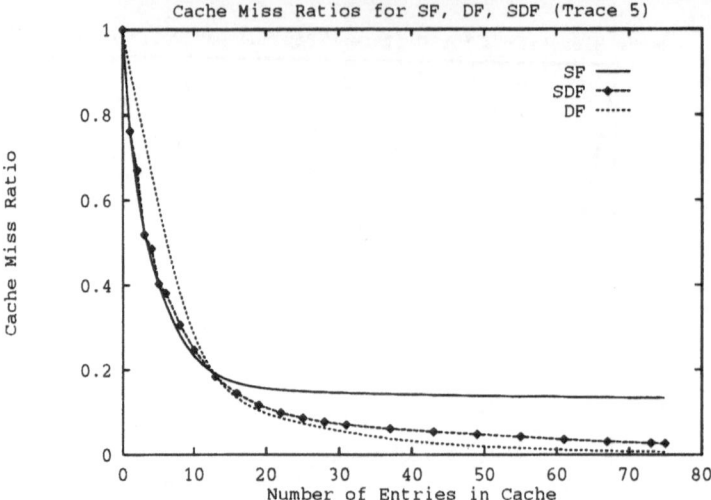

Figure 11. Cache miss ratios for SF, DF, SDF (Trace 5).

As expected, MIN outperforms all other strategies for all cache sizes, but since it looks ahead in the reference string, it cannot be implemented in a real system. Results for MIN nevertheless show that there is room for improvement in implementable policies. Although LRU is the best among the implementable policies considered, it is still sub-optimal. Here, for example, at a cache size of 20 entries, LRU has miss ratios that are 65% higher than those of the optimal strategy (0.096 versus 0.058).

The second question is what fetch policy to use for the cache (i.e., what information to put in the cache, when). The three fetch policies (SF, DF and SDF) were simulated with LRU as the replacement policy. The results are shown for Trace 5 in Figure 11. Trace 6 and Trace 8 show very similar results.

For all three fetch policies, the miss ratios fall sharply up to a certain cache size (about 15 in this example), and then decrease slowly beyond this point. For SF, the miss ratios become almost constant beyond this cache size, while the other two policies continue to show a gradual decrease.

There are two distinct regions of interest in Figure 11. For cache sizes below 15, the three fetch policies perform similarly, although SF performs marginally better than SDF, and both perform better than DF. The better performance of SF and SDF over DF in this region can be attributed to the fact that source address re-use for small window sizes tends to be higher than destination address re-use, as discussed in Section 3.3. Hence fetching just the source address into the cache for each packet received would produce fewer misses than fetching just the destination address. That is, source addresses are good predictors of future destination address references in the short-term future.

For larger caches (more than 15 entries in this example), it is more important to cache destination addresses than source addresses. Caching only source addresses (the SF policy) results in virtually no improvement in miss ratio as cache size increases. One reason for this behaviour is the fact that there are many source hosts that send very few packets. In Trace 5, for example, there are 107 hosts that send exactly one packet during the trace, and 200 hosts that send fewer than five packets. There are also some hosts that appear as destinations in the trace, but never as sources. Since SF does not fetch destination addresses into the cache, no entry for these destination hosts is ever fetched into the SF cache, and each packet going to any of these hosts results in a cache miss. DF and SDF do not face this problem.

The SDF policy provides reasonable (but not optimal) performance over the full range of cache sizes. Caching both source and destination addresses (the SDF policy) is an improvement over SF for large cache sizes, but it still does not perform as well as DF. Caching source addresses as well as destination addresses always hurts the performance for larger caches since all "predicted" destinations occupy cache slots, and some "predicted" destinations are never used. This performance degradation is due in part to lower source address concentration. In other words, source addresses are poor predictors for destination address references in the long-term future. Thus an interconnection device can benefit from address caching, but should use *both* source and destination addresses in its cache, with reasonably short timeouts on idle cache entries to keep unwanted addresses from cluttering the cache.

5. CONCLUSIONS

Packet traffic in packet-switched networks exhibits a phenomenon called network traffic locality, which is similar to the locality phenomenon observed in memory referencing and file referencing behaviour. Two dominant characteristics are present in network traffic locality. First, successive packets on the network tend to be highly related. Second, a small number of hosts tend to account for a large number of the packets transmitted on the network. These characteristics are evident as both short-term and long-term phenomena.

This paper presents four new measures that capture the salient aspects of network traffic locality. These measures are persistence, address re-use, concentration, and reference density. Packet trace data collected in a local area network environment was used to demonstrate and quantify each measure. Little persistence was found in the aggregate network traffic, but address re-use, concentration, and reference density are high, indicating significant temporal locality in packet traffic. Source addresses are good predictors of short-term future references in packet traffic, but poor predictors of long-term references.

The presence of network traffic locality has a number of potential applications, one of which has been studied in this paper. By adding a forwarding table cache to a network interconnection device, it may be possible to reduce the number of forwarding table lookups, thereby reducing the forwarding delay of the device. High concentration values mean that nominally sized caches (e.g., 20 entries) at such devices can give high hit ratios (e.g., 90%). Although this is only one example, it does suggest that the possibilities of exploiting network traffic locality should be investigated further.

One possible direction for future research is to characterize network traffic locality at the individual host level, rather than for aggregate network traffic. Such a locality model for a single host would be useful for estimating the presence (or absence) of network locality as networks increase in size and workload. It remains to be seen how important the notion of network traffic locality will be in the design and operation of future high-speed networks.

6. ACKNOWLEDGEMENTS

This research was supported, in part, by the Natural Sciences and Engineering Research Council of Canada (NSERC) under research grants OGP0121969 and OGP0003707. Financial support was also provided by the University of Saskatchewan.

REFERENCES

[1] L. Belady, "A Study of Replacement Algorithms for a Virtual-Storage Computer", *IBM Systems Journal*, Vol. 5, No. 2, pp. 78-101, 1966.

[2] L. Breslau, D. Estrin, and L. Zhang, "Exploiting Locality to Provide Adaptive Routing of Real-Time Flows in Global Internets", Proceedings of the 4th IEEE COMSOC International Workshop on Multimedia Communications, April 1992.

[3] R. Bunt and J. Murphy, "The Measurement of Locality and the Behaviour of Programs", *Computer Journal*, Vol. 27, No. 3, pp. 238-245, March 1984.

[4] P. Denning, "Working Set Model for Program Behavior", *Communications of the ACM*, Vol. 11, No. 5, pp. 323-333, May 1968.

[5] P. Denning, "Virtual Memories", *Computing Surveys*, Vol. 2, No. 3, pp. 153-189, September 1970.

[6] D. Feldmeier, "Improving Gateway Performance with a Routing-Table Cache", Proceedings of IEEE INFOCOM '88, pp. 298-307, March 1988.

[7] N. Gulati, "Local Area Network Traffic Locality: Characteristics and Application", M.Sc. Thesis, Department of Computational Science, University of Saskatchewan, July 1992.

[8] R. Jain, "Characteristics of Destination Address Locality in Computer Networks: A Comparison of Caching Schemes", DEC-TR-592, February 1989.

[9] R. Jain and S. Routhier, "Packet Trains – Measurement and a New Model for Computer Network Traffic", *IEEE Journal on Selected Areas in Communications*, Vol. SAC-4, No. 6, pp. 986-995, September 1986.

[10] S. Ling, R. Bunt, and D. Eager, "Characterising Client-Server Workload on an Ethernet", Proceedings of CMG '91, Nashville, Tennessee, pp. 59-69, December 1991.

[11] M. Lorence and M. Satyanarayanan, "IPwatch: A Tool For Monitoring Network Locality", *Operating Systems Review*, Vol. 24, No. 1, pp. 58-80, January 1990.

[12] S. Majumdar, "Locality and File Referencing Behaviour: Principles and Applications", M.Sc. Thesis, Department of Computational Science, University of Saskatchewan, August 1984.

[13] D. Mills, C. Boncelet, J. Elias, P. Schragger, and A. Jackson, "Highball: A High Speed, Reserved-Access Wide Area Network", Technical Report 90-9-3, Department of Electrical Engineering, University of Delaware, September 1990.

[14] J. Mogul, "Network Locality at the Scale of Processes", Proceedings of ACM SIGCOMM '91, Zürich, Switzerland, pp. 273-284, September 1991.

[15] C. Song and L. Landweber, "Optimizing Bulk Data Transfer Performance: A Packet Train Approach", Proceedings of ACM SIGCOMM '88, Stanford, California, pp. 134-145, August 1988.

CHARACTERIZATION OF THE TRAFFIC
AT THE OUTPUT OF A DQDB MAN

M.Ajmone Marsan[1], G.Albertengo[1], R.Lo Cigno[1],
M.Munafò[1], F.Neri[1], A.Tonietti[2]

[1] Dipartimento di Elettronica
   Politecnico di Torino
   Corso Duca degli Abruzzi 24
   10129 Torino – Italy

[2] CSELT
   Via G.Reiss Romoli 274
   10148 Torino – Italy

The first trials of the Broadband Integrated Services Digital Network (B-ISDN) will probably aim at business users, for whom connectionless data services are a major concern. Thus, the statistical characterization of the data traffic produced by the interconnection of Local and Metropolitan Area Networks (LANs and MANs) through B-ISDN is of paramount importance for the design of the first B-ISDN services, and of the gateways between LANs and MANs on one side, and B-ISDN on the other. The characterization of the traffic produced by a LAN has already been tackled by several authors. In this paper we provide a first characterization of the traffic at the output of a MAN by using a simulation approach.

INTRODUCTION

The evolution of digital networks towards high speeds is becoming a reality: the first MAN (Metropolitan Area Network) prototypes are being installed and tested, and laboratory prototypes of the first ATM (Asynchronous Transfer Mode) switches for applications in B-ISDN (Broadband Integrated Services Digital Network) are being implemented.

Although the MAN concept evolved from the commercial success of many LAN (Local Area Network) products, it is now clear that on one side MANs will provide geographical extension to LAN-type services, but on the other side they will also constitute the first wave of B-ISDN. This means that one of the first problems that have to be faced in the provision of B-ISDN services is the interconnections of MANs.

This is far from being trivial, mainly due to the fact that the MAN architecture and services were designed as an extension to a metropolitan area of the LAN architectures and services, whereas the B-ISDN architecture and services — ATM in particular — were designed as an upgrade of the ISDN services and architectures. As a consequence, while MAN services are connectionless, ATM services are connection-oriented.

Several researchers already tackled the problem of the interconnection of MANs through B-ISDN. The proposals that appeared in the literature [1, 2, 3, 4, 5, 6, 7, 8] can be classified as follows.

1. The connectionless MAN traffic is transported by special ATM connections established between all the MAN/B-ISDN gateways. Such connections can be of various nature.

   - They can be *permanent,* in the sense that they are allocated identifiers and bandwidth resources. This means that a permanent overlay network exists within B-ISDN for the transport of MAN traffic, so as to transparently support connectionless services through dedicate connections. This approach has the advantage of requiring no set-up procedures for the transfer of a connectionless data unit, but it may lead to a significant waste of bandwidth, and to the allocation of a large number of identifiers for the connections linking all pairs of MAN/B-ISDN gateways. This alternative appears to be viable only if the traffic injected into the ATM network by MAN/B-ISDN gateways is high, and if the number of such gateways remains reasonably low.

   - They can be permanent, but "sleeping," in the sense that they are allocated identifiers, but very little (or no) bandwidth. The latter is provided on demand, when the need arises, through a negotiation with the ATM network control facilities. This alternative reduces the waste of bandwidth, but has no impact on the number of identifiers necessary for the establishment of the overlay network between all gateway pairs. Furthermore, a negotiation phase is introduced, with the associated delays.

   - They can be established *on demand* between MAN/B-ISDN gateways. This means that connections are typically established for the transport of each single connectionless data unit. This alternative solves the problem of the waste in bandwidth and identifiers, but it introduces a very heavy load on the network control facilities, and possibly unacceptable set-up delays for the connectionless traffic.

2. The connectionless MAN traffic is handled within B-ISDN by some "connectionless servers". In this case a permanent overlay network exists within B-ISDN for the transport of MAN traffic, but this overlay network only links connectionless servers, whose number can be much smaller than the number of MAN/B-ISDN gateways. In order to allow the connectionless traffic to reach the first connectionless server, it is possible to either use permanent connections, or allocate them on demand. The same strategy can be applied from the destination connectionless server to the destination MAN/B-ISDN gateway. This alternative tries to reduce the inefficient use of bandwidth and identifiers by allocating a smaller number of connections that should therefore be more heavily utilized. Obviously, the price

that has to be paid is the need for the connectionless servers, and the associated routing and control functions.

3. Since both the permanent allocation of connections of adequate bandwidth, and the establishment of new connections on demand have obvious disadvantages (in the former case, resources are permanently allocated, but possibly severely underutilized; in the latter case, the connection establishment delay needed for each connectionless data unit request can be a serious impairment to the provision of connectionless services), a different alternative was also proposed [8]. It is named "bandwidth advertising", and it consists in the permanent allocation of minimum bandwidth connections, whose capacity is dynamically increased according to the needs, without however renegotiating it, but using the available bandwidth advertised by the ATM network. This scheme can work both for the direct connection of MAN/B-ISDN gateways, and for the interconnections of the connectionless servers. This alternative can have a positive effect on the bandwidth efficiency of the connections used for the provision of connectionless services, but it may require a nonnegligible overhead for the advertisement of the available bandwidth.

The relative merits of the different possible alternatives were qualitatively identified by several authors, but have not yet been quantified, because their quantification largely depends on the assumptions about the connectionless traffic to be carried through the ATM network. This requires a characterization of the traffic at the output of a MAN, since this is the traffic that constitutes the load of the connectionless service in the ATM network. In this paper we try to provide a first quantitative approach to the characterization of the traffic at the output of a MAN. Our study is performed by simulating a set of MAN configurations, and by measuring the traffic characteristics at the MAN/B-ISDN gateway input. The natural choice for the MAN architecture to be simulated was provided by the IEEE 802.6 standard for MANs, known with the name DQDB (Distributed Queue Dual Bus) [9].

THE SYSTEM CONFIGURATION

As we mentioned above, our goal is to quantitatively characterize the connectionless traffic to be carried through B-ISDN for the interconnections of MANs. In order to do this we simulate several DQDB MAN configurations, and measure the statistical characteristics of the traffic at the output of the MAN, i.e., at the input of the MAN/B-ISDN gateway. For the sake of simplicity, we always locate the MAN/B-ISDN gateway at one of the two ends of the MAN, i.e., the gateway is co-located with one of the DQDB headends, as shown in Fig. 1. This configuration permits the complexity of the simulation experiments to be greatly reduced.

The number of MAN nodes is denoted by N; $N - 1$ are the normal MAN users, while the N-th node comprises a normal user *and* the gateway.

Only the MAC protocol effect is considered within the MAN, but this is believed to be the one having the major impact on the output traffic characteristics.

The MAN *input traffic* for each one of the N normal users is assumed to be Poisson with given rate. This appears to be reasonable, since the typical MAN attachment should be either a LAN, or some other device multiplexing the traffic of many individual end users.

The packets transmitted by the users on the MAN (more precisely, the service data units associated with the MA_DATA.request primitives issued by the LLC sublayers of the user protocol piles) are either taken to be of fixed length, or to obey a bimodal

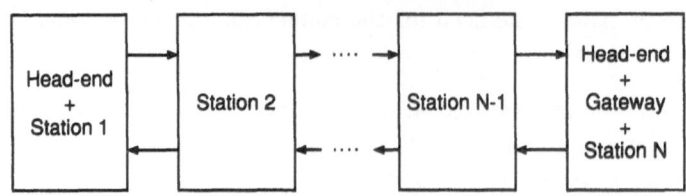

Figure 1: The considered MAN configuration with the gateway at one extreme

distribution, the latter being probably more realistic [10], but the former allowing a simpler interpretation of some performance parameters. Packets at the MAN input are segmented into an appropriate number of cells that then access the QA (Queue Arbitrated) slots on the channel, according to the DQDB protocol. It is assumed that no isochronous traffic is present in the MAN, i.e., no prearbitrated (PA) slots are used.

Each packet transmitted on the MAN has a specific destination, that can be either a local MAN address or a remote MAN address. In the former case, the packet remains in the MAN, and is delivered to one of the users of the local MAN (internal traffic). In the other case, the packet must be delivered to a user of a remote MAN, and thus must be locally delivered to the MAN/B-ISDN gateway (external traffic). The external traffic constitutes a fraction α of the total traffic in the MAN, and is obviously always directed to the gateway. The internal MAN traffic constitutes a fraction $1 - \alpha$ of the total traffic in the MAN, and is assumed to be uniform, i.e., each source user selects the destination of its transmissions according to a uniform distribution spanning the remaining $N - 1$ local MAN users. This means that each packet originated from user i, and remaining in the MAN, is addressed to user j, with $j \neq i$, with probability $1/(N - 1)$.

This assumption, together with the location of the gateway at one of the two MAN extremes, has the effect of producing a trapeze-shaped traffic along the bus that leads to the gateway. Indeed, on the bus leading to the gateway (bus A), user i generates a traffic

$$\rho_{Ai} = \alpha + \frac{(1 - \alpha)(N - i)}{N - 1}$$

Instead, on the other bus (bus B), user i generates a traffic

$$\rho_{Bi} = \frac{(1 - \alpha)(i - 1)}{N - 1}$$

Observe that, altogether, each user generates the same amount of traffic that was normalized to unit value ($\rho_i = \rho_{Ai} + \rho_{Bi} = 1$, $\forall i$). This behaviour is graphically illustrated in Fig. 2.

Note that Fig. 2 illustrates only the portion of traffic generated by the local MAN users, neglecting the traffic originated by remote MAN users, that is transported by B-ISDN to the local gateway, and injected into the local MAN. Under symmetry assumptions, the total exogenous traffic that the gateway injects in the local MAN equals α. Thus, the traffic on the two busses is symmetric, contrary to what appears in Fig. 2.

The MAN/B-ISDN gateway receives streams of cells that form packets addressed to remote MAN users. This means that the gateway must forward these informations to B-ISDN. Whether the packets must be reassembled or not, before forwarding the information to B-ISDN, depends on the gateway implementation.

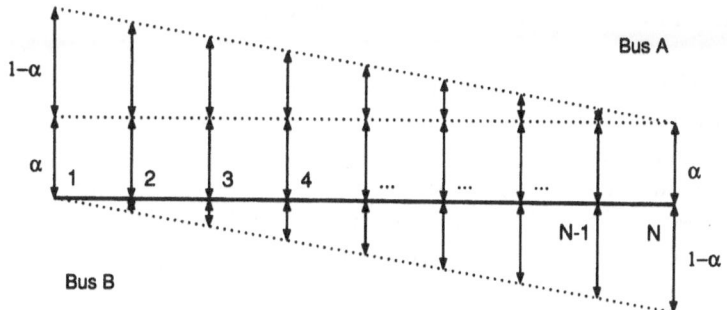

Figure 2: Graphical illustration of the traffic patterns on the two busses

PERFORMANCE INDICES

Several performance indices are obtained from the simulation. They are briefly described here, together with their relevance.

Distribution of the cell interarrival time at the gateway – This is the distribution of the number of slots between two consecutive arrivals at the gateway of cells to be forwarded to B-ISDN. A value 1 means that two consecutive slots carry cells for the gateway. The value of this parameter relates to the data acquisition speed at the gateway.

Distribution of the packet head interarrival time at the gateway – This is the distribution of the number of slots between two consecutive arrivals at the gateway of cells that carry the first segment of a packet. A value 1 means that two consecutive slots carry the first segment of two different packets to be forwarded to B-ISDN. This parameter measures the packet interdeparture time distribution from the MAN, referring to the packet head. Referring to the packet head is necessary when slots are forwarded to B-ISDN *without* packet reassembly.

Distribution of the packet tail interarrival time at the gateway – This is the distribution of the number of slots between two consecutive arrivals at the gateway of cells that carry the last segment of a packet. This parameter measures the packet interdeparture time distribution from the MAN, referring to the packet tail. Referring to the packet tail is necessary when slots are forwarded to B-ISDN *after* packet reassembly.

Statistics on the number of simultaneous receptions at the gateway – The simulator measures the average and 99-th percentile of the number of packets being at any given time received at the gateway (i.e., the average and 99-th percentile of the difference between the number of packet heads and packet tails to be forwarded to B-ISDN). These parameters measure the number of reassembly machines used at the gateway.

Packet length distribution upon arrival at the gateway – This is the distribution of the number of slots between the reception at the gateway of the head and the tail of a given packet. This parameter (by comparison with the distribution of the number of cells produced by the segmentation of a packet) provides an estimate of the spread of the packet due to the multiplexing on the MAN with transmissions by other users (in other words, this is a metrics of the implicit traffic shaping provided by the DQDB MAC protocol.)

THE SIMULATION INPUT PARAMETERS

Several DQDB MAN configurations were simulated, and the performance indices described in the previous section were measured. We summarize here the characteristics of the simulation experiments.

As we already mentioned, the MAN/B-ISDN gateway was always located at one of the two extremes of the MAN, as shown in Fig. 1. The number of MAN nodes was taken to be either 15, or 30, or 90, with equal spacing over a distance of about 15 km in the first two cases, and about 150 km in the third case. This means that in the simulations the parameter N can take either the value 15, or 30, or 90. Each node has a buffer of 125,000 bytes to store incoming packets. When this buffer is full, incoming packets are lost. The data rate on each bus was always taken to be 150 Mbit/s. The slot size is taken to be 53 bytes, 44 of which can be actually used to carry user data. This means that, when the bus length is 15 kilometers, about 20 slots are at any given time propagating along each bus (the speed of the signal on the bus is taken to be $2 \cdot 10^8$ m/s). Users have active insertions on the two busses: the processing time at each user (node traversal delay) is assumed to be equivalent to 50 bit times.

The MAN input traffic, for each one of the N normal users, was assumed to be Poisson, as already mentioned, with a rate set to values such that the normalized overall traffic generated by the local users in the MAN, ρ, equals six different values: 0.12, 0.48, 0.96, 1.067, 1.133, 1.2 (where a unit traffic is defined as the one that saturates one of the two busses, hence equal to 150 Mbit/s). Note that these traffic values refer to the net user-generated traffic, excluding the overhead in each cell. Thus, a packet comprising a given number of bits is first filled so that the number of bits is a multiple of 32, then integrated with a 32-bit address field, and a 64-bit reassembly field. The resulting frame is divided into segments whose length cannot exceed 44 bytes (possibly with a waste in the last segment); then, a 9-byte header is added to each segment to form a cell. The motivation for the selection of those values is as follows. An overall traffic load 0.12 resulted from a "reasonable" estimate of the possible traffic in a business district of the city of Rome [11]. Since this traffic load was too low to observe interesting behaviours on the MAN, it was scaled by factors 4 and 8, thus producing the loads 0.48, and 0.96. Furthermore, the behaviour close to the MAN saturation point was considered to be of interest. This led to the choice of the traffic values 1.067, where the three configurations with 15, 30, and 90 users are stable; 1.133, where only the configuration with 15 users is stable; and 1.2, where all three configurations are in overload conditions.

The packets transmitted on the MAN are either taken to be of fixed length, equal to 10,000 bits (producing 29 cells), or to obey a bimodal distribution, where 60% of the packets have length 10,000 bits, and the other 40% have length 512 bits (producing 2 cells).

Half the packets have a local MAN address, and the other half have a remote address. This means that half the packets transmitted on the MAN are internal traffic, and half are external traffic, and must be locally delivered to the MAN/B-ISDN gateway. Hence, the parameter α that was introduced in the previous section always assumes the value $\alpha = 0.5$. This assumption produces net loads of the bus leading to the gateway equal to $\frac{1}{2}(1 - \alpha)\rho + \alpha\rho$, i.e., equal to 0.09, 0.36, 0.72, 0.8, 0.85, 0.9, respectively.

NUMERICAL RESULTS

Fig. 3 shows results referring to the distribution of the cell interarrival time at the gateway, in the case of 30 users, and bimodal packet lengths, for the six considered load values.

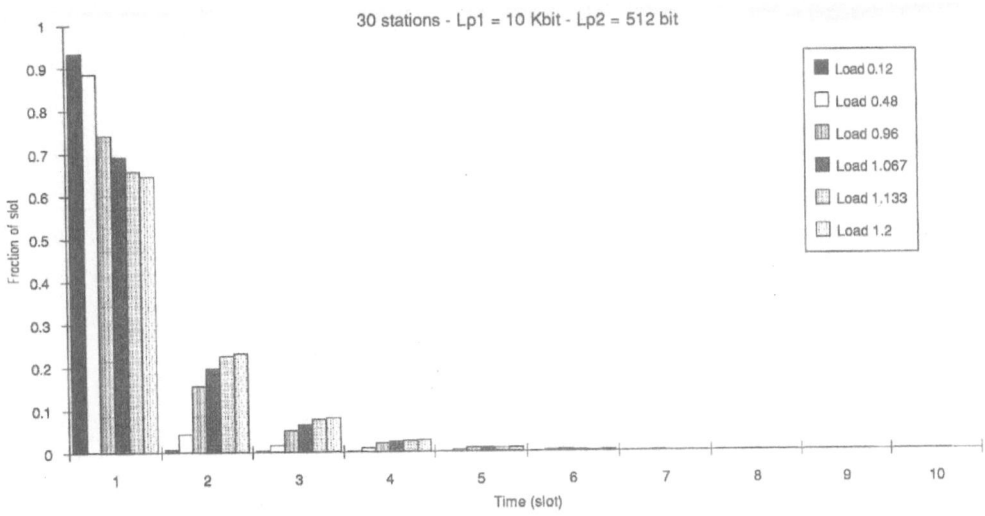

Figure 3: Distribution of the cell interarrival time at the gateway expressed in number of slots in the case of 30 users, and bimodal packet lengths, for the six considered load values

We can observe that, as expected, for low loads the probability of arrivals in consecutive slots is close to unity, and decreases for increasing load. At the same time, the probability of arrivals spaced of 2, 3, 4, 5 slots grows. The probabilities of slot arrivals spaced by 6 or more slots is negligible. One important consideration for the design of the MAN/B-ISDN gateway arises from this performance figure: however light the load of the MAN may be, slots arrive at the gateway in clusters, so that the gateway receiver must be capable of working at the full speed of the MAN. Results remain quite similar for a different number of users, and packet length distribution.

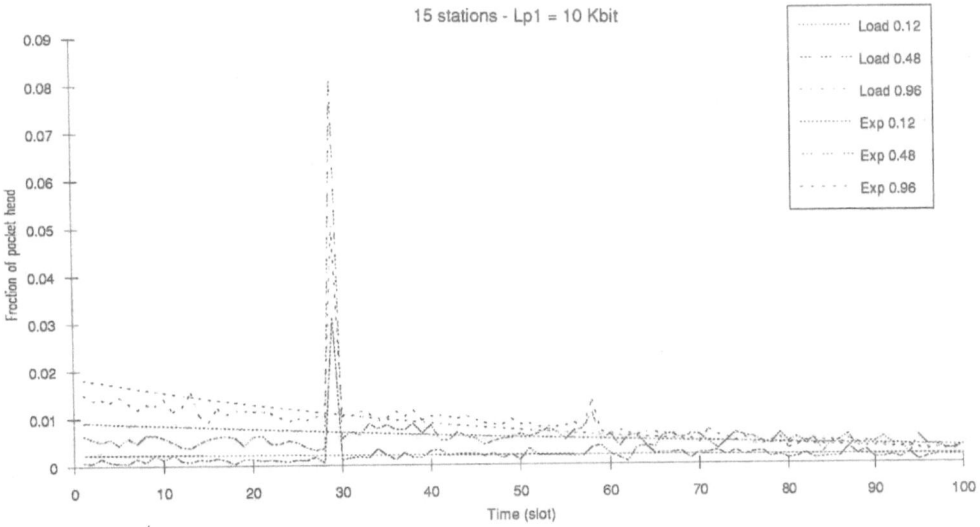

Figure 4: Distribution of the packet head interarrival time at the gateway expressed in number of slots for light loads in the case of 15 users and constant packet length. Exponential distributions with the same mean of the distributions obtained with simulation are also plotted.

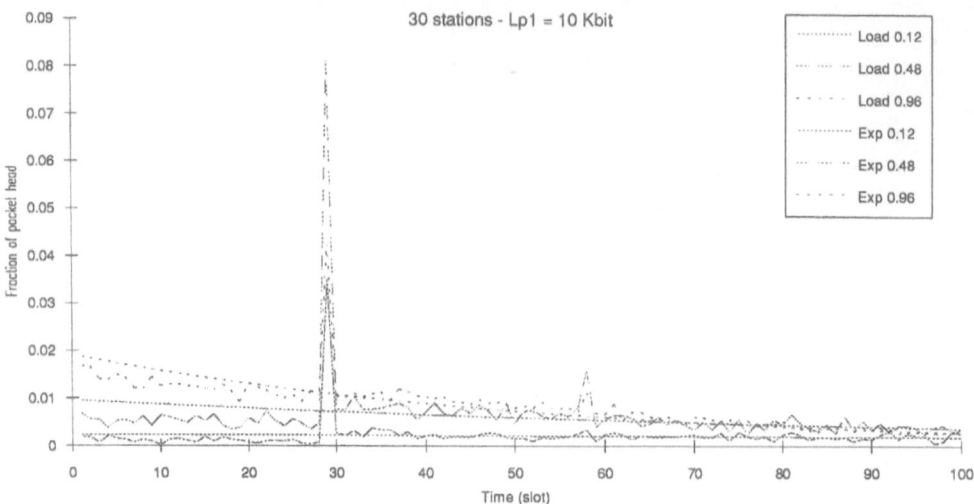

Figure 5: Distribution of the packet head interarrival time at the gateway expressed in number of slots for light loads in the case of 30 users and constant packet length. Exponential distributions with the same mean of the distributions obtained with simulation are also plotted.

Figures 4 and 5 show the distribution of the packet head interarrival time at the gateway for the three load values 0.12, 0.48, and 0.96, i.e., for the three lower loads, in the case of 15, and 30 users, and constant packet length. Fig. 6 shows the same distribution in the case of 90 users for the three load values 0.48, 0.96, and 1.067. The peaks corresponding to the packet length in slots are clearly visible (at 29 slots). Furthermore, we can also see lower peaks at 58, i.e., at twice the packet length in slots. These values are due to two events.

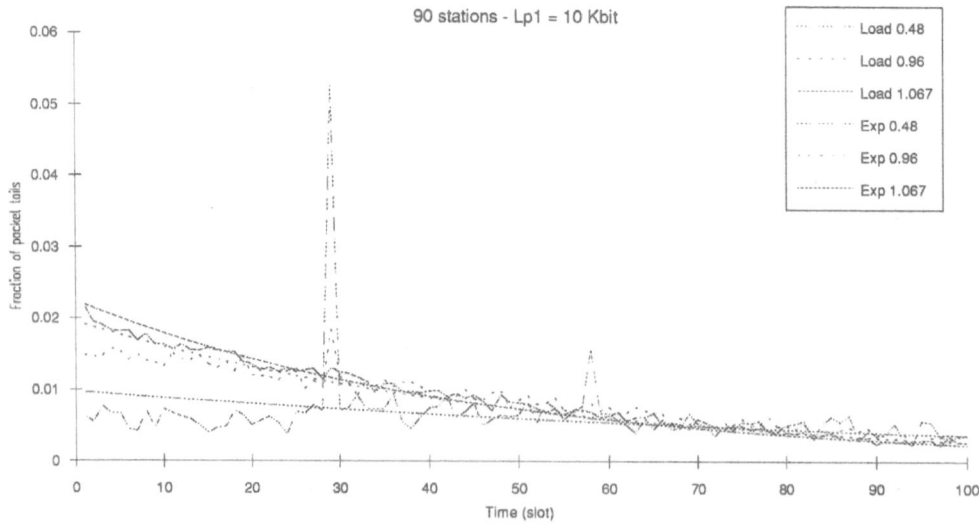

Figure 6: Distribution of the packet head interarrival time at the gateway expressed in number of slots for light loads in the case of 90 users and constant packet length. Exponential distributions with the same mean of the distributions obtained with simulation are also plotted.

1. The successive transmissions of packets buffered at different users: a packet head interarrival time equal to 29 slots may be generated when a downstream user finds the bus busy due to a transmission by one upstream user, and must wait until the end of the upstream packet, before starting its own transmission. This behavior is due to the peculiar characteristics of DQDB, that allows users starting transmission when the bus is empty, to hog the whole bandwidth, when propagation delays are large. A packet head interarrival time equal to 58 slots may be generated in a similar case, when two upstream users prevent the downstream user from transmission.

2. The successive transmissions of packets buffered at the same user: a packet head interarrival time equal to 29 slots may be generated by two consecutive transmissions, when no interference from other users is experienced. A packet head interarrival time equal to 58 slots may be generated by three consecutive transmissions.

This also explains why the largest peak value is obtained at medium load (0.48): at lower loads the probability of having several packets queued for transmission at the same time is lower, and at higher loads the probability of transmitting with no interference from other users is lower.

When bimodal packet length distributions are used, peaks are visible at the packet length values (2 and 29 slots), at twice the packet length values (4 and 58), and at the value resulting from the overlapping of a short and a long packet (31). In this case we only report the results for the case of 30 users, in Fig. 7.

Next we consider the three higher loads. The peaks corresponding to the packet lengths in slots are still clearly visible, but a substantial mass of the probability distribution is now present also for other values. Again, we only report the results for the case of 30 users, in Fig. 8.

It is interesting to observe that, except for the peaks, the distribution shape closely resembles an exponential, both at light and at heavy loads, as can be easily seen

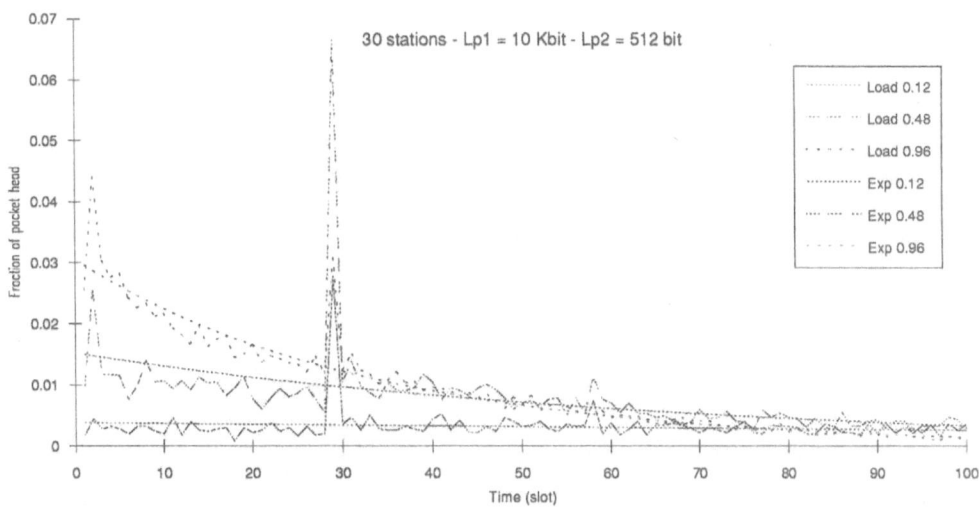

Figure 7: Distribution of the packet head interarrival time at the gateway expressed in number of slots for light loads in the case of 30 users and bimodal packet length. Exponential distributions with the same mean of the distributions obtained with simulation are also plotted.

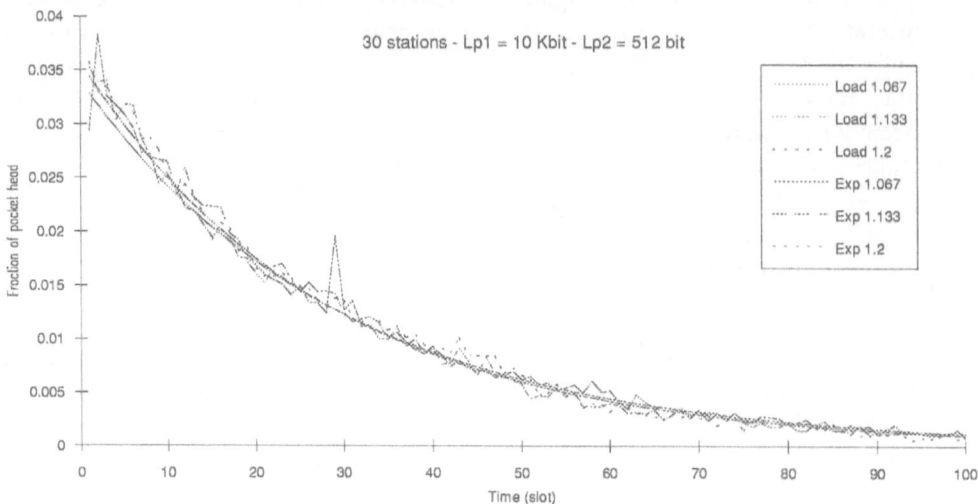

Figure 8: Distribution of the packet head interarrival time at the gateway expressed in number of slots for heavy loads in the case of 30 users and bimodal packet length. Exponential distributions with the same mean of the distributions obtained with simulation are also plotted.

from the figures, where exponential distributions with the same mean of the curves obtained by simulation are also plotted. The exponential matches the experimental results particularly well after the peak, whereas the experimental results are somewhat smaller than the exponential just before the peak. This confirms the previous statement that the peak is generated by consecutive transmissions resulting from close arrivals.

Figure 9 shows the distribution of the packet tail interarrival time at the gateway for the three load values 0.12, 0.48, 0.96, i.e., for the three lower loads in the case of

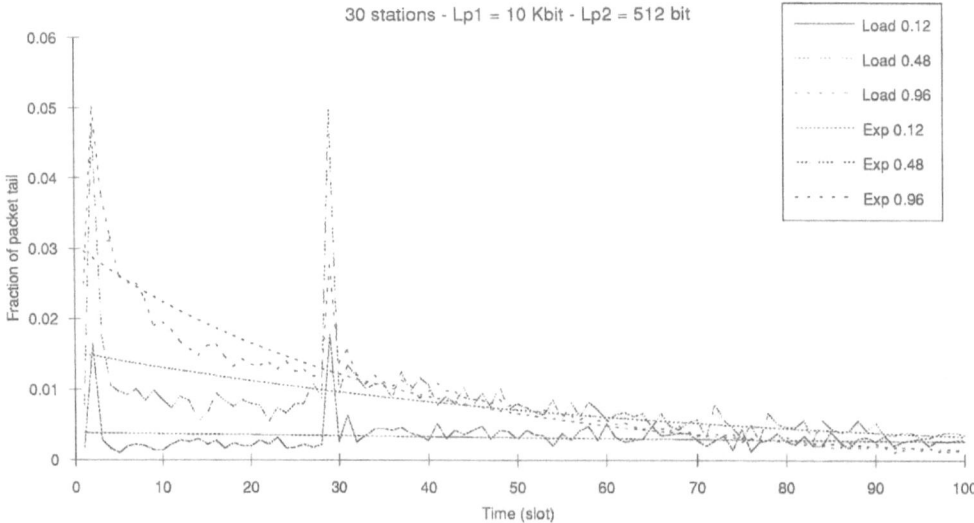

Figure 9: Distribution of the packet tail interarrival time at the gateway expressed in number of slots for light loads in the case of 30 users and bimodal packet length. Exponential distributions with the same mean of the distributions obtained with simulation are also plotted.

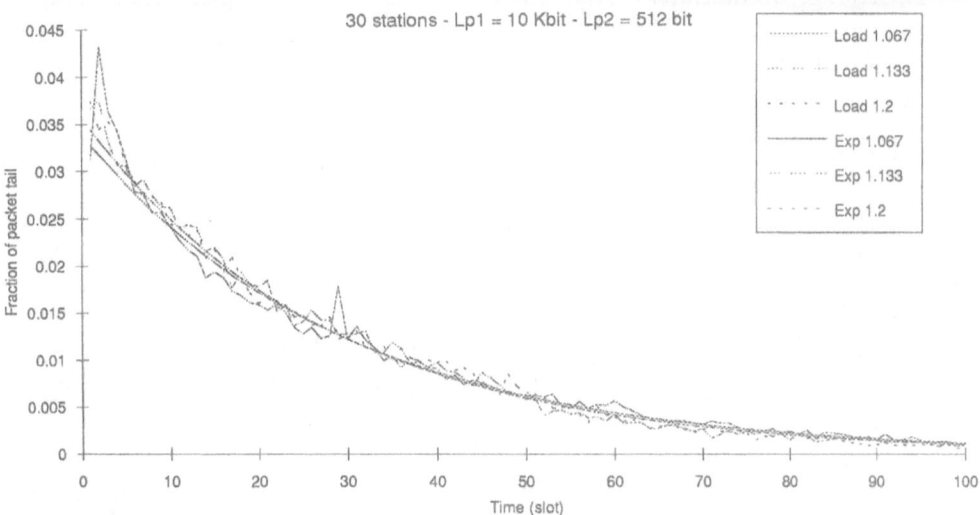

Figure 10: Distribution of the packet tail interarrival time at the gateway expressed in number of slots for heavy loads in the case of 30 users and bimodal packet length. Exponential distributions with the same mean of the distributions obtained with simulation are also plotted.

30 users and bimodal packet length distribution. Once more, the peaks corresponding to the packet lengths in slots are clearly visible, but the distributions now show more significant contributions between the peaks than the equivalent distributions measured with respect to the packet heads.

Figure 10 shows the distribution of the packet tail interarrival time at the gateway for the three load values 1.067, 1.133, 1.2, i.e., for the three higher loads in the case of 30 users and bimodal packet length distribution. The curves are rather similar to the ones in Fig. 8. Figure 11 shows the distribution of packet tail interarrival time at the

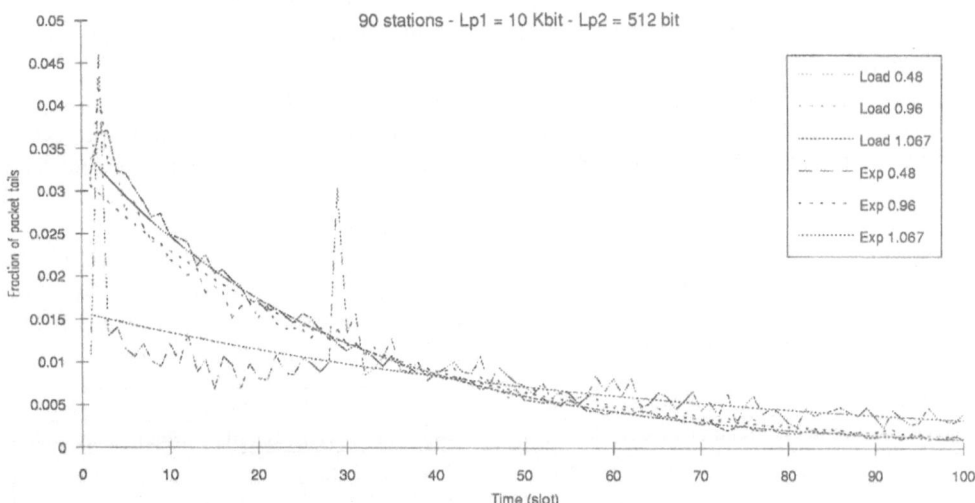

Figure 11: Distribution of the packet tail interarrival time at the gateway expressed in number of slots for loads 0.48, 0.96, and 10.67, in the case of 90 users and bimodal packet length. Exponential distributions with the same mean of the distributions obtained with simulation are also plotted.

Table 1: Average overall interarrival time of packets directed to the gateway (Input Mean), and mean and standard deviation of the packet heads interarrival time measured at the MAN/B-ISDN gateway, for the six considered load values, in the case of constant packet length, for 15, 30 and 90 users

Input Traffic	Input Mean	15 Users		Input Mean	30 Users		Input Mean	90 Users	
		Mean	St.dev.		Mean	St.dev.		Mean	St.dev.
0.12	421.19	421.18	387.58	408.62	408.58	377.94	400.37	400.68	372.34
0.48	108.60	108.51	106.24	104.09	103.74	103.78	102.21	102.16	100.70
0.96	54.10	53.92	48.79	52.45	52.27	47.25	51.11	51.11	47.68
1.067	48.86	48.72	42.88	47.31	47.18	42.93	46.28	46.15	43.69
1.133	46.11	45.99	40.89	44.72	44.58	42.16	44.23	44.05	43.58
1.2	45.42	45.25	41.79	45.08	44.92	43.04	44.61	44.61	43.96

gateway for the three loads 0.48, 0.96, and 1.067, in the case of 90 users and bimodal packet length. Once more, the curves are rather similar to the ones in Fig. 8.

Also in this case, except for the peaks, the distribution shape closely resembles an exponential, both at light and at heavy loads, as can be easily seen from the figures, where exponential distributions with the same mean of the curves obtained by simulation are also plotted.

Since the results in Figs. 4–11 indicate that it is possible to approximately characterize the traffic at the MAN output with the superposition of a negative exponential and some pulses, it is interesting to compare the means and standard deviations of the negative exponential distributions characterizing the input and output traffics.

In Table 1 we present the mean and standard deviation of the measured interarrival time at the MAN/B-ISDN gateway of packet heads, while in Table 2 the results for the packet tails are reported, for the six considered load values, in the case of constant packet length, for 15, 30, and 90 users. These values are compared with the packet generation rates measured at the output of the simulated generators. The similarity in the values of the average and standard deviation confirms the impression that the simulation results are closely matched by an exponential distribution.

Similar results can be observed in Tables 3 and 4, that refer to the case of bimodal packet length distribution.

The remaining results refer to the packet length distribution upon arrival at the gateway, and to the average and 99-th percentile of the number of simultaneous receptions at the gateway. Since considering all combinations of the parameters would lead

Table 2: Average overall interarrival time of packets directed to the gateway (Input Mean), and mean and standard deviation of the packet tails interarrival time measured at the MAN/B-ISDN gateway, for the six considered load values, in the case of constant packet length, for 15, 30 and 90 users

Input Traffic	Input Mean	15 Users		Input Mean	30 Users		Input Mean	90 Users	
		Mean	St.dev.		Mean	St.dev.		Mean	St.dev.
0.12	421.19	421.15	387.02	408.62	408.59	377.02	400.37	400.62	371.09
0.48	108.60	108.52	103.10	104.09	103.69	100.73	102.21	102.14	99.07
0.96	54.10	53.91	44.57	52.45	52.25	43.11	51.11	50.97	45.22
1.067	48.86	48.71	39.92	47.31	47.17	40.08	46.28	46.15	42.69
1.133	46.11	45.99	39.43	44.72	44.58	41.88	44.23	44.12	43.40
1.2	45.42	45.26	41.75	45.08	44.93	43.11	44.61	44.50	44.13

Table 3: Average overall interarrival time of packets directed to the gateway (Input Mean), and mean and standard deviation of the packet heads interarrival time measured at the MAN/B-ISDN gateway, for the six considered load values, in the case of bimodal packet length distribution, for 15, 30 and 90 users

Input Traffic	Input Mean	15 Users		Input Mean	30 Users		Input Mean	90 Users	
		Mean	St.dev.		Mean	St.dev.		Mean	St.dev.
0.12	280.08	279.89	279.44	260.24	260.20	257.03	253.37	253.34	248.85
0.48	67.55	67.43	66.68	65.77	65.63	63.17	63.75	63.37	61.32
0.96	33.98	33.84	32.00	32.82	32.67	31.25	31.89	31.74	30.43
1.067	30.44	30.32	29.44	29.51	29.39	28.63	28.76	28.63	27.97
1.133	28.77	28.68	28.81	28.06	27.94	27.95	27.78	27.63	27.22
1.2	28.37	28.23	29.41	27.98	27.84	27.87	27.89	27.74	27.29

Table 4: Average overall interarrival time of packets directed to the gateway (Input Mean), and mean and standard deviation of the packet tails interarrival time measured at the MAN/B-ISDN gateway, for the six considered load values, in the case of bimodal packet length distribution, for 15, 30 and 90 users

Input Traffic	Input Mean	15 Users		Input Mean	30 Users		Input Mean	90 Users	
		Mean	St.dev.		Mean	St.dev.		Mean	St.dev.
0.12	280.08	279.94	279.55	260.24	260.24	256.95	253.37	253.31	248.85
0.48	67.55	67.43	66.13	65.77	65.64	63.16	63.75	63.37	62.28
0.96	33.98	33.84	31.89	32.82	32.67	31.03	31.89	31.74	30.98
1.067	30.44	30.32	29.61	29.51	29.39	28.72	28.76	28.63	28.47
1.133	28.77	28.68	29.22	28.06	27.94	28.29	27.78	27.67	27.48
1.2	28.37	28.24	29.65	27.98	27.85	27.88	27.89	27.76	27.22

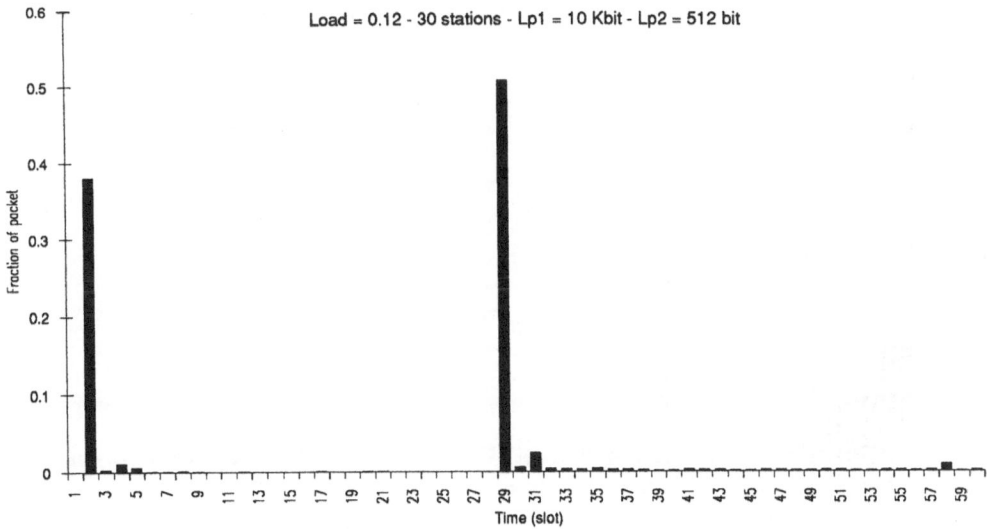

Figure 12: Distribution of the packet length upon arrival at the gateway expressed in number of slots for load = 0.12 in the case of 30 users and bimodal packet length.

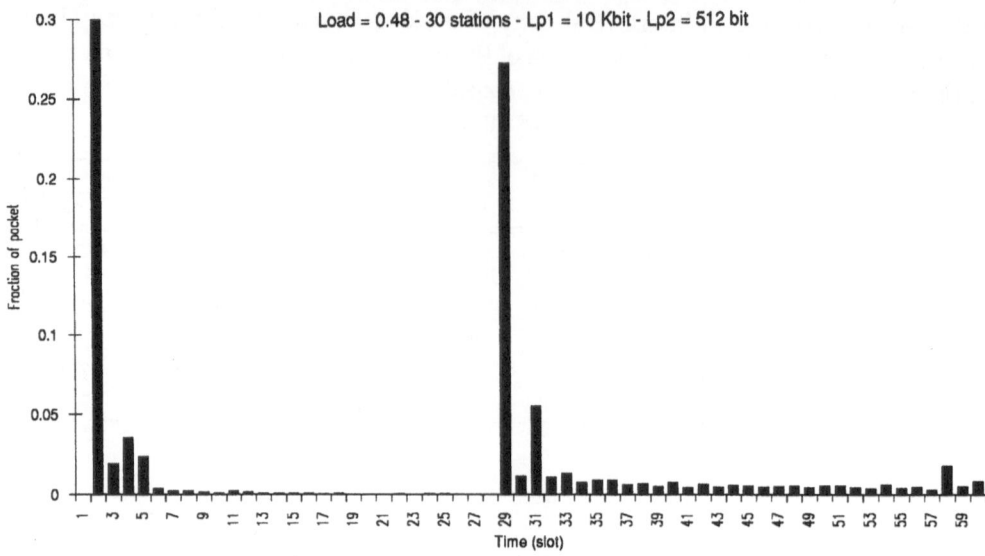

Figure 13: Distribution of the packet length upon arrival at the gateway expressed in number of slots for load = 0.48 in the case of 30 users and bimodal packet length.

to an excessive number of figures, we focus on the case of 30 users and bimodal packet length distribution. Fig. 12 shows the results for the load value 0.12. In this case the packet length upon arrival at the gateway is with very high probability equal to the original packet length in slots (2 slots with probability 0.4, and 29 slots with probability 0.6; note how the percentages are closely maintained). The only other relevant peaks are at twice the packet length (i.e., 4 and 58), or at the length resulting by the superposition of a long and a short packet (i.e., 31). In this case the average number of simultaneous receptions at the gateway is much less than one, and the 99-th percentile is equal to 1.

If we increase the load to 0.48, we obtain the results in Fig 13. Now, both packet types arrive at the gateway with no interference with probability close to 0.3, but, since

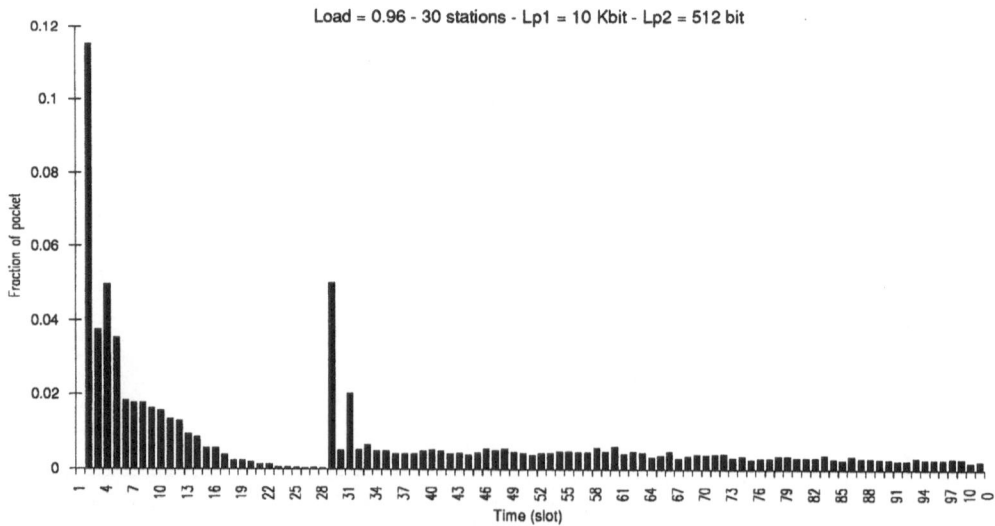

Figure 14: Distribution of the packet length upon arrival at the gateway expressed in number of slots for load = 0.96 in the case of 30 users and bimodal packet length.

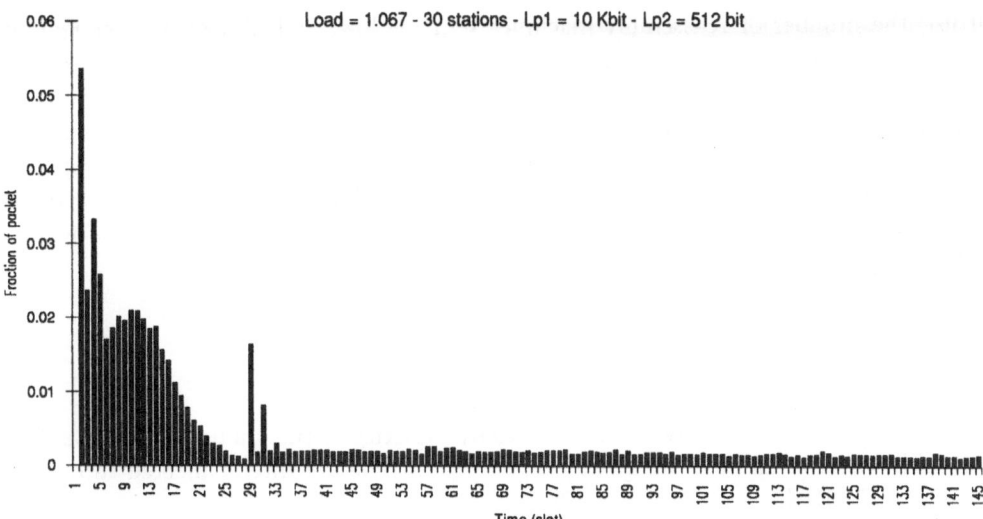

Figure 15: Distribution of the packet length upon arrival at the gateway expressed in number of slots for load = 1.067 in the case of 30 users and bimodal packet length.

the longer packets were originally more numerous, this means that the "no interference" probability for short packets is higher, as intuitively expected. Moreover, the peaks at 4, 31, and 58 can still be observed. In this case the average number of simultaneous receptions at the gateway is close to 0.4, and the 99-th percentile equals 2.

Further increasing the load to 0.96, we obtain the results in Fig 14. Now, long packets arrive at the gateway without interference with very low probability, and the tail of their distribution is quite long. Instead, the tail referring to short packet is practically ended before the value 29. In this case the average number of simultaneous receptions at the gateway is close to 2.3 and the 99-th percentile equals 9.

With a load value equal to 1.067, we obtain the results in Fig 15. We can see that the behaviour is still such that the "no interference" peaks are clearly visible, but their values have become very small, and the distribution tails have become extraordinarily long. At this load the average number of simultaneous receptions at the gateway is close to 4.8, and the 99-th percentile equals 14.

Since the 99-th percentile of the number of simultaneous receptions at the gateway is particularly useful for the design of the gateway hardware (it measures the number of reassembly machines that are simultaneously active in the gateway), we collect the results for this parameter in Table 5. It can be observed that up to moderate bus load

Table 5: 99-th percentile of the number of simultaneous receptions at the gateway in the cases of 15, 30, and 90 users, and of constant and bimodal packet length, for the six considered load values

Total load	Bus load	15 nodes constant	15 nodes bimodal	30 nodes constant	30 nodes bimodal	90 nodes constant	90 nodes bimodal
0.12	0.09	1	1	1	1	1	1
0.48	0.36	2	2	3	2	3	3
0.96	0.72	7	7	10	9	15	16
1.067	0.8	9	9	14	14	32	31
1.133	0.85	11	11	21	21	57	60
1.2	0.9	12	12	20	21	60	64

(0.36), the number of reassembly machines remains low, and almost independent of the number of nodes in the network. When congestion is approached, the number of reassembly machines gets close to the number of nodes in the network, if the number of nodes is low (12 reassembly machines with 15 nodes). Instead, when the number of nodes is larger, the number of reassembly machines simultaneously active remains significantly less than the number of network nodes (21 reassembly machines with 30 nodes, and 64 with 90 nodes). This is due to the limited multiplexing allowed by the DQDB protocol during the transmission of one packet.

CONCLUSIONS

We have quantitatively studied the characteristics of the traffic at the output of a DQDB MAN in an environment where several MANs are interconnected through a wide area B-ISDN.

The configuration of the system that was simulated was based upon a possible scenario in a business district of Rome; since the resulting load was low, the system was analized also for higher loads, up to saturation, in order to study the gateway requirements to insure proper operation under any condition.

Several performance figures were obtained, ranging from the cell interarrival time distribution at the gateway, to the spreading of the packet length upon arrival at the gateway due to the interleaving of the traffic generated from different users on the MAN. All the figures show that the traffic at the gateway maintains the key characteristics of the input traffic, at least until the MAN is brought to saturation, so that packets begin to be lost due to the limited dimension of the user buffers.

Two important parameters for the design of the MAN/B-ISDN gateway emerged from the simulation results: the required gateway speed, and the number of necessary reassembly machines within the gateway (if packets must be reassembled before trasmission on the B-ISDN). Results show that the gateway should be able to work at the full speed of the MAN, even if the MAN itself is lightly loaded. On the contrary, at reasonable loads, a small number of reassembly machines (less that half the number of users) seems sufficient for the gateway operations.

ACKNOWLEDGMENTS

This work was performed in the framework of a research contract between Politecnico di Torino and CSELT (Centro Studi e Laboratori Telecomunicazioni).

References

[1] K.Yamazaki, Y.Ikeda, "Connectionless Cell Switching Schemes for Broadband ISDN," XIII International Switching Symposium, Stockholm, Sweden, May 1990

[2] W.R.Byrne, H.J.Kafka, G.W.R.Luderer, B.L.Nelson, G.H.Clapp, "Evolution of Metropolitan Area Networks to Broadband ISDN," XIII International Switching Symposium, Stockholm, Sweden, May 1990

[3] J.P.Quinquis, A.Lespagnol, J.Francois, G.Gastaud, "Data Services and LANs Interconnection Using ATM Technique," XIII International Switching Symposium, Stockholm, Sweden, May 1990

[4] E.Tirtaatmadja, R.A.Palmer, "The Application of Virtual Paths to the Interconnection of IEEE 802.6 Metropolitan Area Networks," XIII International Switching Symposium, Stockholm, Sweden, May 1990

[5] A.Biocca, G.Freschi, A.Forcina, R.Melen, "Architectural Issues in the Interoperability Between MANs and the ATM Network," XIII International Switching Symposium, Stockholm, Sweden, May 1990

[6] M.De Prycker, "Impact of Data Communication on ATM," ICC '89, Boston, MA, USA, June 1989

[7] T.Van Landegem, R.Peschi, "Managing a Connectionless Virtual Overlay Network on Top of an ATM Network," ICC '91, Denver, CO, USA, June 1991

[8] P.Crocetti, G.Gallassi, M.Gerla, "Bandwidth Advertising for MAN/ATM Connectionless Internetting," INFOCOM '91, Bal Harbor, FL, USA, April 1991

[9] IEEE P802.6 - Metropolitan Area Networks, *Distributed Queue Dual Bus (DQDB) Subnetwork of a Metropolitan Area Networks (MAN)*, Draft Standard - Version D15, October 1990

[10] R.Gusella, "The analysis of diskless workstation traffic on an Ethernet," University of California, Berkeley, Internal Report UCB/CSD 87/389

[11] A.Ciccardi, A.Tonietti, "Scenario di Traffico per Reti MAN," CSELT Technical report, November 19, 1991, in Italian

INTERWORKING FUNCTIONS EFFECT ON BANDWIDTH ALLOCATION

FOR LAN-BISDN INTERCONNECTION

I.S.Venieris, J.-A.Sanchez-Papaspiliou, and M.E.Anagnostou

National Technical University of Athens
Electrical and Computer Engineering Department
Computer Science Division, Telecommunications Laboratory
GR-15773, 9 Heroon Polytechniou St., Zographou, Athens, Greece

Abstract

The interconnection of remote Local Area Networks (LANs) through an Asynchronous Transfer Mode (ATM) based Broadband Integrated Services Digital Network (BISDN) is considered as an imminent interworking application. Major aspects to be considered regarding this application are processing speed of the interworking unit (IWU) for mapping LAN frames to ATM cells and ATM bandwidth allocation for LAN traffic transfer. In this paper we show that these factors are highly correlated. Through an accurate LAN user model, we quantify bandwidth and buffer requirements for the LAN-BISDN interconnection application, taking into account the smoothing effect in the LAN traffic injected in BISDN by the inter-working functionality. This becomes more emphatic under time-consuming, low-cost, firmware-based implementations of the interworking protocols, which however are not bottle-necks in the context of the interconnection of low-speed existing LANs with BISDN.

1. INTRODUCTION

A major issue of current concern in the telecommunication community is the interconnection of networks following different technological and operational principles. The reason is mainly due to the plethora of existing networks with diverse characteristics and size limitations, which cannot be abandoned regarding the cost of introducing a totally new universal network. In this respect, interworking is the most efficient method for achieving integrated communication capability in a diverse environment [1]. From the telecommunication service user perspective interworking should be performed transparently, that is, the involving communication procedures, when passing from one network to the other, should look the same as the communicating users were belonging to the same network.

Interconnection of heterogeneous networks is enabled through the development of appropriate interworking protocols and resource management mechanisms. In general sense the interworking protocols should provide conversion of PDUs (Protocol Data Units) among networks implementing different protocols. The layer in which interworking protocols operate depends on the level of compatibility of the interconnected networks, as well as on the specific communication scenario. The most demanding case of interworking requiring complex protocol functionality is observed when users of heterogeneous networks want to communicate. In the case of similar but remote network users, the interworking protocol functions are restricted to those required for the transfer of information through an intervening Wide Area Network (WAN). In this paper we consider this case of interworking, for the application of ETHERNET LAN interconnection through an ATM based BISDN.

LAN interconnection is considered as a very demanding application due to the wide exploitation of computer networks in business environment. The need to extend LANs in larger geographical areas can be seen in the light of the user requirement for access and retrieve of remotely hosted information. Through the use of virtual circuit based backbone WANs, remote LAN user communication is achieved in a cost effective manner as opposed to the use of leased lines. Furthermore interconnected LANs of the same enterprise may form a private network superposed to the WAN backbone, which can be administered and managed by a particular operating authority [2]. It is true that BISDN provides all those functions necessary to support the LAN interconnection application [3], [4], [5]. Since in the case considered in this paper the BISDN is used as a transit network for LAN traffic the related interworking protocols perform solely adaptation of the ETHERNET MAC PDUs to ATM PDUs. The protocol stack of the IWU performing relay of information between LANs and BISDN follows to the CCITT view for connectionless data service transfer in BISDN [6], [7], [8]. We will focus on the interworking mechanism and protocol issues in Section 2.

The most simple resource management method for LAN interworking applications is to allocate WAN bandwidth equal to the LAN capacity. This mechanism however results in poor utilization of the WAN resources and additionally in unfair charging. The reason is twofold: The LAN users are typical Variable Bit Rate (VBR) sources with bursty characteristics. On the other hand, one cannot ignore the smoothing effect of the IWU intervention to the LAN traffic injected to the WAN. To take into account both factors contributing in the proper definition of the bandwidth value allocated for LAN interconnection, a LAN user model is required together with the IWU protocol processing times. Both are essential for dimensioning the buffers of the IWU performing the protocol conversion functions [9].

To efficiently emulate the traffic injected by a LAN to a BISDN backbone more sophisticated models than the ones currently used in the literature are required. The Poisson assumption and the fixed size of LAN frames are clearly inadequate for the objective of IWU dimensioning and BISDN bandwidth definition. This is the view expressed in [10], [11] even for compound Poisson arrivals. Towards an accurate representation of a LAN user, a user model is developed. The model recognizes the multiplicity of services available to LAN users and effectively represents the bursty periods of a LAN source. Furthermore a general distribution describing the length of LAN frames is introduced.

The paper is organized as follows. In Section 2 the LAN-BISDN interworking scenario is presented and the functionality of the IWU is defined. Section 3 covers the modelling part of this study. User, LAN, IWU models are presented and protocol processing times are extracted through a definition of procedural parameters. In Section 4 we present numerical results based on simulations adopting the models of section 3 for a range of traffic values. Within this section the IWU dimensioning is performed and the BISDN bandwidth required for the specific interworking application is extracted. This is in turn used for dimensioning of a policing mechanism operating in the entrance to the BISDN. Conclusions are summarized in Section 5.

2. LAN INTERWORKING THROUGH ATM NETWORKS

The interconnection of existing LANs following the IEEE 802.x standards [19] is considered as an important application for the forthcoming ATM based BISDN. From the

BISDN standpoint, the interconnection application complies to the "support of broadband Connectionless (CL) data service on BISDN" as exemplified in CCITT draft recommendation I.364 [8]. In this specification, connectionless data (the case of LAN traffic) is handled by specific designated nodes of the ATM network able to support Connectionless Service Functions (CLSFs). CLSFs are interconnected through Virtual Paths (VPs) which are set-up and released by management or signalling procedures depending on the CL traffic load volume. CLSF nodes terminate CL protocols and route information based on layer 3 addresses. Depending on the underlying ATM Adaptation Layer (AAL) protocols, information need not be reassembled in the CLSF nodes but instead routed on the fly. This is achieved through a set of labels present in each segment of a message, which are mapped to addresses and outgoing VP Identifiers (VPIs). The CL protocols of the CLSF need not be the same with the LAN CL protocols. Hence protocol conversion is performed in the LAN-BISDN IWUs for LAN interconnection applications. Options regarding label translation and investigation of the IWU protocol complexity for both cases of protocol encapsulation and/or concatenation have been thoroughly studied in [3], [5], [9].

CLSFs are not the only way for the support of "broadband CL data service". In case of LAN communication with well-known traffic volume, semi-permanent ATM VPs may be used to interconnect LANs. These VPs are then administered by management procedures performed by the operator of a private enterprise network. Provisions for such inter-connection mechanisms are included in CCITT recommendations I.211 and I.327 [6], [7]. LAN information is routed within ATM by VP cross-connects and CL protocols are terminated outside the BISDN. This simplicity in information routing does not require the wide number of labels per segment as it is the case when using CLSFs. Simple AAL protocols may be employed in the IWUs for mapping LAN frames to cells and vice versa. Layer 3 addresses are mapped to VPIs while the Virtual Channel Identifier (VCI) of the ATM cell may be used to code other LAN protocol related information that will be interpreted only in the IWUs. Several scenarios for VCI/VPI allocation supporting point-to-point or point-to-

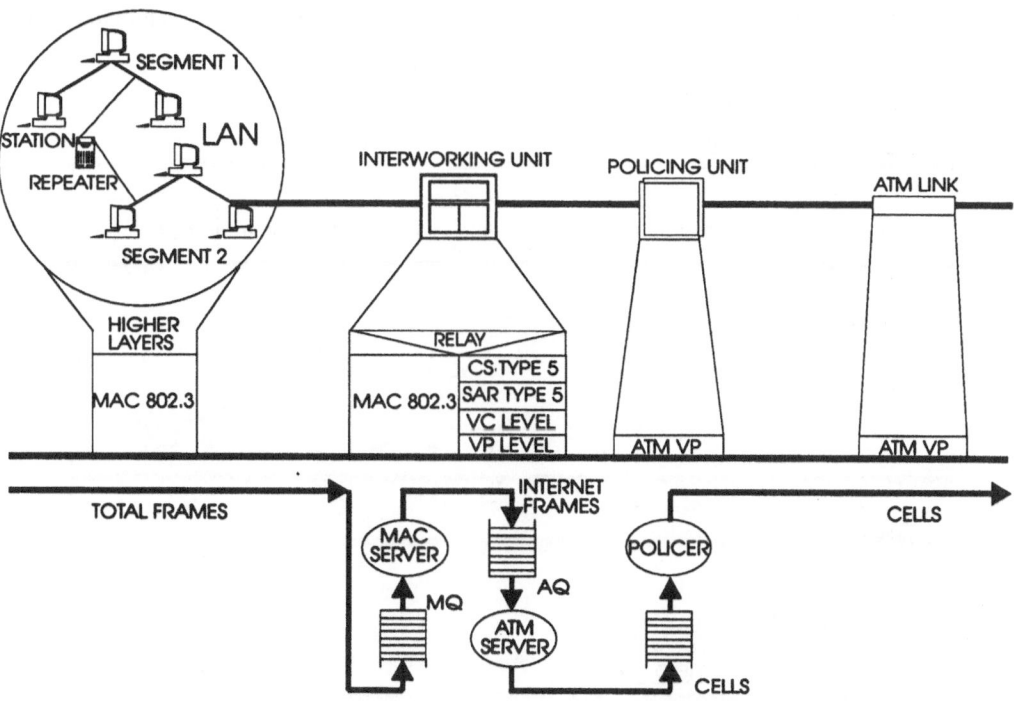

Figure 1. Physical configuration - Queueing model

multipoint connections are presented in [3], [12] all putting their emphasis on relaying addressing problems while reducing the size of the headers.

In this paper, we concentrate on the indirect provision of CL data service support by BISDN for LAN interconnection. CLSFs are not considered. The physical set-up of the configuration studied is shown in figure 1. Remote ETHERNET LANs are assumed to communicate through an intervening ATM network. The IWU performs encapsulation of MAC frames to AAL PDUs after checking the Destination Address (DA). It acts as a filtering bridge that forwards only frames with not local DAs. In the most general case the relay protocol of the bridge (denoted as R in figure 1) could perform a protocol translation functionality among different MAC protocols so that interconnection of heterogeneous LANs (i.e. LANs following different MAC standards of the IEEE 802.x series) is supported. Such unified MAC protocols are presented in [12].

The AAL protocols used in the IWU for this interconnection application are the AAL type 5 protocols currently under standardization in CCITT [13]. The AAL PDUs are illustrated in figure 2. AAL type 5 is decomposed in a Convergence Sublayer (CS) and a Segmentation and Reassembly Sublayer (SAR). In contrast to SAR type 4 devoted to operate on CLSFs, SAR type 5 performs segmentation of Common Part CS PDUs (CPCS PDUs) to 48 octet segments dispensing with the need of the 4-octet SAR PDU headers and trailers. This results in higher bandwidth utilization [14], [15]. Information is protected by a 32-bit CRC operating on a CPCS PDU basis. The CPCS PDUs are padded so that the SAR SDU size is always a multiple of 48 octets. After segmentation the ATM header is assigned. VPI/VCI values depend on the label allocation scheme employed and in all cases enable delivery of cells to the appropriate destination IWU. It is possible that more than one LAN share the same VP towards a given destination and more than one VPIs are allocated to a LAN for different destinations [16].

In our study we assume that the total internet traffic of an ETHERNET LAN is injected into a single VP of the ATM network after passing through the IWU. The IWU architecture (see figure 1) is based on two functions: the LAN interface (MAC Queue (MQ) and server) and the ATM interface (AAL+ATM Queue (AQ) and server). The LAN interface stores all frames that have been transmitted on the LAN in the MQ and looks for the destination MAC

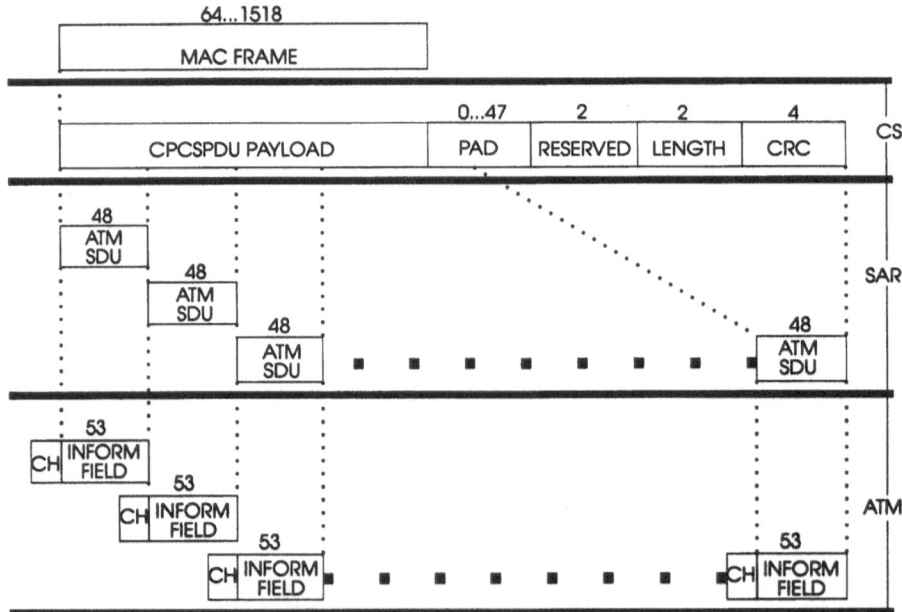

Figure 2. AAL type 5 protocols.

address to determine whether that frames are internet or intranet. Intranet frames are discarded and internet frames are forwarded to the ATM interface. The ATM interface stores internet frames in the AQ and provides AAL type 5 and ATM functionality. Internet frames are segmented into fixed-size cells and forwarded over the defined VP. In the following sections we investigate the required IWU buffer space and determine the VP bandwidth value for LAN interconnection.

3. MODELLING

3.1. LAN Traffic Characteristics

One of the events that cannot be foreseen with accuracy is the human volition to do something at a given time. We can verify the appearance of this conjecture in the shape and the characteristics of the traffic in an ETHERNET cable. This traffic shows an enormous variability on time scales ranging from milliseconds to months with relatively low mean utilization and high burstiness. The dominant traffic variability is in the range of milliseconds but very low frequency variability is not negligible. This seems to be an intrinsic quality of data traffic due to LAN and application changes. LAN environments have changed dramatically in the past few years due to technological advances in computer communication and software engineering. Therefore one must consider both low frequency and high frequency components, since data traffic does not have a predictable level of busy period traffic. Traffic "spikes" ride on longer-term "ripples" that in turn ride on still longer term "swells" as spelled out in [11], [17]. All these features make LAN traffic to be an important test of the performance of any high-speed integrated-service network.

3.1.1. LAN User Source Model and Microscopic Parameters. Typical data traffic models provide little variability at time scales longer than the duration of packet trains. This property constrains theoretical analysis and simulations. The user traffic model presented in this paper exhibits the required properties. It is an application of a new source model whose details are given in appendix A. The total traffic transmitted in the LAN that is the aggregation of the traffic generated by the users after passing the LAN MAC protocol displays characteristics similar to real LAN traffic.

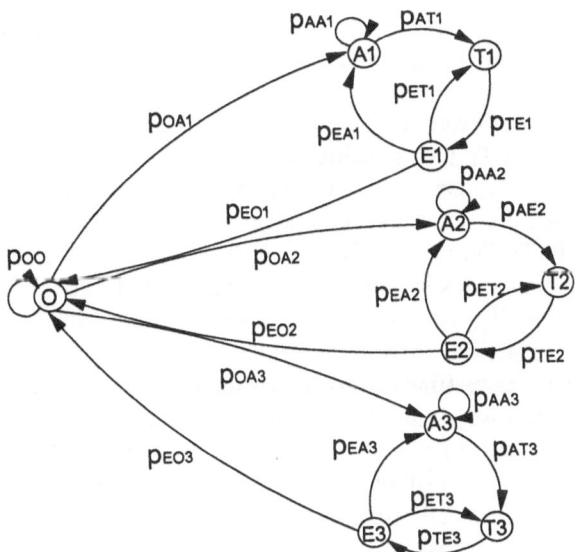

Figure 3. LAN user source model.

Table 1. Microscopic parameters

Notation	Description
$P_{ET1}, P_{ET2}, P_{ET3}$	Transition probabilities from states E_1 to T_1, E_2 to T_2, E_3 to T_3.
$P_{EO1}, P_{EO2}, P_{EO3}$	Transition probabilities from states E_1, E_2, E_3 to O.
$P_{OA1}, P_{OA2}, P_{OA3}$	Transition probabilities from state O to states A_1, A_2, A_3.
$P_{AA1}, P_{AA2}, P_{AA3}$	Transition probabilities from states A_1, A_2, A_3 to themselves.
pdf_{Ti}, pdf_{Ei}	Distributions for the time spent in states T and E

The model provides a simple way for characterizing the behavior of a telecommunication user and is able to describe the telecommunication session by adopting a multilevel view starting from the user activity and concluding with the traffic profile of the telecommunication service. The model accommodates user transitions from one service to another and respects the different duration of calls depending both on the specific service as well as the particular user. Finally from the traffic engineering point of view the model fully describes services with widely accepted traffic parameters; i.e. peak transmission rates, mean transmission rates, activity ratio, call rates, duration of an active period, burst rate in an active period, interburst gap time between bursts belonging to the same active period, number of bursts in an active period, duration of a burst period, frame transmission rate in a burst period, interframe gap time between frames belonging to the same burst period, number of frames transmitted in a burst period.

The model is general enough to fit a variety of different services by appropriately varying the values of the parameters defining it. It is simple enough because it is described by a small number of parameters and the defining parameters make a good intuitive sense by being closely related to actual physical quantities. In addition the model is easily amenable to analysis and it can be easily implemented. The application of this model for a LAN user is depicted in figure 3.

The model involves a semi-Markov chain. The time spent in states O and A_i (i=1,2,3) is geometrically distributed. The time spent in states T_i is defined by the CSMA/CD protocol, and the time spent in states E_i is defined by the frame transmission time. The microscopic parameters of the model that we must define are the independent transition probabilities of the chain and the distributions for the time spent in states T_i and E_i. Table 1 lists the microscopic input parameters. In the model we can see that three types of cycles are formed. These cycles are formed from the transition edges and each one hides a different conception that will be revealed. The small cycles are formed from the transition edges of states E_i to states T_i and backward. The intermediate cycles that include the small ones are formed from the transition edges of states A_i to states T_i, then transition edges from states T_i to E_i and finally transition edges from states E_i back to states A_i. The big cycles that include the small and the intermediate ones are formed from the transition edges from state O to states A_i, then transition edges from states A_i to T_i, T_i to E_i and finally from states E_i back to state O.

States E_i are the only states in which traffic is generated (frames). The time spent in the other three kinds of states T_i, A_i, and O represent silent periods (time units). The structure of the model enables these periods to play the role of the time regulators for the traffic generated in states E_i. Hence rates (frames/time units) or frequencies or macroscopic parameters of different scales can be easily introduced to the microscopic parameters of the model if we define the duration of the silent periods.

The total time spent in a small cycle, when one of these is reached, represents a burst period of a service, since the small cycles are translated into the highest rates or frequencies in the model. A burst period consists of consecutive frame transmissions separated by the interframe gap time (the time spent in states T_i). The total time spent in an intermediate cycle, when one of these is reached, represents an active period of a service, where traffic is generated with some features depending on the silent periods in states A_i and T_i. The intermediate cycles are translated into intermediate rates or frequencies in the model. An active period consists of consecutive bursts separated by the interburst gap time (the time spent in states A_i). The time spent in state O represents the inactive period of the source and regulates the call rates of the services - the smallest rates or frequencies in the model.

The existence of more than one big cycle that includes intermediate and small cycles is indispensable for the representation of a source that generates traffic with different features over different time periods. A LAN terminal that supports applications, processes and services such as: file transfer and access protocols, graphical applications, word processing, electronic mail, remote data base access, etc. is a source of this type. We categorize these services in three general classes, therefore we introduce three big cycles in the model, each one generating traffic with different features.

As already mentioned traffic is generated only in states E_i and each transition to these states from states T_i, since this is the only way, defines the transmission of a frame. As we can see there are no transitions from states E_i to themselves. So after the transmission of a frame in state E_i there are three different options:

a) Transition to state T_i. This means that the burst period will continue with the transmission of another frame since the only way out from state T_i is state E_i. The time spent in states T_i is the interarrival time regulator.

b) Transition to state A_i. This means that the burst period has finished but not the active period since the only way out from state A_i is state T_i and finally state E_i. The time spent in states A_i is the interburst time regulator.

c) Transition to state O. This means that the active period of a service is finished and the source will sleep for a while. The time spent in state O is the intercall time regulator.

The microscopic input parameters are easily defined by the transformation equations given in appendix A, for the general model, using as input the macroscopic parameters of a given source. We define the macroscopic parameters for the LAN user source by a qualitative analysis based on measured LAN traffic [11], [17]. Therefore the LAN model must be very close to real avoiding simplifications leading to wrong results.

3.1.2. LAN Model. For the ETHERNET LAN we use the 10BASE5 medium specification of IEEE 802.3 [18], [19]. Complying to the 802.3 standard, the LAN parameters to be used in our performance study are recapitulated in table 2. We assume that our source model lies in the Data Encapsulation component of the ETHERNET LAN functional architecture. Hence it generates directly MAC frames, each one consisting of a MAC header (destination address, source address, data length indicator), trailer (frame check sequence) and an information field (the LLC packet). The MAC frames of a particular LAN user access shared medium using the mechanism and procedures specified in the IEEE 802.3 standard.

3.1.3. Macroscopic Parameters. Observed Traffic in an ETHERNET LAN [11], [17] and measurements in our campus show that for the total traffic the mean monthly utilization exceeds 15% of the network capacity with daily peak hours exceeding 30% and peak minutes exceeding 50%. For the internet traffic that is important for our study, the mean daily utilization fluctuates in the level of 1% with peak traffic minutes in 14% and seconds in 20%. The frame length follows the distribution depicted in figure 4 with an average value of 430 bytes. The interarrival times of frames in the medium are smaller than 5msec with a probability of 95% for 15% mean utilization. Obviously most of the assumptions used in previous studies regarding network configurations and workload characteristics are no longer valid [20].

Table 2. Specific LAN parameters

Description	Value	Units
Number of segments	2	
Coaxial cable propagation velocity	0.85c	m/sec
Transceiver cables propagation velocity	0.75c	m/sec
Transceiver cables length	25	m
Transceiver transmit path delay	0.35	microsec
Transceiver receive path delay	0.5	microsec
Transceiver collision path delay	0.5	microsec
Encoder delay	0.1	microsec
Decoder delay	0.3	microsec
Preamble generation delay	6.4	microsec
Preamble removal delay	6.4	microsec
Carrier sense delay	0.1	microsec
Repeater path delay	0.5	microsec

On the basis of the data given above we qualitatively resort to the macroscopic parameters of table 3. The exact meaning of macroscopic parameters is given in appendix A. These values are sufficient for the definition of the microscopic parameters based on the transformation equations given in appendix A. The qualitative analysis for deriving the macroscopic parameters is now presented.

The transmission time of a frame is analogous to the frame length and follows the same distribution as the frame length. This distribution is the distribution of the time spent in states E_i and is the same for all the services. The transmission rate (10Mbits/second) and the mean frame length values define the macroscopic parameters CD_i.

The distribution for the time spent in states T_i is defined by the interaction of the CSMA/CD protocol and the traffic the sources generate. Therefore it is dependent on the traffic volume and shape and as such its definition becomes complex. However the parameters CT_i (average gap time between two successive frame transmissions) that we really need, are defined as the sum of three components: (1) the minimum interframe gap time defined in the 10BASE5 specification, (2) the time needed to transmit the preamble and the start byte also defined in the 10BASE5 specification and (3) the average time collisions delay a frame transmission. The average time that a collision experiences is dependent on the one way propagation delay in the specific ETHERNET LAN channel, the jam time complying to the

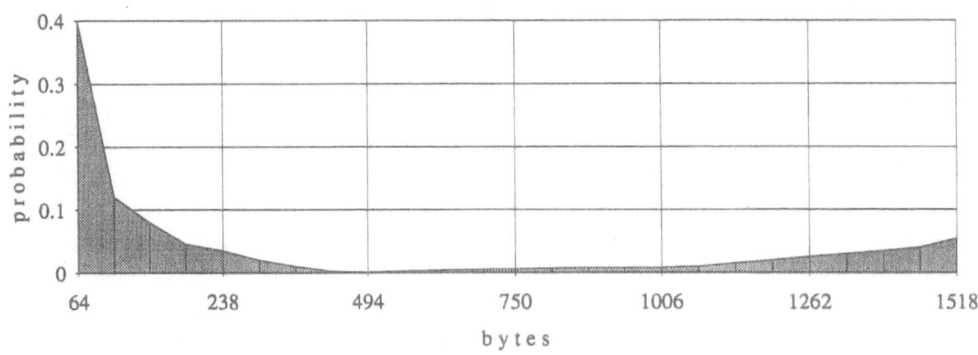

Figure 4. Frame length distribution.

Table 3. Macroscopic parameters

Notation	Description	Service 1	Service 2	Service 3	Units
CD	Transmission time of a frame	344	344	344	microsec
CT	Interframe gap time	18 for 15% utilization	18 for 15% utilization	18 for15% utilization	microsec
C	Number of frames belonging to the same burst period	1	2	3	
BT	Inter-burst gap time	2	0.2	0.002	sec
AD	Activity period duration	60	60	60	sec

10BASE5 specification and also components (1) and (2). For our LAN configuration it becomes equal to 26.7 microseconds. Finally the number of collisions before a successful transmission is dependent on the traffic volume and is given by [22].

The C_i values are defined by the peak rates in time intervals from 1ms to 10ms, under the condition the model achieves the minimum interarrival times of frames in the medium as defined by the CSMA/CD protocol and consequently maximum allowed rates by the same protocol in these time intervals (high frequencies). The BT_i values are defined by the interarrival times of frames in the medium, under the condition the model (considering also the interaction with the CSMA/CD protocol) produce a distribution where 95% of its values are lower than 5msec (intermediate frequencies). The definition of AD_i values accommodates the requirement of monotonically increase of dispersion throughout a time span of five orders of magnitude. The AD_i values together with the activity ratio (AR) define the time that the source will remain in a state (low frequencies). The activity ratio together with C_i and BT_i values define the peak transmission rates of the source. This explains why an AR=1/15 has been chosen. Finally equal call rates for the three services are assumed.

With these parameters the sources generate traffic with an average rate of 3.18 frames/second. When the source is active generates 0.49 frames/second in service 1, 9.96 frames/second in service 2 and 142 frames/second in service 3. Figure 5 illustrates the peak rates appearing in the total and internet LAN traffic of our model. A typical instance of LAN total and internet traffic is shown in figure 6. In figure 7 the distribution of interdeparture times of frames from the ETHERNET LAN is depicted and figure 8 gives the index of dispersion for the total and internet traffic. Similar figures have been extracted by [11] and [17] using real-time measurements. This further validates the model capability to describe real traffic situations through careful selection of input parameters.

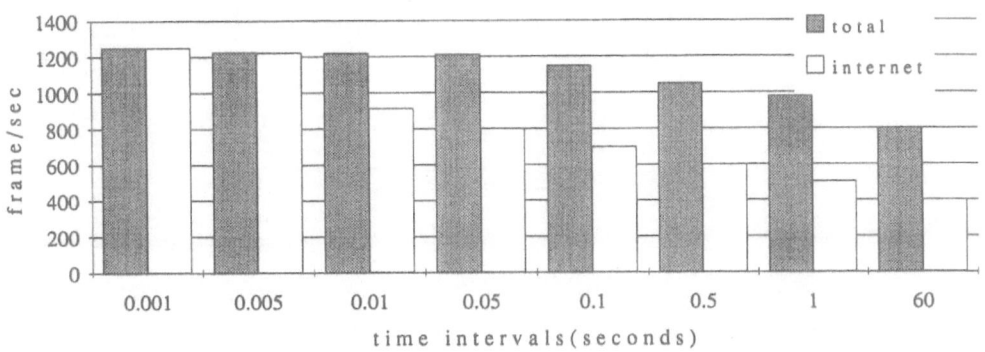

Figure 5. Peak rates in ETHERNET traffic.

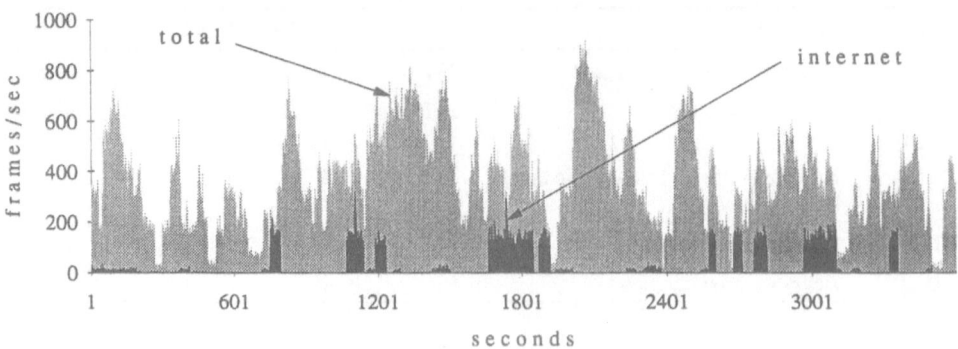

Figure 6. ETHERNET traffic.

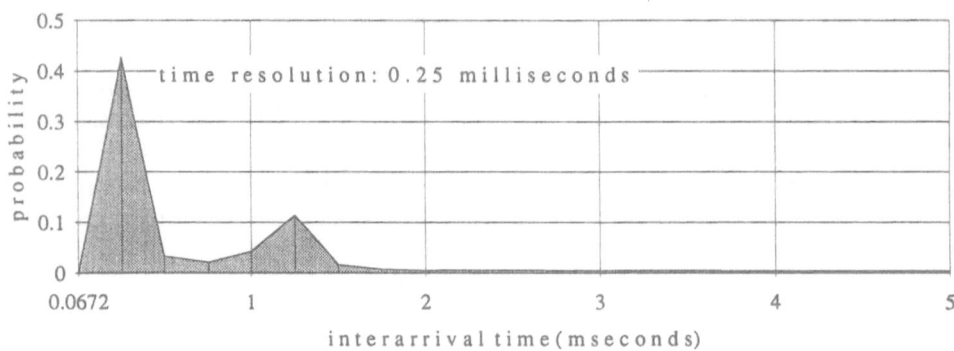

Figure 7. Interdeparture distribution in ETHERNET traffic.

3.2. ATM Interworking Unit

The ATM IWU as presented in figure 1 consists of two distinct parts. The LAN related part implements the MAC protocol functionality while the ATM related part is devoted to AAL and ATM protocol operations. IWU's processing speed is not considered as a bottleneck in the interconnection of low speed existing LANs with ATM. This means that instead of using fast hardware implementations we could benefit from the longer processing times experienced in firmware-based solutions. These cost-effective implementations lead to a drastic smoothing of the bursty LAN traffic that implies that less bandwidth is allocated in ATM. For the processing times calculated below we show the trade-off between buffer space requirement and bandwidth savings in ATM.

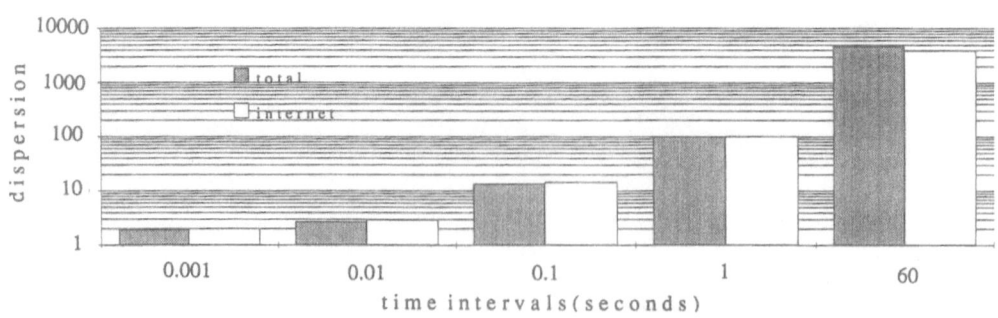

Figure 8. Index of dispersion in ETHERNET traffic.

Table 4. MAC protocol processing time components

Notation	Description	Value
A	Waiting time until the delivery of the first byte of the frame from the Physical Layer to the Receive Link Management component. This time includes the transceiver receive path delay, the decoding delay and the preamble removal delay	7.6 microseconds
B	Buffering time of the frame in the MQ	16*framelength*a
C	Processing time devoted to octet boundary alignment and identification of the frame length	100*a
E	Processing time for addressing	4000*a
D	Inspection time for the frame check sequence	34*framelength*a

3.2.1. MAC Protocol Processing Time. For the estimation of the MAC protocol processing times experienced in the ATM bridge we use the 10BASE5 specification of the IEEE 802.3 and the ETHERNET LAN functional architecture. The process involves all MAC layer functions up to and including those of the Receive Link Management component functionality. The processing time per frame is further decomposed in 5 components (A-E) each one implementing a group of the MAC protocol functions. The implementation of a functional group depends on the frame length as well as on the frame destination (internet or intranet). If the frame length is not acceptable (smaller than 64 bytes-a collision has happened) then the process includes the components A, B, C. If the frame length is acceptable (between 64 and 1518 bytes) but the frame destination is local the process includes the components A, B, C, D. If the frame length is acceptable and the frame destination is not local then the process includes all the components (A-E).

The processing times for the all the components, except A that depends on the 10BASE5 specification, are presented in table 4 and are calculated based on a microprocessor implementation involving the complexity of the processing and a parameter(a) characterizing the microprocessor speed.

3.2.2. AAL and ATM Processing Time. The processing time at the ATM part of the IWU is calculated for the AAL type 5 and ATM protocol functionality outlined in section 2. The processing time of an internet frame (the process in the convergence sublayer, the segmentation and reassembly sublayer and the ATM layer) includes the components presented in table 5. The processing times for the all the components are calculated based on a

Table 5. IWU processing time components

Notation	Description	Value
F	Buffering time of the MAC frame AQ	16*framelength*a
G	Time for adding the PAD	16*(x*48-framelength)*a
H	Time for adding the length indicator	32*a
I	CRC computation and CRC appending time	34*(framelength+x*48+8)*a
J	Segmentation time	3*x*a
K	Time for cell header assignment	80*a
L	Cell transmission time to the Physical layer	848*x*a

microprocessor implementation involving the complexity of the processing, a parameter(a) characterizing the microprocessor speed and x equal to the integer part of

$$\frac{framelength + 48}{48}.$$

4. NUMERICAL RESULTS AND DISCUSSION

The average internet frame delay from the time the source generates it to the time the first cell of the frame leaves the ATM IWU is decomposed in the following components: LAN delay, MAC delay, AAL+ATM delay. The LAN delay consists of the waiting time of frames due to the CSMA/CD protocol plus transmission and propagation delay. The MAC delay component is the waiting time in the MQ plus the MAC processing time in the IWU. The AAL+ATM delay is the waiting time in the AQ plus the processing time for accomplishing AAL and ATM layer protocol functionality.

In figure 9 we obtain the MQ buffer size as a function of the channel utilization for two different firmware implementations of the MAC protocol (a=100nanoseconds and a=50nanoseconds) and for a frame loss limited to 10^{-4}. Channel utilization increases by adding new sources to the LAN; i.e. the user model parameters, except CT_i, are kept constant throughout our experiment. The frame loss experienced in the MQ buffer is due to the total LAN traffic and is observed under persistent bursty periods. We noticed that further expansion of the buffer sizes, while increasing dramatically the waiting time variance, does not contribute significantly in reducing the loss probability even further. The same result has been observed for real traffic [11]. This is a validation of the accuracy of the user model developed here. In figure 9 we observe that the MQ buffer dimension for the slow implementation and for LAN utilization higher than 20% is over 1Mbytes for the above QoS (10^{-4}) and increases dramatically while the LAN utilization increases. If the MQ buffer is limited to 1Mbyte then the frame losses are 1% of the total frames transmitted for 20% of LAN utilization and reach the level of 25% for 50% of LAN utilization. For the fast implementation (a=50nanoseconds) the buffer size is reduced dramatically, however in expense of increased bandwidth as will appear in the policer dimensioning and losses higher than 10^{-4} appear only for LAN utilization above 50% and buffer dimension limited to 1Mbyte.

Regarding the AQ buffer (figure 10) we have to observe that now the frame loss probability depends both on the total traffic as well as on the proportion of the internet traffic and persistent bursty periods. We assume that the MQ buffer dimension is limited to 1Mbyte

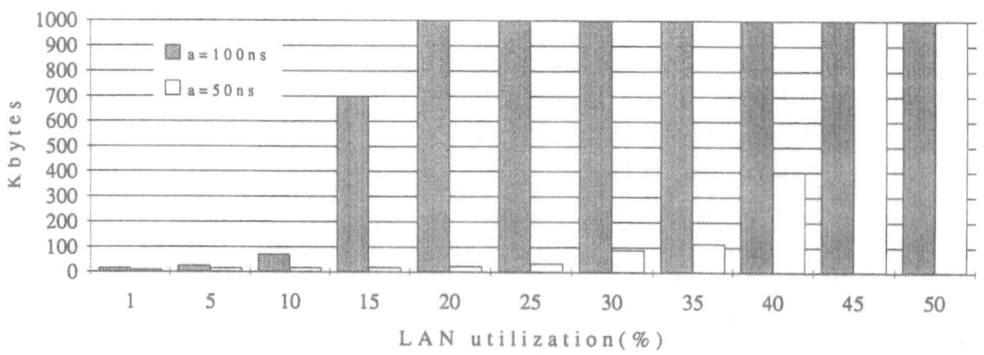

Figure 9. MQ buffer dimensioning.

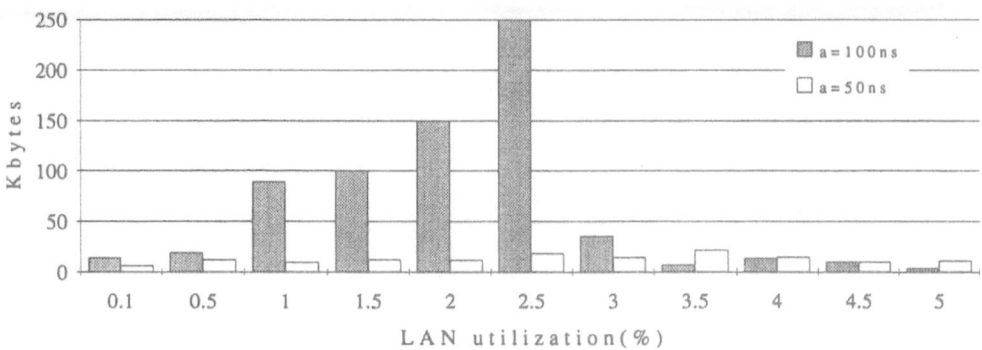

Figure 10. AQ buffer dimensioning.

and that the internet traffic is 10% of the total traffic transmitted in the LAN. Results show that traffic injected to the AQ is smoothed as it passes the first component (MAC server) since the internet traffic is 10% of the total traffic and the AAL+ATM server processing time is above 3 times the MAC server processing time. We also observe that for high LAN utilization the AQ buffer dimension demanded decreases dramatically. This happens for two reasons. The contention interval increases for high utilization of the LAN since the number of collisions increases before a frame transmission (0.01 collisions before a transmission for 1% LAN utilization but 0.70 collisions before a transmission for 50% LAN utilization) and the probability to find active sources transmitting internet frames alone in the channel is close to zero. Finally we observe again that for the fast implementation the buffer size is reduced dramatically, however in expense of increased bandwidth.

The most interesting result of our study concerns the impact of the interworking devices to the burstiness of the aggregate LAN internet traffic when they are firmware implemented. Results show that traffic is smoothing progressively when passing from one component to the other towards the ATM network. The traffic seems to lose some of his high and low frequencies and the IWU could be seen as a filter for the traffic generated by the LAN. This observation is essential for determining the bandwidth allocated to VPs interconnecting remote LANs.

To clearly demonstrate this issue we consider that a policing unit operating according to the Jumping window principles [20] is positioned in the entrance of the ATM network. We first inject in the policing unit the cell stream resulting from the IWU processing operation for the two different firmware implementations. We then compare the policing unit performance under a cell stream traffic that is the internet frame stream at the output of the LAN

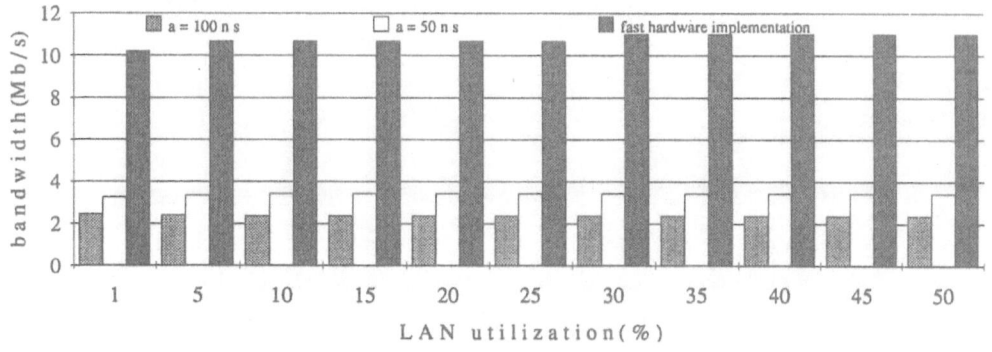

Figure 11. Bandwidth allocation (Decrementing interval: 0.01s).

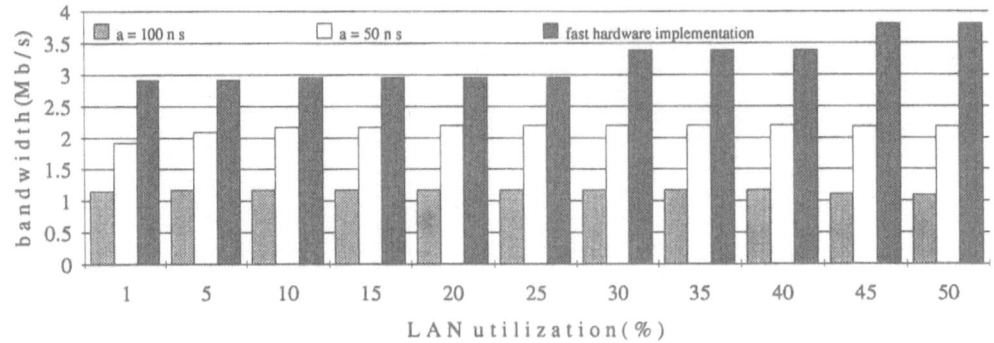

Figure 12. Bandwidth allocation (Decrementing interval: 0.1s).

segmented in cells with the same AAL type 5 protocol (i.e. IWU's processing times and buffering delays are ignored) assuming a loss of information equal to that experienced in the MQ and the AQ. This is equal to a fast hardware implementation with no buffering. Figures 11, 12 and 13 show the required VP bandwidth for the above cases of traffic and three different policer implementations (different decrementing intervals). We observe that the VP bandwidth demand decreases significantly in the case of firmware implementation of the IWU. Due to the presence of a slow IWU and for a window of 10mseconds the bandwidth demand is 5 times lower (a=100nseconds) or 3.5 times lower (a=50nseconds) than the fast hardware implementation. The required bandwidth is less even for a very slow policing unit with a window interval equal to 1second.

5. CONCLUSIONS

Existing LAN traffic models are not adequate for dimensioning the buffers of IWUs as well as for defining the ATM bandwidth allocated for remote LAN interconnection. Through the use of an accurate LAN user model, we have been able to evaluate the impact of the LAN traffic burstiness to IWU dimensioning. Given the IWU protocols for the LAN interconnection application we have investigated the contribution of specific implementations both to buffer dimensioning and ATM bandwidth allocation. The result is that for certain low traffic applications where IWU speed is not a bottleneck, firmware developments can be employed. The trade-off is then between the IWU buffer size and the ATM bandwidth reserved for LAN interconnection application. It should be stressed however, that for fast

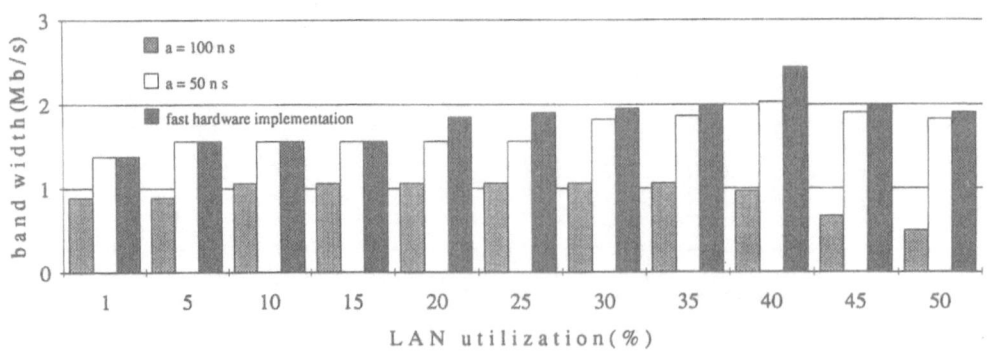

Figure 13. Bandwidth allocation (Decrementing interval: 1s).

IWUs and under a cell loss objective, the unpredictable peaks of LAN traffic result in much higher bandwidth demand than the expected. This would then necessitate the use of a spacer at the ATM entrance that should be taken into account in the techno-economical evaluation procedure of the implementation alternatives. Hence the problem for interconnection of existing LANs is formulated in the following question, slow IWUs with buffers or fast IWUs with spacers?

APPENDIX A - THE SOURCE MODEL

The general representation of the new source model is like the one in figure 3 but with i medium cycles. Let p_{T_i} be the probability to return to state T_i if we are in state T_i or equally the probability to have another small cycle of service i or equally to have a transmission followed by a silent period (time in state T_i) in service i. It is obvious that this probability is equal to the transition probability from state E_i to state T_i. Therefore $p_{T_i} = p_{ET_i}$ and the probability to have s small successive cycles is given by

$$p_{T_i}^{(s-1)} * (1 - p_{T_i}) = p_{ET_i}^{(s-1)} * (1 - p_{ET_i}) \text{ with average time equal to } \frac{1}{(1 - p_{ET_i})}$$

But in each small cycle we have a transmission. Consequently the probability to have s small successive cycles is equal to the probability to have s transmissions of cells in a burst. Thus given the macroscopic parameters C_i we can define the microscopic parameters p_{ET_i} by

$$p_{ET_i} = 1 - \frac{1}{C_i} \text{ since } C_i = \frac{1}{(1 - p_{ET_i})}$$

Let pdf_{T_i} be the probability density function of the interarrival time between two successive cells belonging to the same burst of service i where CT_i is the average value of the distribution as mentioned before. Then this pdf will be the pdf of the time spent in state T_i. Since the only way-out from states T_i are states E_i the transitions from states T_i to states E_i is

Table 6. Notation

Notation	Description
i	(i=1,2,...,n) number of services represented by the source
CD_i	Average transmission time of a cell of average size (average if the cells have variable sizes) of service i in the given media
C_i	Average number of cells belonging to the same burst period of service i.
CT_i	Average gap time between two successive cells belonging to the same burst period of service i (intercell gap time)
CR_i	Average cell transmission rate (cells/time units) in a burst period of service i
BD_i	Average burst period duration of service i
B_i	Average number of bursts belonging to the same active period of service i
BT_i	Average gap time between two successive bursts belonging to the same active period of the service i (interburst gap time)
BR_i	Average bursts rate (bursts/time unit) in an active period of service i
AD_i	Average activity duration of service i
CAR_i	Average call rate of service i
AR	Average activity ratio of the source or probability that the source is active
ATR	Average transmission rate of the source

sure in average time equal to CT_i. If the time spent in state T_i is geometrically distributed the microscopic parameters p_{TT_i} are given by

$$p_{TT_i} = \frac{1}{(1-C_i)}$$

Let SCT_i be the time needed to complete a small cycle of service i. This time is the sum of CD_i and CT_i. Therefore

$$SCT_i = CD_i + CT_i$$

$$CR_i = \frac{1}{SCT_i}$$

$$BD_i = \frac{C_i}{CR_i} = C_i * SCT_i = C_i*(CD_i + CT_i)$$

and given a combination of the macroscopic parameters CD_i, CT_i, SCT_i, CR_i, BD_i, C_i the microscopic parameters p_{ET_i} can be also defined as

$$p_{ET_i} = 1 - \frac{1}{C_i} = 1 - \frac{1}{BD_i * CR_i} = 1 - \frac{(CD_i + CT_i)}{BD_i}$$

Let p_{A_i} be the probability to return to state A_i if we are in state A_i or equally the probability to have another intermediate cycle of service i or equally to have a burst period followed by a silent period (time in state A) of service i. This probability is given by

$$p_{A_i} = p_{EA_i} + p_{EA_i} * p_{ET_i} + p_{EA_i} * p_{ET_i}^2 + ... + p_{EA_i} * p_{ET_i}^m + ... = p_{EA_i} \sum_{m=0}^{\infty} p_{ET_i}^m = \frac{p_{EA_i}}{(1-p_{ET_i})}$$

Thus the probability to have k intermediate successive cycles in service i is equal to

$$p_{A_i}^{(k-1)} * (1 - p_{A_i}) \text{ with average time equal to } \frac{(1-p_{ET_i})}{p_{EO_i}}$$

But in each intermediate cycle we have a burst. Consequently the probability to have k intermediate successive cycles is equal to the probability to have k bursts in an active period and obviously

$$B_i = \frac{(1-p_{ET_i})}{p_{EO_i}}$$

Since the probabilities p_{ET_i} have been defined given the macroscopic parameters B_i the microscopic parameters p_{EO_i} can be defined as

$$p_{EO_i} = \frac{(1-p_{ET_i})}{B_i}$$

The transition probabilities p_{EA_i} depend on probabilities p_{ET_i}i and p_{EO_i}. Thus

$$p_{EA_i} = 1 - p_{ET_i} - p_{EO_i}$$

Let pdf_{A_i} pdf_{A_i} be the probability density function of the interburst gap time in service i where BT_i is the average value of the distribution as mentioned before. Then this pdf will be the pdf of the time spent in state A_i. Since the only way-out from states A_i are states T_i the transition from states A_i to states T_i is sure in average time equal to BT_i. If the time spent in state A_i is geometrically distributed the microscopic parameters p_{AA_i} are given by

$$p_{AA_i} = 1 - \frac{1}{BT_i}$$

Let ICT_i be the average time needed to complete an intermediate cycle of the service. This time is the sum of BD_i and BT_i. Therefore

$$ICT_i = BD_i + BT_i = \frac{C_i}{CR_i} + BT_i = C_i * (CD_i + CT_i) + BT_i$$

$$BR_i = \frac{1}{ICT_i}$$

$$AD_i = \frac{B_i}{BR_i} = B_i * ICT_i = B_i * (BD_i + BT_i) = B_i * (C_i * (CD_i + CT_i) + BT_i)$$

Thus given a combination of the macroscopic parameters BD_i, BT_i, ICT_i, BR_i, AD_i, B_i the microscopic parameters p_{EO_i} can be also defined as

$$p_{EO_i} = \frac{1}{C_i * B_i} = \frac{1}{C_i * AD_i * BR_i} = \frac{(BD_i + BT_i)}{C_i * AD_i} = \frac{(1 - p_{ET_i}) * (C_i * (CD_i + CT_i) + BT_i)}{AD_i}$$

Let pdf_O be the probability density function of the inactive periods of the source and CAT the average intercall time of all the services of the source. Since the only way-out from state O are states A_i we will have a transition from state O to a state A_i after an average time equal to CAT. If the time spent in state O is geometrically distributed we can define the microscopic parameter p_{OO} by

$$p_{OO} = 1 - \frac{1}{CAT}$$

Let $AD = \sum_i AD_i$ and by definition $AR = \frac{AD}{AD + CAT}$ then p_{OO} can be defined also as

$$p_{OO} = 1 - \frac{AR}{AD * (1 - AR)}$$

Using the above information the call rates of the services are now given by

$$CAR_i = \frac{k_i}{(AD_i + CAT)} \text{ where } k_i \text{ are greater than zero.}$$

Let $SCR = \sum_i CAR_i$. Then the probability the source to be at service i at a given time is given by

$$p_{S_i} = \frac{CAR_i}{SCR_i}$$

Assuming a general distribution in state O the microscopic parameters p_{OA_i} become equal to the probabilities p_{S_i}. Thus $p_{OA_i} = p_{S_i}$ and for a geometrical distribution these parameters are given by

$$p_{OA_i} = (1 - p_{OO}) * p_{S_i}$$

By definition the average transmission rate of a small cycle of service i is

$$CR_i = \frac{1}{(CD_i + CT_i)}$$

The average transmission rate of an intermediate cycle of service i can be defined as

$$\frac{CR_i * BD_i}{ICT_i} = CR_i * BD_i * BR_i = C_i * BR_i = \frac{BR_i}{(1 - p_{ET_i})}$$

The average transmission rate of the source in the active periods can be defined as

$$\sum_i p_{S_i} * C_i * BR_i = \sum_i \frac{p_{S_i} * C_i}{(BD_i + BT_i)} \text{ where } i=1,2,...,n$$

Finally the average transmission rate of the source is

$$ATR = AR * \sum_i \frac{p_{S_i} * C_i}{(BD_i + BT_i)} \text{ where } i=1,2,...,n$$

REFERENCES

[1] *IEEE Select. Areas Commun.*, Special Issue on Heterogeneous Computer Network Interconnection, Vol.8, No.1, Jan.1990.

[2] *IEEE Network Mag.*, Special Issue on LAN Interconnection, Vol.8, No.5, Sep.1991.

[3] I.S.Venieris, E.N.Protonotarios, G.I.Stassinopoulos, R.Carli, "Bridging Remote CL LANs/MANs Through CO ATM Networks," Vol.15, No.7, Sep.1992.

[4] N.Kavak, "LAN Interconnection over BISDN," *in Proceeding of Interworking 1992*, Bern, Nov.1992.

[5] T.Hadoung, I.S.Venieris, E.N.Protonotarios, N.Tat, "LAN, MAN, Frame Relay and ATM: A Taxonomy of Interworking Units," *in Proceedings of Interworking 1992*, Bern, Nov.1992.

[6] CCITT SG XVIII, "Rec.I.211 - BISDN Service Aspects," TD 42, Geneva, June 1992.

[7] CCITT SG XVIII, "Draft Rec.I.327 - BISDN Functional Architecture," TD 16, Geneva, June 1992.

[8] CCITT SG XVIII, "Rec.I.364 - Support of Broadband Connectionless Data Service on BISDN," TD 58, Geneva, June 1992.

[9] I.S.Venieris, J.D.Angelopoulos, G.I.Stassinopoulos, "Efficient Use of Protocol Stacks for LAN/MAN-ATM Interworking," to appear *in IEEE J. Select. Areas Commun.*, Special Issue on Networks in the Metropolitan Area, 1993.

[10] T.Y.Mazraani, G.M.Parulkar, "Performance Analysis of the ETHERNET Under Conditions of Bursty Traffic," *in Proceedings of IEEE Globecom 1992*.

[11] H.J.Fowler, W.E.Leland, "Local Area Network Traffic Characteristics with Implications for Broadband Network Congestion Management," *IEEE J. Select. Areas Commun.*, Vol.9, No.7, Sep.1991.

[12] G.I.Stassinopoulos, I.S.Venieris, R.Carli, "ATM Adaptation Layer and IEEE LAN Interconnection," *in Proceedings of the 15th IEEE LCN 90*, Minneapolis, Oct.1990.

[13] CCITT SG XVIII, "Rec.I.362,3 - BISDN ATM Adaptation Layer (AAL) Functional Description, Specification," TD 57, TD 60, Geneva, June 1992.

[14] K.Kerhof, A.VanHalderen, "AAL type 5 to support the Broadband connectionless data bearer service," *in Proceedings of Interworking 1992*, Bern, Nov.1992.

[15] G.I.Stassinopoulos, I.S.Venieris, "ATM Adaptation Layer Protocols for Signalling," *Comput. Networks ISDN Systems*, Vol.23, No.4, 1992.

[16] R Bubenik, M.Gaddis, J.DeHart, "Communicating with Virtual Paths and Virtual Channels," *in Proceedings of IEEE Infocom 1992*, Vol.2, May 1992.

[17] W.E.Leland "LAN Traffic Behaviour from Milliseconds to Days," *7th ITC Seminar*, NJ, Oct.1990.

[18] W.R.Franta, I.Chlamtach, *"Local Networks-Motivation, Technology and Performance,"* Lexington Books, Oct.1982.

[19] ANSI/IEEE Standards 802.2 ("Logical Link Control"), 802.3 ("Carrier Sense Multiple Access with Collision Detection"), 1985.

[20] J.F.Shoch and J.A.Hupp, "Measured performance of an ETHERNET local network," *Commun. ACM*, Vol.23, No.12, Dec.1980.

[21] E.P.Rathgeb, "Policing Mechanisms for ATM Networks-Modelling and Performance Comparison," *7th ITC Seminar*, NJ, Oct.1990.

[22] D.Bertsekas, R.Gallager, *"Data Networks,"* Prentice-Hall International, 1987.

MULTI-CHANNEL TOKEN RING, INTRODUCTION AND THROUGHPUT ANALYSIS

Kenneth J. Christensen
Franc E. Noel

LAN/WAN Systems Council
Department H28, Building 002
IBM Corporation
Research Triangle Park, NC 27709

ABSTRACT

If traffic on a large LAN segment can be localized, increased user throughput can be achieved by dividing the LAN segment into smaller interconnected segments. Each localized segment contains a group of communicating stations. If traffic cannot be localized, increasing the data-rate of the LAN is the typical solution to providing increased user bandwidth. An alternative solution is architecting multiple parallel LANs to increase bandwidth. This paper proposes a technique for increasing the bandwidth of a LAN without changing the data rate; namely, by utilizing a number of independent channels operating in parallel and at the same data rate. This technique is applied to token ring and is named Multi-Channel Token Ring (MCTR). Driving forces for MCTR are; preserving existing hardware and software as much as possible, being IEEE 802.5 standards compliant, and offering increased reliability. MCTR can be implemented without changing work-station interfaces or LAN protocols. A means of precisely computing peak single station throughput on a token ring LAN is developed. This throughput analysis is then applied to MCTR.

1. INTRODUCTION

One of the key performance measures for a LAN user is throughput. As the number of active users on a single LAN segment increases, throughput decreases for each user. The effects of decreasing throughput can be seen by users as slower access to network services, for example, increased client/server file transfer times.

By dividing a single large (many user) LAN segment into smaller (few user) segments, user throughput can be increased. This is done by localizing groups of communicating users to individual segments. Often it is not possible to localize user traffic, for example if much of the traffic is between a majority of users and a few specific gateways or file servers. If it is not possible to localize user traffic, then increasing the total available bandwidth is necessary. One way to do this is by upgrading all of the LAN users to a higher data-rate LAN (e.g., upgrading from 16-Mbps token ring to 100-Mbps FDDI). Another method of increasing user throughput is to offer each user access to parallel LAN segments with a combined bandwidth greater than the bandwidth of the original single channel LAN.

Applications driving the need for additional bandwidth include multimedia and image-based client/server computing. However, the demand for greater overall LAN bandwidth does not mean that every station on a LAN requires access to the full bandwidth. Individual client work-stations are likely to require less bandwidth than a file server or gateway. A parallel LAN offers the capability to scale the amount of bandwidth available to a single station. A parallel LAN also offers the possibility of preserving the existing investment in LAN adapter cards, cabling, protocol software, and application software.

The remainder of this paper is organized as follows. Section 2 describes parallel token ring networks as described in the literature. This includes Multi-Channel Token Ring (MCTR) proposed by the authors of this paper. Section 3 contains a throughput analysis of token ring. A means of computing maximum throughput for a single station is described. Section 4 extends this analysis to MCTR. Following section 4 are the conclusions, acknowledgments, and references.

2. PARALLEL TOKEN RING NETWORKS

Figure 1 shows a multiple token ring network as proposed by Chen and Bhuyan [ChBh88]. In [ChBh88] multiple token ring is intended as a reliable, expandable, and cost effective multiprocessor interconnect architecture. In the multiple token ring network each user is physically attached to B independent token rings. Each token ring operates independently and asynchronously.

Three station access mechanisms are proposed by [ChBh88] for multiple token ring, they are :

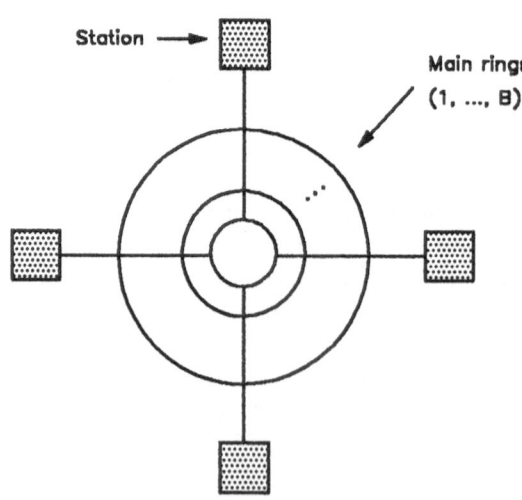

Figure 1 - Multiple token ring

1. Separate queues for each of the B main rings with simultaneous transmissions.

2. Single queue with simultaneous transmissions.

3. Single queue with single transmission.

At the expense of a complex access mechanism, options 1 and 2 give a single user access to the entire bandwidth of the B main rings. In [BhGY89] the multiple token ring with single queue and single transmission, option 3, is further considered. For this access mechanism, when a station has a frame queued for transmit the station monitors all rings for the first available token. When a frame is transmitted, the transmitting station also strips the frame from the ring on its return. Hence, when a station is transmitting on one ring it cannot transmit on another ring until its first transmission is complete.

In [ChBh88] the mean waiting time (in a station queue) and mean transfer time of a frame is analytically solved for the three access mechanisms of multiple token ring. For the analysis, a Poisson arrival process of frames into the station queue is assumed. An assumption is made that token arrivals at a station are uniformly distributed across the B main rings. However, simulation results showed that tokens tend to cluster together instead of being distributed independently across the B main rings. In [MFMK92] service and waiting times in parallel token rings, similar to the multiple token ring with single queue and simultaneous transmission of [ChBh88], are evaluated. Compared to a single channel token ring of similar bit rate, a parallel token ring can yield performance advantages due to a decreased mean and variance in wait time for tokens. A model is developed which separates the influence of mean and variance of token wait time on frame transfer time and frame wait time in the queue. An analytical expression to handle the case of non-dispersed (i.e., traveling together) tokens is proposed. An analysis to determine the optimum number of parallel token rings to minimize mean transfer time for a given load and number of stations is presented in [VaRa92]. The analysis is for a multiple token ring with single queue and single transmission. The results show that the best ratio of number of rings to number of stations is in the range of 0.2 to 0.5 depending on the load.

Figure 2 shows a Multi-Channel Token Ring (MCTR) as proposed in [Noel91] and [ChNo92]. In MCTR existing single channel token ring stations are attached to a concentrator. The concentrator provides access to multiple independent parallel token rings with a single queue, simultaneous transmission access mechanism.

The basic functions of the MCTR concentrator, shown in Figure 3, are to multiplex frames from a single ring to multiple rings and de-multiplex frames from multiple rings to a single ring. The MCTR concentrator is very similar to a multiport bridge. Frames incoming to the concentrator from the multiple main rings are forwarded to the station lobe ring(s) on which stations with matching destination addresses reside. Frames incoming to the concentrator from the station lobe rings are forwarded to the first available ring in the multi-channel main ring. The MCTR concentrator contains a "learning function" to learn where stations (i.e., Media Access Control or MAC addresses) reside. Outgoing frames queued in the MCTR concentrator are transmitted on the first available token on any one of the B main rings. Transmitting on the first available token optimizes the total bandwidth utilization of the multiple main rings and guarantees proper sequential delivery of frames to the destination station.

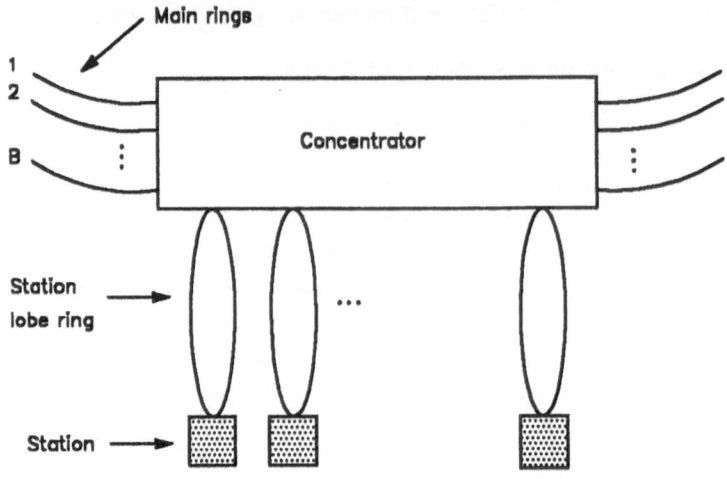

Figure 2 - Multi-Channel Token Ring (MCTR)

The basic data flow of an MCTR concentrator is as follows. Assume a station with MAC address A is transmitting to a station with MAC address B. Assume also that stations A and B are attached to different concentrators. Then, the following occurs:

1. Station A captures a token on its station lobe and transmits a frame addressed from A to B.

2. The station lobe MAC in the concentrator recognizes that this frame is not from itself (i.e., does not have its unique MAC address as a source address) and thus copies the frame.

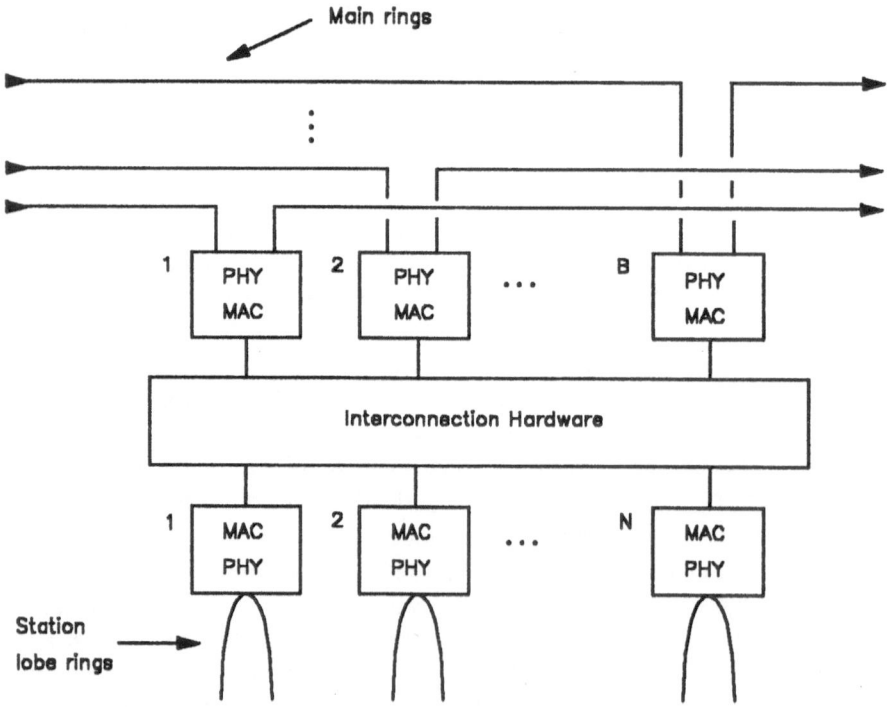

Figure 3 - MCTR concentrator

3. The frame is now queued for transmit onto the MCTR.

4. The first token that is available on any channel is captured and the frame is transmitted.

5. A downstream concentrator recognizes that the address B is one of its attached stations and thus copies the frame.

6. The frame is now queued for transmit onto the station lobe for station B.

7. A token is captured on station B's station lobe and the frame is sent to station B.

8. Station B receives the frame from station A.

Step 5 requires that each of the B MCTR MAC's be able to recognize and copy frames destined to any of the stations attached to the concentrator. As stations insert into their station lobes, the concentrator "learns" their addresses. The data flow described above does not require store-and-forward operation between the lobe ring and main rings. This is due to an identical bit rate (i.e., 16-Mbps) for lobe rings and the separate main rings. Cut-through operation between lobe rings and main rings can minimize the latency of the MCTR concentrator.

A B channel MCTR concentrator with N ports requires B + N MAC and PHY implementations. A MAC and PHY implementation is the VLSI hardware that directly implements the IEEE 802.5 Media Access Control (MAC) and Physical (PHY) layer protocols. Connecting the B channel MAC and PHY implementations and N station lobe MAC and PHY implementations is high-speed interconnection hardware (e.g., a high-speed bus or crossbar switch).

MCTR uses existing token ring LAN adapters in the work-station. Thus, MCTR preserves the large investment made in token ring LAN adapters, lobe cabling, and software. MCTR supports IEEE 802.5 [*IEEE*88] standard token ring connections between concentrators. This makes MCTR a fully standards compliant solution for increased bandwidth using parallel LANs. MCTR offers high reliability between concentrators. If one of the B rings fail, the remaining B - 1 rings can continue to operate with no impact to connectivity. Frame loss, due to channel loss or other causes, is handled by higher layer communications protocols. MCTR also offers scalability of bandwidth. MCTR allows basic access to 1/B of the total bandwidth. Less bandwidth per user is possible by increasing the number of attached stations to a lobe ring (i.e., there can be one or more stations attached to each lobe ring). Increased bandwidth is possible in two ways. By having a single station attach to a concentrator with multiple lobe rings or by being directly attached to the B main rings.

In the following sections, a means of computing single station throughput of MCTR is developed. The authors believe that throughput is the critical performance measure for LAN users, hence the emphasis in this paper is on throughput (more than on the arrival characteristics of tokens or end-to-end frame delays which are typically small compared to end-station processing delays).

3. THROUGHPUT ANALYSIS OF TOKEN RING

The bit rate of token ring is 16-Mbps, however the data throughput rate is less than 16-Mbps due to framing overheads and token rotation delays. The data throughput of a station in bits per second, S, is computed as

SD	AC	FC	DA	SA	Info	FCS	ED	FS	Gap

Length of frame fields:

SD = 1 byte
AC = 1
FC = 1
DA = 6
SA = 6
Info = Variable length
FCS = 4
ED = 1
FS = 1
Gap = 5

Figure 4 - IEEE 802.5 frame format

$$S = \frac{\text{Number of data bits in a frame}}{\text{Total time to send a frame}}. \tag{1}$$

More specifically,

$$S = \frac{L}{T_{frame} + T_{overhead} + T_{phy} + T_{other\_xmits}}, \tag{2}$$

where

L = Frame length in bits defined as the following IEEE 802.5 frame fields, $AC + FC + SA + DA + Info$ (see Figure 4).

T_{frame} = Time to transmit L bits given a bit rate of R ($R = 16 - Mbps$ for this paper).

$T_{overhead}$ = Time to transmit overhead fields in a frame defined as, $SD + FCS + ED + FS + Gap$ for a total of 12 bytes overhead per frame.

T_{phy} = Physical layer delay for a token rotation. This is a function of the cable length and number of stations. The computation of T_{phy} is described following Figure 4.

T_{other_xmits} = Time for other stations to transmit frames (i.e., $T_{frame} + T_{overhead}$ for each station transmitting a frame on a given token rotation).

The delay caused by other transmitting stations, T_{other_xmits}, is very difficult to compute. This is a stochastic variable that depends highly on the arrival distribution of frames at other stations. Sethi and Saydam [SeSa85] approximate the number of stations with frames queued to transmit (per token rotation) as binomially distributed. In [SeSa85] frame arrivals at a station are assumed to be Poisson.

The physical layer delay, T_{phy}, is calculated as follows. Define the following delay components:

292

$T_{monitor}$ = Delay from the active monitor on the ring. This delay component includes a 24 bit fixed token buffer delay and (for 16-Mbps Token-Ring) a 16 bit elastic buffer delay.

$T_{station}$ = Station delay for repeating a frame or token. For 16-Mbps Token-Ring this is approximately 1.5 bit delays per station.

T_{cable} = Propagation delay in copper cable. This is approximately 5.0 nanoseconds per meter.

Then, for a ring with M stations, each with L_{cable} meters of cable,

$$T_{phy} = M(T_{station} + (L_{cable}T_{cable})) + T_{monitor}. \tag{3}$$

For example, for a one station ring with zero meters of cable, T_{phy} = 2.6 microseconds. For a 256 station ring with 100 meters of cable per station, T_{phy} = 154.5 microseconds.

Equation (2) computes throughput assuming Early Token Release (ETR) is enabled. In ETR a transmitting station releases a token immediately following transmission of its frame. With no ETR, the transmitting station must have stripped the frame header (i.e., $SD + AC + FC + DA + SA$) before releasing a token. Thus, ETR can only occur for the case where T_{phy} is greater than $T_{frame} + T_{overhead} + T_{header}$ (T_{header} is the time to transmit the 15 bytes of the header). For this case and with ETR not enabled,

$$S = \frac{L}{T_{phy} + T_{phy} + T_{other\_xmits}}. \tag{4}$$

Using equation (2) the maximum single station throughput can be computed. This is the maximum attainable throughput a single station can sustain for a given frame length and token ring size (i.e., number of stations and cable length). For a 16-Mbps token ring, Figure 5 shows maximum throughput as a function of frame size for a one station "zero length" ring, a 128 station ring with 100 meters of cable between stations, and a 256 station ring also with 100 meters of cable between stations. Figure 5 assumes ETR is enabled and that T_{other_xmits} = 0.

Comparing the single station throughput curves for small and large frame sizes (in Figure 5), it can be seen that the percentage of throughput reduction (as a function of the number of stations) is greater for small frame sizes. A large frame size reduces the effect of adding new stations since $T_{frame} >> T_{phy}$ for this case.

4. THROUGHPUT ANALYSIS OF MCTR

For MCTR with B main rings attaching K concentrators, the maximum single station throughput is the minimum of the lobe ring throughput, S_{lobe_ring}, and total B main rings throughput, S_{main_ring}. S_{lobe_ring} is computed using equation (2) with the number of stations equal to two (i.e., one each for the attached station MAC and the concentrator MAC assuming one station per lobe ring), or M = 2. S_{main_ring} is computed as,

$$S_{main\_ring} = \frac{BL}{T_{frame} + T_{overhead} + T_{phy} + T_{other\_xmits}}. \tag{5}$$

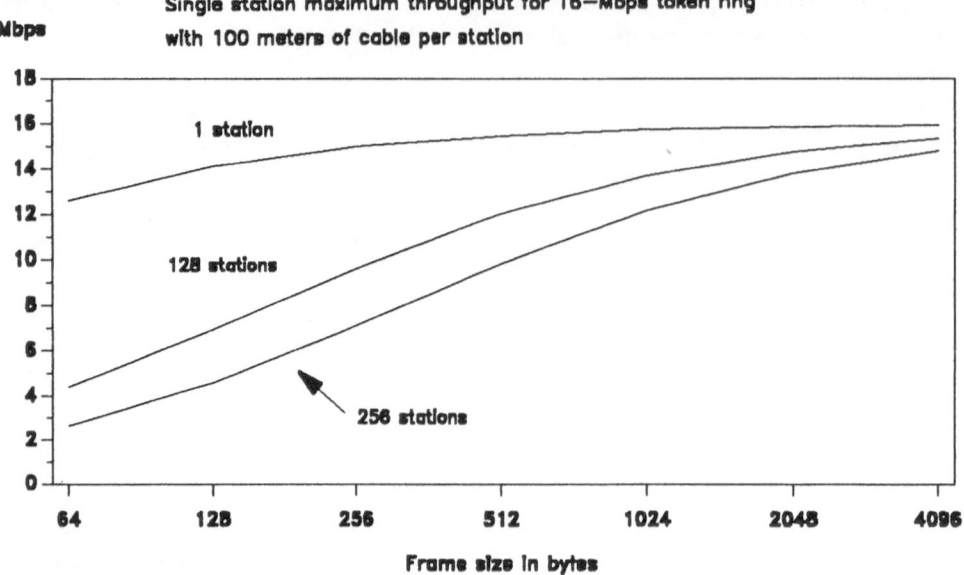

Figure 5 - 16-Mbps single station maximum throughput

Using $M = K$ to compute T_{phy} for equation (5), the maximum single station throughout for MCTR, S_{MCTR}, can be computed,

$$S_{MCTR} = \min(S_{lobe\_ring}, S_{main\_ring}). \tag{6}$$

For B greater than two, S_{lobe_ring} will usually be less than S_{main_ring}. Thus, $S_{MCTR} = S_{lobe\_ring}$ for most cases.

As described by equation (5), an MCTR with B 16-Mbps token ring channels has a greater throughput capacity than a single 16-Mbps token ring. For a large number of

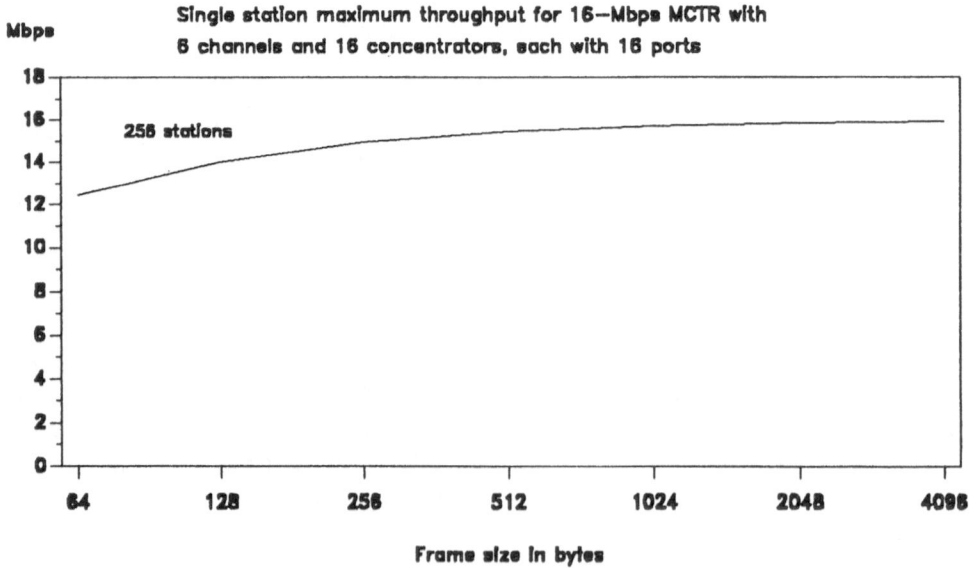

Figure 6 - 16-Mbps MCTR single station maximum throughput

stations, MCTR also yields a higher single station maximum throughput than does a single 16-Mbps token ring segment. This is can be seen by comparing figure 5 with figure 6. Figure 6 shows maximum single station throughput for a 256 station, 6 channel MCTR with 16 concentrators connected by 100 meters of cable. Each concentrator is assumed to have 16 attached station. Figure 6 assumes ETR is enabled and T_{other_xmits} = 0. From Figures 5 and 6, for a 64 byte frame size and 256 stations, S = 2.66-Mbps for a single channel 16-Mbps token ring and S = 12.43-Mbps for an MCTR (with 6 channels and 16 stations per concentrator). This higher single station throughput is due to the decreased T_{phy} of the MCTR implementation compared to single token ring. For MCTR the tokens on the station lobe rings experience only two station delays (the station itself and the concentrator) and one cable length. For a single token ring segment, each token experiences the full delay of M stations and their respective cable lengths.

5. CONCLUSIONS

Multi-Channel Token Ring (MCTR) demonstrates a method of using parallel token rings to increase LAN bandwidth. MCTR was designed to preserve existing investments in work-station adapters, cabling, and software while increasing both overall bandwidth and reliability. A means of computing single station throughput of a token ring station was developed. This analysis demonstrates the impact of physical layer delays on small frame throughput. This analysis is then applied to MCTR. The analysis shows that MCTR, compared to a single channel token ring, both increases overall bandwidth (due to its B main ring channels), but can also in some cases increase single station small frame throughput due to reduced physical layer delays.

ACKNOWLEDGMENTS

The authors would like to acknowledge Dr. Arne Nilsson, Franc Noel's Ph.D. Committee chairman, for his guidance with the original work in [Noel91]. The authors would also like to acknowledge the helpful comments from the anonymous referees.

REFERENCES

[BhGY89]   Bhuyan, L., Ghosal, D., and Yang, G. "Approximate Analysis of Single and Multiple Ring Networks." *IEEE Transactions on Communications*, Vol. 38, No. 7, July 1989.

[ChBh88]   Chen, C. and Bhuyan, L. "Design and Analysis of Multiple Token Ring Networks." *Proceedings of IEEE INFOCOM*, 1988.

[ChNo92]   Christensen, K. and Noel, F. "Parallel Channel Token Ring Local Area Networks." *Proceedings of the 17th Conference on Local Computer Networks*, 1992.

[IEEE88]   IEEE Standard 802.5. "Local Area Networks: Token Ring Access Method and Physical Layer Specifications." IEEE 802.5X-88/55, 1988.

[*MFMK*92] Mukkamala, R., Foudriat, E., Maly, K., and Kale, V. "Modeling and Analysis of High Speed Parallel Token Ring Networks." *Proceedings of the 17th Conference on Local Computer Networks*, 1992.

[*Noel*91] Noel, F. "Dynamic Resource Allocation Using Wavelength Division Multiplexed Channels in Token-Ring Local Area Networks." Ph.D. Thesis, North Carolina State University, 1991.

[*SeSa*85] Sethi, A. and Saydam, T. "Performance Analysis of Token Ring Local Area Networks." *Computer Networks and ISDN Systems*, Vol. 9, No. 3, March 1985.

[*VaRa*92] Vasko, D. and Rajsuman, R. "An Analysis and Simulation of Multiple Token Ring Networks." *Proceedings of the 17th Conference on Local Computer Networks*, 1992.

Parallel Networking: Improving Capacity and Multi-traffic Performance Together

E.C. Foudriat, K. Maly, R. Mukkamala and C.M. Overstreet

Computer Science Department
Old Dominion University
Norfolk, VA 23529
foudr_e@cs.odu.edu

Abstract

The research reported in this paper shows that parallel token ring networks are more effective in handling multiple traffic conditions than equivalent capacity single or bridged networks. This is in addition to the generic capacity increase obtained by added links. This capability is obtained because parallel networks can use message load distribution and joint token control policies not available to or effective for single link networks. The improvement is as much as a 7 fold decrease in message wait time for high priority (real-time and access time critical) traffic at high network loads. At the same time, regular message traffic is supported equally, if not better, for parallel rings than for other configurations. As a result, parallel ring networks are more suitable for handling multiple traffic (combination of synchronous and asynchronous) conditions. Further, this performance increase is attained without augmenting the token ring media access protocol by adding separate access mechanisms for synchronous and asynchronous traffic. The paper presents data for a two ring parallel network performance for a number of message distribution and token control policies. The best policies are then compared against single and bridged network performance. Finally, data is presented for a range of policy and network operational parameters in order to demonstrate that performance gains for parallel networks occur over a wide range of conditions.

I. Introduction

A. Statement of the Problem

There are two reasons for replacing a network with an upgraded, more modern one. The first is to increase network capacity in order to alleviate overloading. For most of today's networking problems this justification is most applicable. The second justification is that the media access protocol of the system in place does not effectively support multiple traffic types, especially when the network is operating at loads near its capacity[1]. With the advent of multimedia and peer-to-peer activity becoming more relevant, the reason may much more applicable in the near future.

---

[1] By multiple traffic, we mean a combination of traffic some of whose delivery (access plus travel) time is critical and others which, although not as time critical, must be delivered regularly so that the user does not view the network as inhibiting his operations. Synchronous, i.e., voice, video and real-time control traffic are examples of the former category; asynchronous traffic, such as file server data, as an example of the latter. For this paper, we categorize the former as high priority and the latter as regular messages, respectively.

Local Area Network Interconnection, Edited by R.O. Onvural
and A.A. Nilsson, Plenum Press, New York, 1993

One solution which satisfies both reasons above is to install a modern, higher capacity network with a protocol designed to support multimedia traffic. It requires new hardware and software. However, installing a new network system can be traumatic as well as costly, and it could create problems if some of the computers in a heterogeneous collection may not have such hardware and/or software available.

A second solution which may satisfy the overload problem is to split the network into sub-networks and connect them via bridging. This eliminates much of the trauma accompanying change over to a new network and for many users is completely transparent. It works assuming that network use is effectively separable, i.e., that a distribution of users is available where those on a link mainly communication with others on the same link.

However, bridging is not completely painless and it does not effectively address the multimedia problem. It may create operating restrictions on the broad utility of the overall network. For instance, problems arise concerning placement and coupling, compounding network management problems. Who should be closely connected and where should servers be placed to minimize the number of network hops? Should the interconnection be multiple bridges or bridge hubs serving multiple links? For example, some file servers in the Sun network serve multiple Ethernets creating a mini-hub. However, the hub node now must serve as a bridge as well as a file server, reducing its capacity to provide effective file service during heavy loads. Also, peer-to-peer operations by those assigned to work cooperatively can create variable, irregular loads on the network. Reconfiguring nodes to different links may be required each time a new job is started or as jobs change with new people assigned. If the bridging solutions is extended to where there are a number of links, real-time performance can deteriorate significantly between "distant" nodes when messages traverse a number of bridges.

However, bridging does not solve multiple traffic type problems (other than reducing load) unless the original network supports them effectively. Future networks must effectively support integrated traffic, i.e., both high priority and regular traffic [17]. If nodes are singly connected, then the network protocol itself must provide this support. It is well accepted that the IEEE802.3, CSMA/CD protocol is not effective for synchronous traffic, especially at high loads [19, 20]. Token ring, IEEE802.5, protocols and slotted rings have similar problems [21]. To solve these problems, many researchers have suggested dual access protocols. For busses, DRAMA [19] is typical of a dual access protocol mechanism. It has a repetitive frame with two cycles per frame. The first cycle allows nodes to reserve virtual channels and the second uses CSMA/CD contention for access. For token ring, a node sending high priority access does so by eliminating regular message traffic [20]. FDDI-II uses a 125 μsec. frame with reserved channels for isochronous traffic and the free channels support token ring operations [18]. Other dual access protocols are applicable for slotted rings [21] and adaptive protocols, a form of dual access, are used when heavy loading affects operational performance [17].

B. Solution Using Parallel Networks

In this paper, we propose an effective alternative mechanism for upgrading network performance. The solution uses parallel networks where each computer is effectively connected to two or more network links simultaneously. The parallel networks use single link protocols either as they presently exist or with minor modifications. This significantly reduces the trauma associated with switching to an entirely different system. Obviously, connecting to two links simultaneously requires some hardware changes. Suffice to say, the minimum for two link parallelism would be to add an additional receiver or a second network card to each node. This allows all nodes to receive on both links and to talk on one or more links. A more sophisticated system would replace the single link port with a single card which could listen on both links and talk, more simply on one link at a time, or, with more complexity, on both simultaneously. This eliminates bridging and routing between the networks. Some changes in software are required but would be considerably less than replacing the entire network system.

Significant enhancement accrue when parallel links exist. First, capacity is increased, thereby reducing overloading. Second, parallelism improves the response time between computers in comparison to a the bridged system since the "distance" effects of bridging are reduced or eliminated. Parallelism also enhances reliability.

In this paper, we show an additional advantage — parallel token ring networks can significantly improve the capability of the network system to handle multiple traffic types. This is above and beyond the increase afforded by distributing load over the parallel networks, i.e., above the generic increase in capacity. By improved capability, we mean high priority and regular data are transferred with either the same or reduced time and with greater predictability. Also, the networks links work in consort to support multiple traffic types without adding additional access protocol mechanisms.

The approach taken in this paper is to examine a range of strategies afforded by message assignment, token use and token rotation control policies for a parallel network consisting of two token rings. For an example of token use, one ring uses an exhaustive submittal policy and the other a target token rotation time (TTRT) policy[2]. Message assignment policies could give a high priority message to the ring where the token is expected next based upon some method of estimating next token arrival time. The next regular message could be assigned to the ring with the smallest queue length. Further, high priority messages can be given preferred access over regular messages. As the number of parallel channels increase, the mix of strategies grows rapidly.

For this paper, we limit the investigation to two parallel token rings. Why? Even with only two rings the number of combinations is quite large. Second, in previous work [10, 11] discussed in the next subsection, factors which influence token arrival for multiple rings are distinctly different from those for single rings. Also, we consider only two classes of traffic, high priority and regular messages. With only two traffic classes, we can use policies which provide for the predominance of support for one traffic type on one ring. Finally, we are able to measure the results of operations on each ring separately using a computer simulation model and glean further understanding by comparing similarities and differences between each. Thus for two rings, we are able gain some understanding of why different policies provide different performance and are not overwhelmed by the shear number of policy combinations which can exist.

In the paper, we discuss possible policy strategies in the next section and then, using a simulator, compare the performance for various strategies to those of single ring operation under identical conditions. We discuss briefly the simulator system, then present network performance comparisons. After showing the performance improvement, we vary the policy and network operational parameters to demonstrate that the performance gains are available over a wide range of conditions.

C. Background

Considerable research relating to parallel networking exists. The great majority is directed toward increasing throughput by eliminating bottlenecks which can occur at or between various levels in the networking protocol. Many use parallelism as a mechanism for improving protocol processing at a node; results can be found in references [1-9]. This work employs parallel processing mainly at the transport/network level to increase throughput and reduce latency at a node. Very little of this work explicitly employs parallel physical channels and when physical parallelism is modelled, only simple assignment policies are considered. Performance gains available are attributed mainly to a generic increase in capacity.

Maly et al. [9, 15] briefly examine assignment strategies at the media access level but do so for the single application transfer problem needing high data rates. The disturbing effects of other network users are minimally modeled. We have, in other papers [10, 11], examined particular aspects of the message assignment and token control problem. Reference 10 shows that there is a significant improvement due to parallel token rings if message assignment at a particular node employs the concept of minimum queue length. Queue length can be measured either by minimum message count or by minimum bit length. These results are confirmed in reference 9 where an *adaptive* assignment is shown to be better than *random* assignment.

---

[2]TTRT is a token control policy used in FDDI [18, 20], for example, where each node is required to release the token after a fixed time based upon its previous release time. Enforcing TTRT guarantees each node an opportunity to send. An exhaustive policy sets the TTRT = ∞.

The work in reference [11] develops an analytical model for token rotation. It shows that when nodes are serviced by multiple rings, the statistical arrivals of tokens at a node is considerably different from that occurring in single rings. It does suggest the potential for using token control and message submittal policies to enhance performance. Others [12 - 14] had looked at multiple token rings. Their work mainly extends single ring token analysis and they do not consider detailed message submittal or token control policies.

To date very little work has been done on parallel network access hardware. One analysis by Maly et al. [9, 15] considers that each FDDI ring has a separate access card but that information available about each ring so that message assignment decisions to each card can be made. Work by Christensen and Noel [16] consider that a node has access to all networks equally. They describe a Multi-Channel Token Ring system and a Media Access Control Supervisor for handling arriving messages. Their work, while it does not consider various message assignment strategies, demonstrates that single card, multiple channel ring access systems are feasible and could be effectively realized with today's hardware technology.

II. Access Strategies for Single and Multiple Parallel Rings

In this section, we describe strategies available for single and two parallel rings networks.

A. Single Ring Strategies

In single token rings, the number of strategies is limited message access at a node can be based upon priority and the ring can enforce a Target Token Rotation Time (TTRT) policy. At a node, higher priority messages are given precedence in the message queue. This is really the only message based strategy available. For the ring, establishing a TTRT is the basic token control strategy. Further, the TTRT can either be enforced for all messages or high priority messages can have access after the TTRT has expired. In the latter case, high priority messages receive the same preferential treatment as synchronous traffic in the FDDI-I ring operation [18]. A third strategy is to enforce a priority on the network as a whole, i.e., only messages of sufficient priority can be submitted [20]. However, since lower priority messages are not submitted during this period, the network does not support integrated traffic simultaneously. Hence, it is not considered as being an effective strategy in a general token ring environment.

B. Parallel Ring Strategies

Strategies for two parallel networks are considerably more diverse. There are message assignment and ring and node operational strategies which can be implement. In many cases, these are independent so that different strategies can be combined. As a result, the total number of strategies available can be quite extensive. In this paper, we examine some of the more obvious ones. We only consider strategies assuming that all nodes are connected to and message rotation direction is the same for both rings.

Message assignment strategies are probably the most diverse. For the research reported in this paper the message assignment strategies investigated include:

1. Random - When the message arrives, it is assigned randomly to a ring. In most cases, the probability of assignment is uniform across the rings.
2. Minimum queue length - When the message arrives, it is assigned to the ring with the minimum number of messages in its message queue. Alternatively, the minimum number of bytes in the ring message queue can be used.
3. Fixed assignment - When the message arrives, it is automatically assigned to a particular ring, probably because of the operational policies supported on that ring.
4. Any ring assignment - The message is not assigned to a ring until it is at the head of the queue. It is then assigned to a ring only after the token for that ring has arrived at the node.
5. Estimated next token arrival - Here, the node examines the messages as they pass. It observes the source node and message length of the last passing message on each ring. It knows where each token was at the time that message was submitted and possibly, when the token will be

released from that node. With this information, it estimates on which ring the next token is expected to arrive.

Note that the message assignments above are independent of priority and size. Thus, high priority messages may use one assignment policy while regular messages may use another.

Ring policies are nearly as diverse. They include:

1. Token rotation policies - When the token arrives at a node, it may be released after a single message (single), released after all queued messages are exhausted (exhaustive), or released based upon a TTRT expiration.
2. Token use policies - When a token arrives at a node, it may be used only if another token is not already at the node. Alternatively, multiple token use at a node may be permitted.
3. Token release policies - The token may be held past the TTRT overflow time to send high priority messages. Alternatively, the TTRT policy may be enforced for all messages.

Like message policies, ring policies may differ on each ring and it is possible for some ring policies to be operational at specific nodes without being uniformly applicable to the total ring. As noted above, node location policies, i.e., messages may be sent via a shortest route using clockwise or counterclockwise message progression are not considered in this paper.

As a result, the combination of message and ring policies for parallel ring networks with two or more rings becomes extensive. In this paper we deal with two rings, and will examine the most reasonable combinations.

III. Simulation Model for Studying Parallel Ring Operational Capabilities

It is obvious that with the policy combinations available, it is impossible analytically to determine and compare there effectiveness. Therefore, computer simulation is used to study parallel token ring operations. In this section, we briefly describe the system in order to provide the reader with an understanding of how the model operates to study the transfer of diverse traffic over a multi-ring network.

Each node and each ring is modeled as an array structure. Each node contains information about ring policies in effect at that node, e.g., simultaneous use of more than one token, and the last token arrival information for each ring. It also has message structures which generate traffic for each type at this node and a linked list which constitutes the message queue. Finally, statistics are collected about token and message activity at this node.

The ring structure is similar. It contains policies which exist for the ring in general, such as rotation direction, TTRT, ring data rate capacity, whether the ring can override the TTRT for high priority packets, etc. In addition, the ring structure keeps statistics about token rotation by maintaining information about token activity at node 1.

Since both nodes and rings can operate individually a separate structure is used to keep statistical data about each node's operation on each ring. Operational and statistical information is accumulated about how a particular ring's token is used by a node and about the number and character of the messages transmitted on this ring by this node. With this overall structure, each node and each ring can be looked at separately, providing significant insight on policy effectiveness differences. The simulation model is able to study nonuniformly distributed loads as well.

Further information is kept about messages. Messages are classified by priority and length. There are three categories for each, high, regular and low priorities and short, medium and long, respectively. Multiple message types can occur at a node and each node can have a different set of conditions for its messages. Message statistics are collected at each node for each type. As a result information is available on how messages perform based upon policies existing at a particular node and as influenced by other nodes and ring operations.

As a result of the manner in which the system is modeled and information collected about individual message types, nodes and rings, the amount and character of the information which is available at the end of a simulation run is both extensive and very versatile. For this paper, we mainly present information about policies and message types as they exist across the network in general.

We have not studied how particular nodes and messages performance under special loading conditions.

IV. Comparison of Operational Policies for Single and Parallel Rings

A. General Performance Features for Single and Parallel Rings

To provide an understanding of how various policies affect ring operation, we first present some general performance information for the two ring parallel network. Figures 1a) - 1c) show the mean wait time for various message policies for two parallel rings operating under identical ring policies. Mean wait time is defined as the time from the arrival of the message until the last byte has cleared the sender[3]. A single message type, regular, is used for all runs. Regular messages have a fixed length of 4K bytes. The capacity for each ring is 5 Mbytes/sec. Load factor is the ratio of total submitted byte rate to the total ring byte rate capacity. Load is distributed uniformly over all the nodes and arrivals are offset from the previous arrival by an exponential distribution with a mean time of $1/\lambda$. There are 10 nodes, each connected to both rings; nodes are assumed to be separated by 20 km. making a total ring distance of 200 km. The justification for selecting this distance is to capture the effects of transit time between nodes.

Figure 1a), Figure 1b) and Figure 1c)[4] show mean wait time for exhaustive, single and a TTRT token policy, respectively, for various message distribution policies for a single message type, regular messages. The TTRT time is set at 10 msec. which is 10 times the minimum rotation time[5]. The exhaustive policy is best for load factors of 0.9, bettering the single message policy by a factor of about 7 and the TTRT policy by about 5. However, below a 0.7 load factors, there is not much difference between the policies since the difference between message assignment policies is small (usually less than a factor of 2). However, it is clear that the policy which assigns the message after the token has arrived is generally superior to the others.

For comparison, Figure 2 shows performance for a comparable single ring for various token rotation control policies. Policies include exhaustive, single message and 4 TTRT times between 4 and 30 times the minimum rotation time. Wait time performance for the single ring is generally similar for the same token policy as for the 2 ring condition as shown in Figures 1a) - 1c). Further, we can conclude from Figure 2 that TTRT assignments below 7 are not effective and those above 30 are nearly equivalent to an exhaustive policy.

One feature shown in Figures 1 & 2 needs further clarification when interpreting the results in these and the remainder figures in this paper. The slope of many of the curves reduces as the load increases. For example, this flattening starts at a load factor of about 0.9 for the curves Figures 1b) and 1c), about 1.0 for Figure 1a) and about 0.8 for the TTRT = 4 curve in Figure 2. We investigated this anomaly, to make sure that it was not caused by errors in the simulation system.

The flattening is caused by overloading the capacity of the network to handle a particular message type. When overloaded, wait time becomes unstable and will tend to increase toward ∞ as long as the overload exist. However, in the simulator since run time is finite, a finite value for wait time is measured. Flattening occurs after overload because measured wait time does not increase as fast as load for the simulation times used thereby causing the curves to flatten out. This affect is further exacerbated visually by plotting wait time on a log scale as commented in footnote 3.

To further identify and illustrate overload occurrences, Figures 3a) and 3b) present results for a single ring with both short (100 bytes), high priority, medium (4000 bytes), regular messages. Load factor is the submitted to total capacity (bytes/sec.) as noted previously and high priority load is 10% of the submitted load. Simulator run time the important parameter. Run times vary from 2

---

[3]Wait time, the time from message arrival till the last bit is placed by the source, is used as the basic measure of performance. It, as opposed to response time, is a better indicator of access protocol performance.

[4]Note: some plots are on a log scale so that performance features can be seen over the whole load range.

[5]TTRT values are in terms of the minimum round trip time. A TTRT = 1 is 1 msec. for a 200 km long network. This form extrapolates to different length networks.

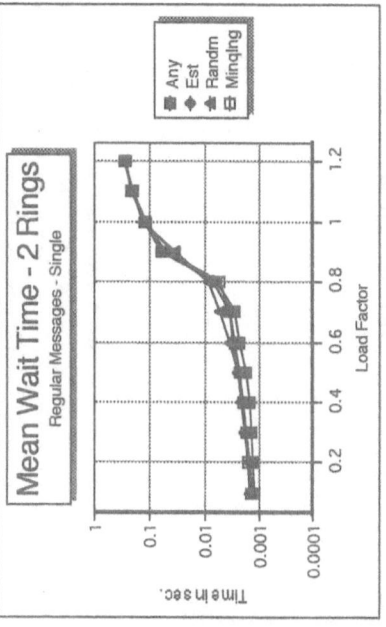

Figure 1a) Performance for Message Distribution Strategies

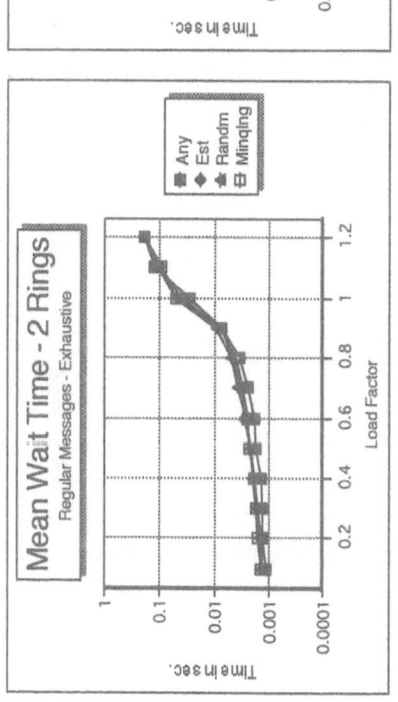

Figure 1c) Performance for Message Distribution Strategies

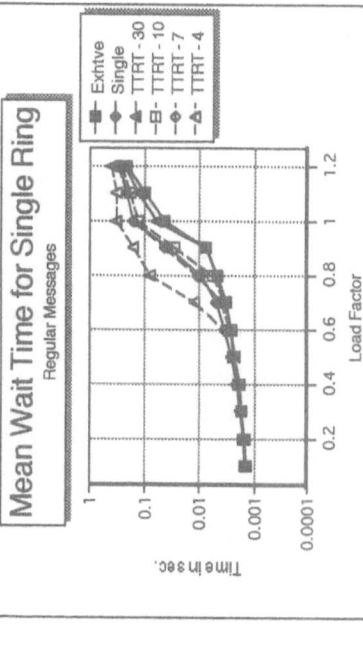

Figure 1b) Performance for Message Distribution Strategies

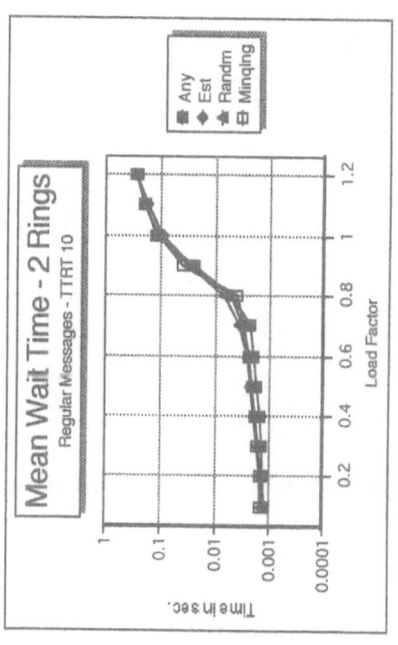

Figure 2. Performance for Single Ring with Various Ring Policies
Conditions: Date Rate - 5Mbytes/sec/ring; Length - 200 km.;
Packet Length - 4K bytes; Rings - 1;
Nodes - 10

Figure 1. Performance of 2 Ring Network for Various Message
Assignment Policies
Conditions: Date Rate - 5Mbytes/sec/ring; Length - 200 km.;
Packet Length - 4K bytes; Rings - 2;
Nodes - 10

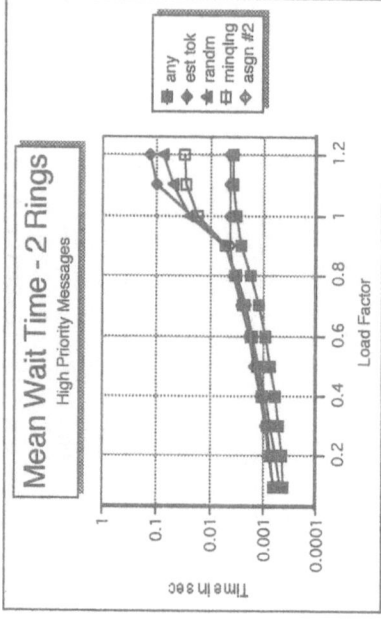

Figure 3a) Performance for High Priority Messages

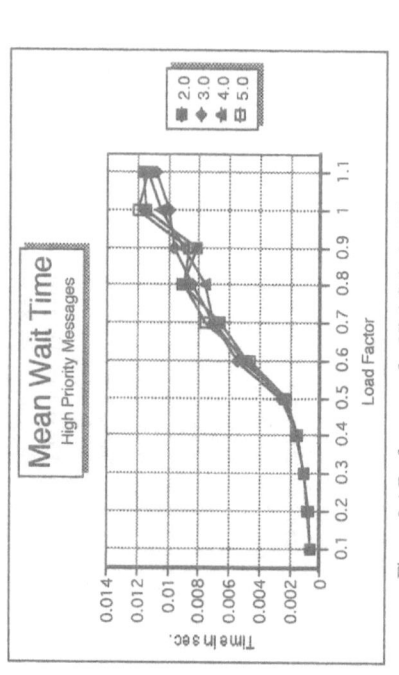

Figure 3b) Performance for Regular Messages

Figure 3 Performance for Single Ring vs Simulator Run Time
Conditions: Date Rate - 5Mbytes/sec/ring; Length - 200 km. ;
Packet Length - 4K bytes; Rings -1;
Nodes - 10;
Token Service Policy - single

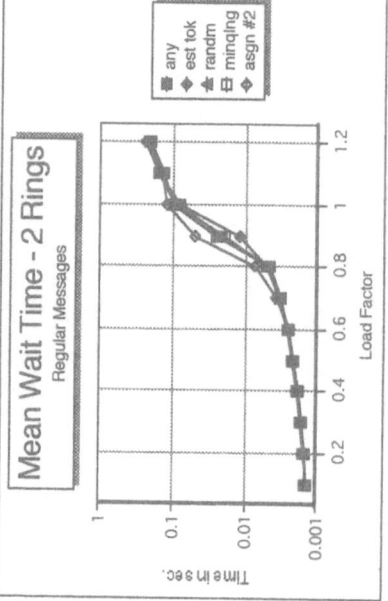

Figure 4a) Mean Wait for High Priority Messages

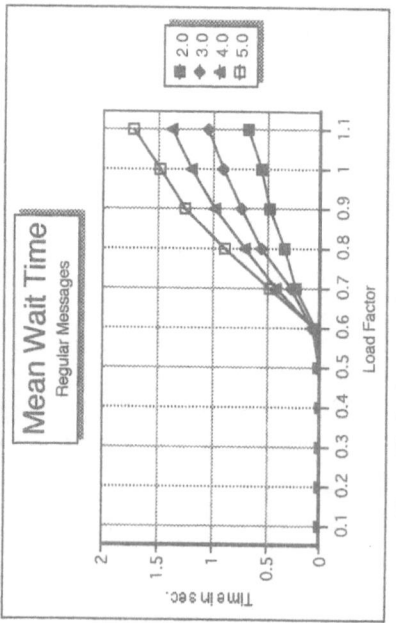

Figure 4b) Mean Wait for Regular Messages

Figure 4 Performance for Multiple Message Traffic for 2 Parallel Rings
Conditions: Date Rate - 5Mbytes/sec/ring; Length - 200 km. ;
Packet Length - 4K bytes; Rings - 2; Nodes - 10
Token Service Policies Ring #1 - Exhaustive; Ring #2 - TTRT = 10.

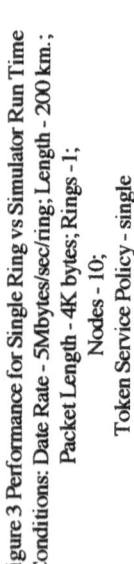

304

to 5 seconds for the actual time the network is in operation and data is collected after a 100 msec. startup time. The high priority messages, Figure 3a) do not overload the network capacity even at a load factor of 1.1 since the there is relatively little change in value due to change in simulation time. However, Figure 3b) shows that regular messages start to overload the network capacity at about 0.6 load factor. Above this load, wait time is unstable as it increases as the simulation time increases. The growth appears to be nearly linear as a function of simulator run time.

Therefore, we must be careful to identify conditions under which overload occurs. Once it does for a particular message type, wait time results are meaningless and should not be considered in comparisons. Also note that the flattening is further exacerbated by plotting wait time on a log scale as commented in footnote 4.

B. Performance Capabilities of Parallel Rings to Support Multiple Message Traffic

In this section we demonstrate the effectiveness of parallel rings to support integrated traffic. Figures 4a) and 4b) present the wait time performance for a 2 ring system with two traffic types, high priority and regular messages. The token policy for each parallel ring is different; for ring #1 — exhaustive and ring # 2 — a TTRT of 10. Figures 5a) and 5b) plot identical runs for a single ring with a TTRT of 10. Load factor, as noted previously, is the ratio of submitted byte rate to the total ring byte rate capacity with high priority message characteristics as noted in Figures 3..

Figure 4a) shows high priority message performance using various message assignment policies for these messages. The best performance is for message assignment policy #4, above. For most load conditions, this policy provides an improvement by a factor of at least 2 - 4 over the other assignment policies. After the total load exceed network capacity, policy #4 still maintains a significant advantage over other policies except direct ring assignment, policy #3. Note that policy #4 provides improved performance because it most effectively uses the arrival of tokens on both rings. Both policies appear to limit the mean wait time to about 6 milliseconds. This is true because both policies #3 and #4 take advantage of the TTRT control policy on ring #2 *and not because they have overloaded capacity*. Another factor which influences high priority message performance is that they can be sent even after the TTRT timer has expired at a node.

Figure 4b) shows that regular messages performance is not appreciably influenced by the high priority message assignment policy used although it is slightly worse at high loads for policy #4. In these runs, regular message assignment is random, policy #1, for all cases.

Figures 5a) and 5b) shows comparable results to those of Figures 4 for a single ring. Setting TTRT =< 10 is fairly effective in supporting high priority messages. Doing so however is disastrous for regular messages where wait time increases by at least a factor of 8. Further overload appears to start for regular messages at 0.9 load factor or less. If TTRT > 10, regular message performance is improved but high priority message performance now deteriorates. Any TTRT setting for a single ring severely compromises the performance results of one or both message traffic types.

From Figures 4) & 5), we conclude that using two parallel rings with different policies is more effective as a means of enabling network performance for multiple traffic types than policies available on a single ring. For the parallel rings, traffic over the load range is more predictable and controllable and less sensitive to operational parameter selection.

Figures 6a) and 6b) further compare the advantage of parallel rings over other configurations — single and bridged networks. The single ring uses a TTRT = 10 with other parameters as noted above. The bridge couples two identical single rings. The bridge performance assumes both networks are equally loaded but is conservative especially at low loads since it assumes no additional wait time due to internal bridge processing. Figure 6a) shows parallel network performance for high priority messages is much better, with up to a 7 fold wait time decrease over the others at high loads. For regular messages performance for both parallel and single rings is equivalent with the exception that the parallel ring system reachs an overload condition for regular messages sooner than for the single ring. The comparisons, Figures 4 - 6, clearly show that a two ring parallel network can handle integrated traffic loads more rapidly and effectively than single or bridged systems.

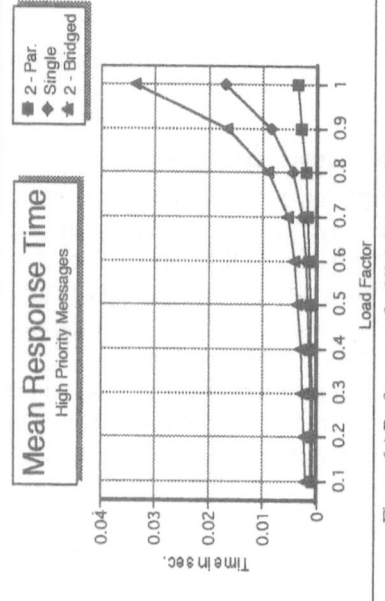

Figure 6a) Performance for High Priority Messages

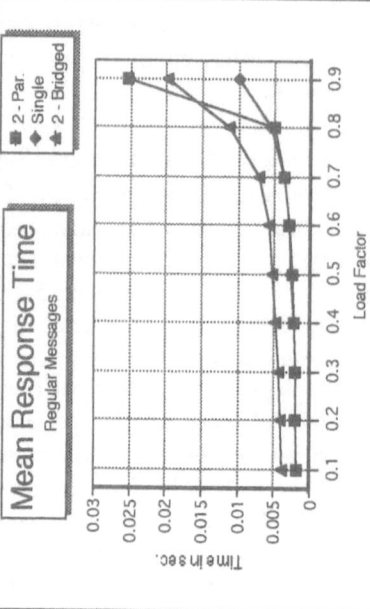

Figure 6b) Performance for Regular Messages

Figure 6 Response Time Comparison for Parallel, Single and Bridged Networks
Conditions: Date Rate - 5Mbytes/sec/ring; Length - 200 km.;
Packet Length - 4K bytes; Nodes - 10
Token Service Policy for 2 Rings: Ring #1 - Exhaustive; Ring #2 - TTRT = 10.
Token Service Policy for 1 & bridged rings: TTRT = 10

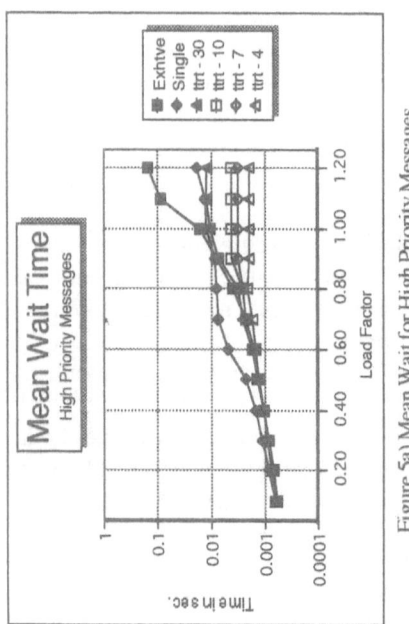

Figure 5a) Mean Wait for High Priority Messages

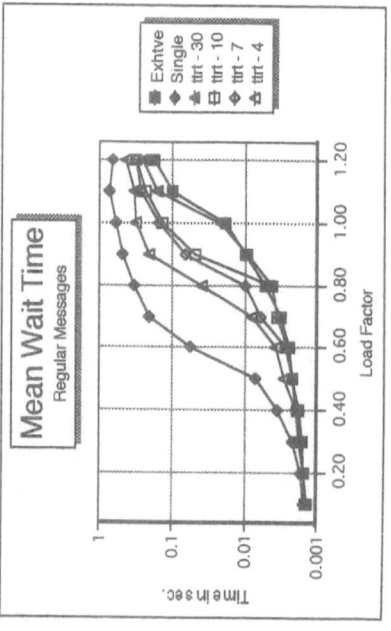

Figure 5b) Mean Wait For Regular Messages

Figure 5 Performance for Multiple Message Traffic for Single Ring
Conditions: Date Rate - 5Mbytes/sec/ring; Length - 200 km.;
Packet Length - 4K bytes; Rings - 1; Nodes - 10

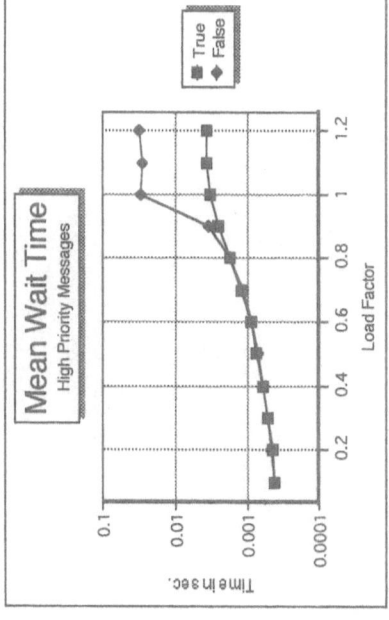

Figure 7a) Performance for High Priority Messages

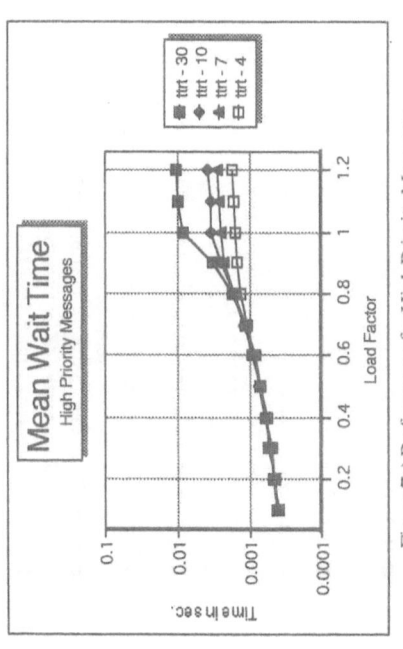

Figure 7b) Performance for Regular Messages

Figure 7 Affect of Target Token Rotation Time Setting for 2 Ring Parallel Network

Conditions: Date Rate - 5Mbytes/sec/ring; Length - 200 km. ;
Packet Length - 4K bytes; Rings - 2; Nodes - 10
Token Service Policies Ring #1 - Exhaustive; Ring #2 - Varied.

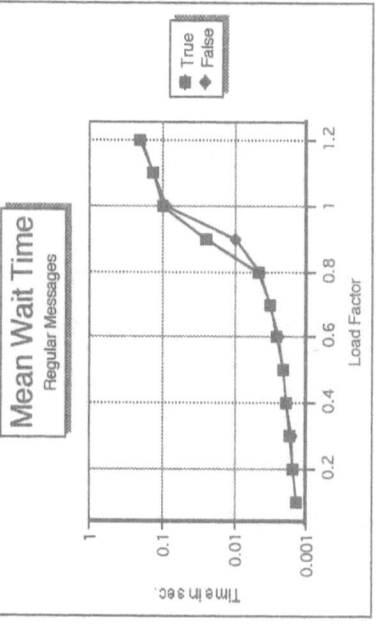

Figure 8a) Performance for High Priority Messages

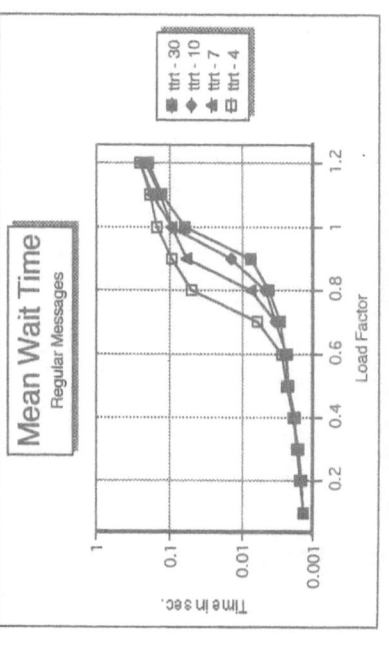

Figure 8b) Performance for Regular Messages

Figure 8 Affect of TTRT Override for 2 Ring Parallel Network

Conditions: Date Rate - 5Mbytes/sec/ring; Length - 200 km. ;
Packet Length - 4K bytes; Rings - 2; Nodes - 10
Token Service Policies Ring #1 - Exhaustive; Ring #2 - TTRT = 10.

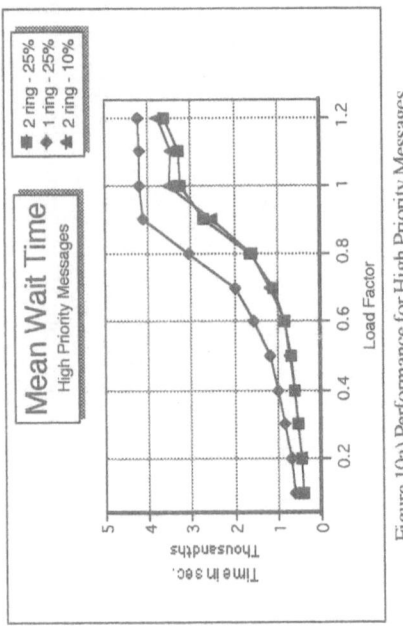

Figure 1(0a) Performance for High Priority Messages

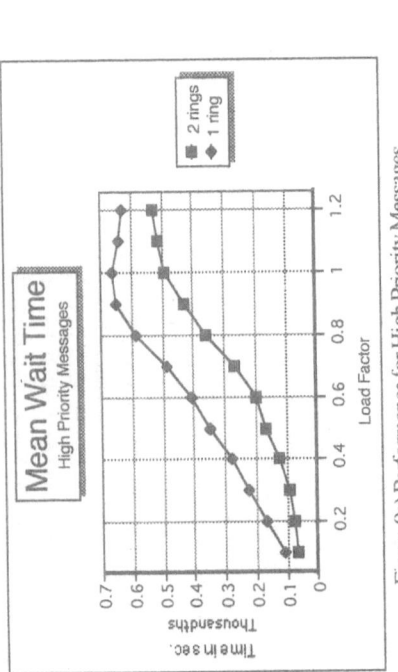

Figure 9a) Performance for High Priority Messages

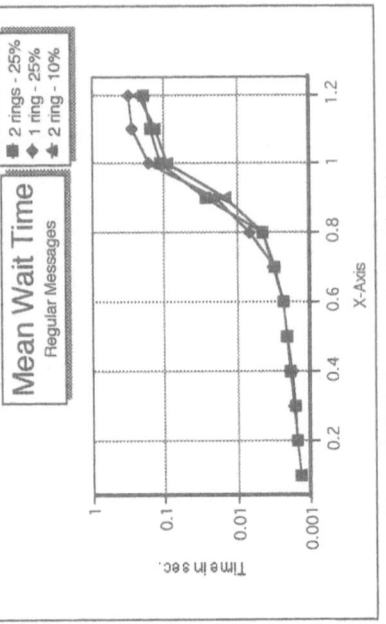

Figure 10b) Performance for Regular Messages

Figure 10. Affect of Increasing High Priority Message Load
Conditions: Date Rate - 5Mbytes/sec/ring; Length - 200 km.;
Packet Length - 4K bytes; Rings - 2; Nodes - 10
Token Service Policies for 2 Ring #1 - Exhtive; Ring #2 - TTRT = 10.
Token Policy for Single Ring: TTRT = 10.

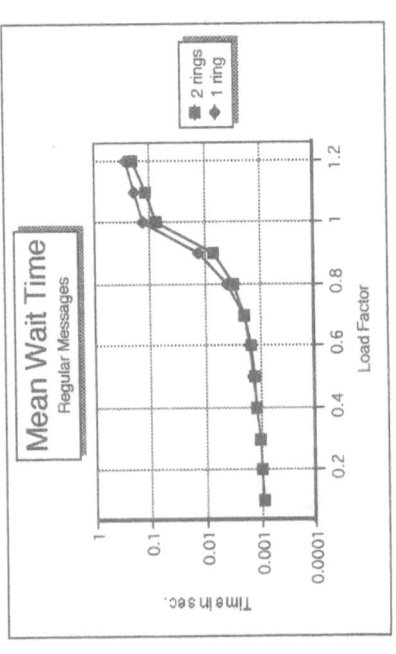

Figure 9b) Performance for Regular Messages

Figure 9 Comparison Between 2 Parallel and Single Ring for 20 Km
Network
Conditions: Date Rate - 5Mbytes/sec/ring; Length - 20 km.;
Packet Length - 4K bytes; Nodes - 10
Token Service Policies for 2 Rings: Ring #1 - Exhaustive; Ring #2 -
TTRT = 10.
Token Policy for Single Ring: TTRT = 10.

C. Study of Parametric Effects on Parallel Ring Network Performance

In the previous section, we established that a parallel two ring network was superior for supporting a combination of high priority and regular type traffic. In this section, we examine the effect of changing control and operational parameters on network performance.

The selection of TTRT, as shown, has a significant affect on performance. The affect of changing TTRT in parallel rings is shown in Figure 7 for the two ring case. Ring #1 uses an exhaustive submittal policy and ring #2 uses a range of TTRTs between 4 and 30 times the minimum token rotation. It is evident from the curves that a TTRT = 10 is a very suitable compromise. The mean wait time for high priority messages at high loads decreases by up to a factor of 3 using a TTRT < 10 while regular message wait time decrease by about a factor of 2 for TTRTs > 10. TTRT changes for the parallel ring system do not cause so drastic a change as for a single ring, Figure 5.

A second factor in TTRT control is to allow high priority messages to be sent even after the TTRT has expired. As noted previously, this policy is used for high priority traffic in FDDI-I systems. Figures 8 show this affect for high priority and regular messages. When token use after TTRT expiration is set to *true* at all nodes, high priority message performance at high loads is much better than when token us is set to *false*. However, regular message wait time is slightly degraded. Since good, predictable high priority message access is very desirable, setting the factor to *true* is recommended.

Figure 9 compares the performance between two parallel rings and a single ring for the condition where the ring length has been reduced from 200 km to 20 km. This case is more typical of a LAN covering a campus; the longer distance represents a metropolitan area network covering a city. Here again, the high priority message performance for the 2 parallel rings is substantially better than for a single ring with equivalent loading and the regular message performance in both cases is about equal.

Another factor which may affect performance is the percentage of high priority load a parallel network can effectively support. Figures 10 compare a 25% high priority load with the nominal 10% load used for previously presented results. The results show a minimal change in performance for the high priority load change while the substantial improvement over single ring performance still remains.

Another factor possibly affecting performance is the use of more than one token simultaneously by a node. Simultaneous use complicates the equipment at the node since multiple transmitters are now required. Figure 11 compares single and multiple token use. It shows negligible performance difference between the two conditions. Therefore, single token use appears to be warranted.

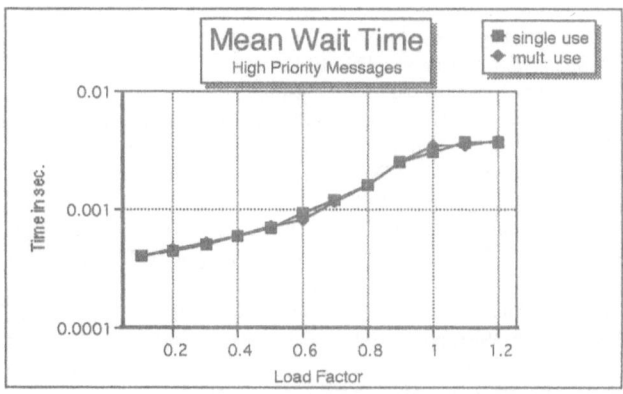

Figure 11 Affect of Single Token Use
Conditions: Date Rate - 5Mbytes/sec/ring; Length - 200 km.;
Packet Length - 4K bytes; Rings - 2; Nodes - 10
Token Service Policies Ring #1 - Exhaustive; Ring #2 - TTRT = 10.

V. Concluding Remarks

The results presented in this paper support the following conclusions for parallel token ring configurations over single or bridged systems:

1. The parallel network supports integrated traffic with a significant reduction in mean wait time.
2. The superior performance of the parallel network is demonstrated over a wide range of operational conditions. Performance is less sensitive to TTRT policies settings.
3. Parallel network system has a much wider range of message assignment and token control policies. Using different policy choices for each parallel ring enables the performance improvement.
4. The policy combination determined as most effective for high priority and regular message traffic are:
   a. The token control policy is for one ring to have its TTRT set to a value needed for effective access for high priority messages and for the second ring to have an exhaustive submittal policy. Token use after TTRT expiration should be set *true*. Simultaneous token use at a node is unnecessary.
   b. The message submittal policy is for high priority messages is to the ring where the token arrives first; for regular messages a number of message submittal policies are acceptable.
5. The performance gains are available without significantly altering the media access protocol for token rings.

From these conclusion, it is evident that parallel token ring networks have the capability and potential to support a much wider range of network operational requirements and quality of service features with less complex access protocols than their single link or bridged counterparts.

We feel that the research presented in this paper represents only the beginning for understanding the potential performance available from using parallel links in networking. First, it deals only with token rings; many networks are bus connected. Although we have shown in previous work that access gains are available for parallel bus networks [19], their ability to effectively support integrated traffic with simple protocols is unknown. Further, there are many other network performance and management requirements which may be amenable to improvement or simplicity of operation by employing parallel token ring or bus networks. In addition, there are many untried control and distribution policies and the effectiveness of extensions beyond two parallel links have not been investigated. Finally, the hardware required to support parallel network access with effective control and message distribution policies must be available and cost effective.

VI. References

1. Maly, K., et al.: "Parallelism for High Speed Networks", *Proc. of IEEE Workshop on Arch. & Implementation of High Performance Communication Subsystems*; Feb. 1992; 6 pp.
2. Zitterbart, M.: "Parallel Protocol Implementation on Transputers," Ibid.
3. La Porta, T. F.: "Design, Verification and Analysis of a High Speed Protocol Implementation Architecture," Ibid.
4. Hoffman, B., et al.: "On the Parallel Implementation of OSI Protocols," Ibid.
5. Netravali, A. N..; Roome, W. D.; Sabnani, K.: "Design and Implementation of a High-Speed Transport Protocol," IEEE Trans. on Comm., Vol. 38; 1990; pp. 2010-2024.
6. Haas, Z.: "A Communication Architecture for High Speed Networking," *Proc. of IEEE Infocom '90*; pp. 433-441.
7. Jain, N., Schwartz, M.; Baskow, T.R.: "Transport Protocol Processing at GBPS Rates," *Proc. of SIGCOMM '90*; pp. 188 - 199.
8. La Porta, T.F.; Schwartz, M.: "Architectures, Features and implementation of High-Speed Transport Protocols," *IEEE Network Magazine*; Vol. 5; 1991; pp. 14 - 22.
9. Maly, K., et al.: "Parallel TCP/IP for Multiprocessor Workstations," *Proc. of 4th IFIP Conference on High Performance Networking*; Liege, Belgium; Dec 1992; pp. C1.1 - C1.16.
10. Mukkamala, R., et al.: "Message Assignment Policies for High Speed Parallel Networks," *Proc. of 1st Int. Conf. on Comp. & Comm.*; 1992, pp. 77 - 81.
11. Mukkamala. R., et al.: "Modelling and Analysis of High-Speed Parallel Token Ring Networks," *Proc. of 17th LCN*, Minn., MN; Sept. 1992; pp. 624 - 632.
12. Vasko, D.A.; Rajsuman, R.: "Analysis and Simulation of Multiple Ring Token Networks," Ibid. ; pp. 649 - 657.

13. Bhuyan, L.N.; Ghosal, D.; Yang; Q: "Approximate Analysis of Single and Multiple Ring Networks," *IEEE Trans. on Computer*; Vol. 38, No. 7; July 1989; pp. 1027 - 1040.

14. Sethi, A.S.; Saydam, T.: "Performance Analysis of Token Ring Local Area Networks," *Computer Networks and ISDN Systems*, Vol. 9; March 1985; pp. 191 - 200.

15. Maly, K.; et al.: "Scalable Parallel Communications," Old Dominion University Technical Report; Feb. 4, 1992

16. Christensen, K.J.; Noel, F.C.: "Parallel Channel Token Ring Local Area Network," *Proc. of 17th LCN* ; Minn., MN; Sept. 1992; pp. 634 - 638.

17. Skov, M.: "Implementation of Physical and Media Access Protocols for High Speed Networks," *IEEE Comm. Magazine*; June 1989; pp. 45 - 53.

18. Ross, G.E.: "FDDI - A Tutorial," *IEEE Comm. Magazine*; Vol. 24; May 1986; pp. 10 - 23.

19. Maly, K., et al.: "Dynamic Allocation of Bandwidth in Multichannel Metropolitan Area Networks," To appear in Computer Networks and ISDN; 1992.

20. Stallings, W. *Data and Computer Communications* 3rd Edition; Chapter 11. Local Area Networks; Macmillan Pub.; NY.

21. Zafirovic-Vukotic, M.; Niemegeers, I. G.; Valk, D. S: "Performance Analysis of Slotted Ring Protocols in HSLAN's," *IEEE JSAC*, Vol. 6; No. 6; July 1998; pp. 1011-1024.

The Journey of SMDS -- A LAN Interconnect Solution

Mahendra Joshi and Fuyung Lai

IBM
500 Park Offices Drive
Research Triangle Park, NC 27709
Internet: joshi@ralvm29.vnet.ibm.com
Internet: fuyung@ralvm29.vnet.ibm.com

Introduction

SMDS is a switched, high-speed, connectionless data service and is based on the IEEE 802.6 metropolitan area network specification [1,2,3]. The SMDS system is composed of several major components: a subscriber-network interface (based on subset of 802.6), a switching system (SS), an inter-switching system interface (ISSI), and an Inter-carrier interface (ICI), as shown in Figure 1. An important characteristic of this structure is its compatibility with Asynchronous Transfer Mode (ATM) which is the building block for cell-switching based BISDN services. Bellcore has defined the Switching System, the Inter-Switching System Interface, and the Inter-Carrier Interface in the Technical References. However, it is clear that the strategic direction is to allow easy migration from SMDS to ATM [7,19].

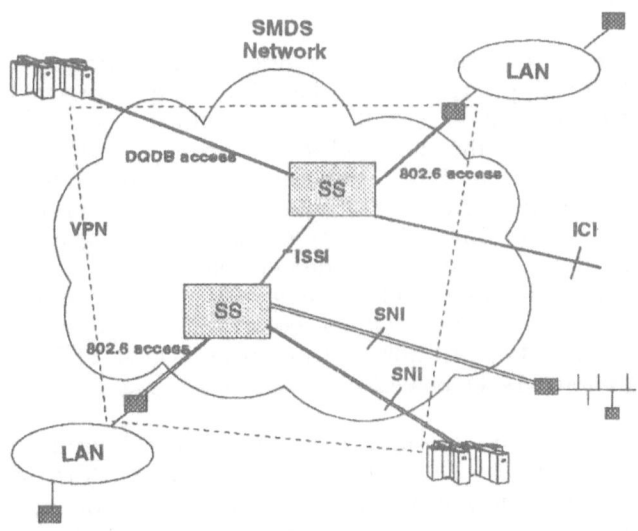

Figure 1. Switched Multi-megabit Data Service

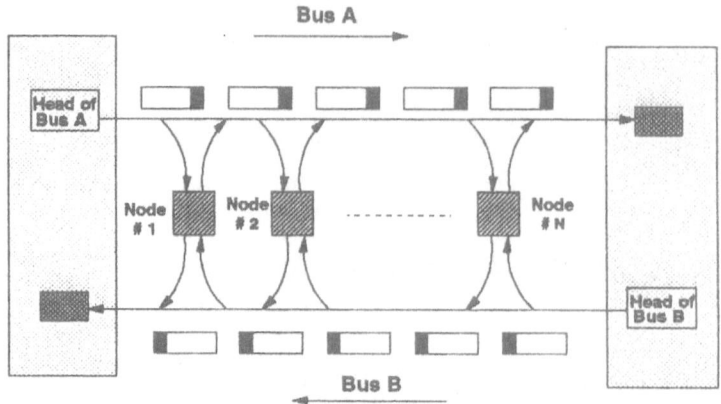

Figure 2. Distributed Queue Dual Bus

Initial SMDS offerings are likely to be confined to traffic within a single local access and transport area (LATA) so that a Bell Operating Company can serve all its SMDS customers within a Metropolitan area [10,18]. Long distance carriers have provided SMDS service at T1 rate for Inter-LATA traffic. However, interoperability between different carriers are expected to be available later. When the interoperability is available, the users will be able to transmit data across wide area networks.

The cell-based IEEE 802.6 MAN standard has been selected as the basis for the user-access protocol to SMDS over T1, T3 and higher speed links. Historically, the dual bus proposal by Telecom Australia, called QPSX, was agreed upon as the basis for a MAN standard and the network, known as Distributed Queue Dual Bus (DQDB), has since become the building block for the IEEE MAN. The DQDB network consists of two unidirectional slotted buses with a collection of end stations attached to both buses, as shown in Figure 2. The stations could transmit or receive data through both buses. Access requests for transmission on the upstream bus are generated over the down stream bus, and vice versa. The 802.6 Standard provides two structures for the bus, a looped-bus or a parallel-bus [20].

Due to the expense of providing dual paths for the looped-bus, SMDS uses 802.6's parallel bus approach rather than the looped bus. There are two access modes supported in DQDB: the queue arbitrated access mode (for non-isochronous traffic) and the pre-arbitrated access mode (for isochronous traffic). However, most of the SMDS implementations only use queue arbitrated access mode. One consideration that have been brought up is the DQDB fairness problem [5]. When the network is running under heavy load condition, the fairness access breaks down. The heavily used nodes closer to the heads of the buses starve the nodes further away. There are many solutions proposed to solve this problem, such as bandwidth reservation schemes [6]. However, on the point-to-point link that provides access to SMDS, DQDB fairness is not a problem. [4,11].

Even though the MAC level interface, called the SMDS Interface Protocol (SIP), is based on IEEE 802.6, the customer premise network (CPN) may or may not use DQDB based network. A customer premise equipment, such as host, bridge, or router, usually uses a channel service unit/data service unit (DSU/CSU) to connect to SMDS at the Subscriber Network Interface (SNI). In the next section, the features and the characteristics of the SMDS are discussed. Following that, the interface protocol between the DSU/CSU and the DTE, called DXI, and a low speed frame based user network interfaces are described.

Specific Features and Characteristics of SMDS

SMDS is a high-speed connectionless or "datagram" packet public switched service. To efficiently serve high speed data applications, Level 3 Protocol Data Units are variable length packets, up to a maximum of 9188 octets. This can avoid the need to segment and reassemble MAC frames for bridging IEEE 802 LANs. The source and destination addresses are based on E.164, the ISDN numbering scheme, to be consistent with the support of SMDS towards the BISDN service. In addition, the other important features and characteristics of SMDS are discussed further. Many of these features are selected by the subscriber through a prior arrangement with the SMDS service provider. However, some features may be activated on demand for selected SMDS data units being transmitted.

SMDS Addresses

SMDS complies with CCITT E.164 addressing scheme and the North American Numbering Plan. Accordingly, a ten digit address followed by a prefix of "1" is recognized. SMDS also provides a feature similar to the multi-casting feature provided by LANs, known as Group Address. When a CPE sends a SMDS data unit to a group address, the SMDS network will send copies of the SMDS data unit to a set of destination addresses previously specified by the group address. A group address can represent up to 128 individual addresses and is distinguished from an individual address based on the address type field contained in the L3PDU header. A specific individual address can be part multiple group addresses

Multiple addresses assigned to SNI

SMDS switch can support upto 16 addresses for each SNI. Some switches support even a much larger number. The subscriber can use these multiple addresses to carry traffic for multiple CPEs, multiple protocols, etc.

Individually addressed SMDS Data Unit Transport

SMDS provides flexibility to address each SMDS data unit (up to 9188 bytes) individually. The CPE specifies both a source address and a destination address for each SMDS data unit and the network transports these data units individually based on the destination addresses.

Local communication for Multi-CPE Access

Since IEEE 802.6 is the basis for SIP, local communication on the multi-CPE access is not precluded. Some of the local communication will be visible to the switch system, however the switching system will distinguish the local communication from the out-going traffic by looking at the destination addresses. The local communication is ignored by the switching system.

Access Classes

SMDS provides DS-1 and DS-3 (in future OC-3) based access paths between CPE and the switch. No access classes are provided for DS-1 access; so the access class for DS-1 is essentially the maximum throughput which can be achieved on DS-1 access after removing the overhead (approximately 1.17

	Access Rate	Sustained Information Rate (SIR)		
OC3	155 Mbps	123.6 Mbps (From ETSI CBDS Draft)		
DS3	45 Mbps	Access Classes	1	4 Mbps
			2	10 Mbps
			3	16 Mbps
			4	25 Mbps
			5	34 Mbps
DS1	1.544 Mbps	1.17 Mbps		
N*64	N*64 Kbps	N*63.6 Kbps		
N*56	N*56 Kbps	N*55.7 Kbps		

Figure 3. Access Class of SMDS service.

Mbps). The SMDS Interest Group (SIG) has agreed on a DXI based low speed interface to support fractional DS1 access.

For DS3-based SMDS access, there is a range of Access Classes. Each access class provides for different traffic characteristics, in other words, it prescribes limits on the sustained information rate and on the burstiness of the information transmitted from the CPE to the SS. There are 5 different classes defined for the DS3 SMDS access. The allowed sustained information rate (SIR) for class 1, 2, 3, 4, and 5 is 4, 10, 16, 25, and 34 Mbps, respectively as shown in Figure 3. Of course, different access class implies different charges for the user. These access classes were specified to meet token ring (4 and 16 Mbs), ethernet (10 Mbs), and full DS-3 rate reduced by SMDS overhead (34 Mbs). An additional access class (25 Mbs) was added to provide a mid-value between 16 and 34 Mbs. Some DSUs allow the CPE to comply with the access class by providing two different clocks: DS-3 rate clock on the network side and an access class matching clock on the CPE side. Access classes permit efficient planning on the part of subscribers as well as service provider.

Credit Manager

The credit manager is an algorithm (similar to Leaky-bucket scheme) implemented by the network as the admission control for user packets to the SMDS. Credit consumption is based on the number of level 3 protocol data units (L3PDU) transferred between CPE and the network. L3PDUs that exceed the access class are discarded by the network. It's up to the end user to recover the lost packet. To avoid unnecessary data loss, it's preferred that users also implement a congestion avoidance and control schemes at the end nodes. Some DSUs are equipped to help this situation by providing two different clocks on the network side and CPE side.

For each direction of traffic follow, a credit manager is maintaining a balance of available credit measured in octets. The parameters that define the operation of each credit manager are chosen at subscription time and are described as follows:

- C_max -- the maximum amount of credit, measured in octets.

- I_inc -- the interval between increments to the credit, measured by counting L2PDUs received or transmitted by SS.

- N_inc -- the number of octets that the credit is incremented by whenever I_inc L2PDUs have been counted.

As an example, every L3PDU that is sent by CPE to the SS, the credit manager compares the length of the information field contained in the L3PDU with the value of the credit. If there is sufficient credit available, the L3PDU is accepted by the SMDS network and the credit is decremented by the length of the L3PDU. The L3PDU will not be delivered and the credit is unchanged if there is insufficient credit. The values of the three parameters control not only the allowed burstiness of the information transfer but also the Sustained Information Rate (SIR), which is the maximum information transfer rate that can be sustained for a long period of time.

Address Screening

SMDS address screening can be used to allow or to restrict the reception of SMDS L3PDUs or the delivery of L3PDUs. There are two types of screening: source screening and destination screening. The screening features can be positioned to discard L3PDUs when addresses do not match the candidate addresses in the screening table. The screens provide a set of allowed or disallowed addresses. By using the source address screening and the destination address screening features in conjunction, a "virtual private network" may be provided. With other addressing features in SMDS, such as group addressing and the ability to assign multiple addresses to a single user interface, a group of end users can form a virtual private network. Meanwhile, an end user could be a member of several groups or several virtual private networks. Both individual address screens and group address screens can be provided. Since there may be more addresses assigned to SNIs than a number of address screens; an address screen can be used by more than one source addresses.

Security

Security and privacy issues for the subscribers are handled in various ways. SMDS traffic is routed over an access path which is assigned to an individual subscriber. SMDS switch validates the SMDS source address for each SMDS data unit to ensure that the source address is indeed assigned to that SNI. Address screening feature (associated with source address and destination address) discussed earlier also enables the subscribers to keep the data exchange within a virtual private network.

SMDS Interface Protocol (SIP)

SMDS Interface Protocol (SIP) defined in Bellcore TR-TSV-000772 is required by Data Terminal Equipments (DTEs) to access SMDS Network. There are three levels of SIP according to the function performed. The level 1 SIP is dependent on the specific physical medium used for transmission and DS-1 and DS-3 specific SIP is described in Bellcore TR-TSV-000773. This physical layer is further divided into two sublayers, the physical layer convergence procedure (PLCP) and transmission system dependent layer (TSD). The SIP level 2 functions include segmentation/reassembly of variable length L3PDU, bit error detection and framing for SIP L3PDU through the use of fixed length L2PDUs. SIP level 3 provides SMDS addressing for the data as well as the ways to detect lost L2PDUs.

Level 3 Protocol Data Unit (L3PDU)

L3PDU and its relationship to L2PDUs is shown in Figure 4. Some highlights of the L3PDU are described next.

- **BEtag** - There are two one octet fields in the header and trailer portion of L3PDU which contain matching beginning-end tags. The value may range 0

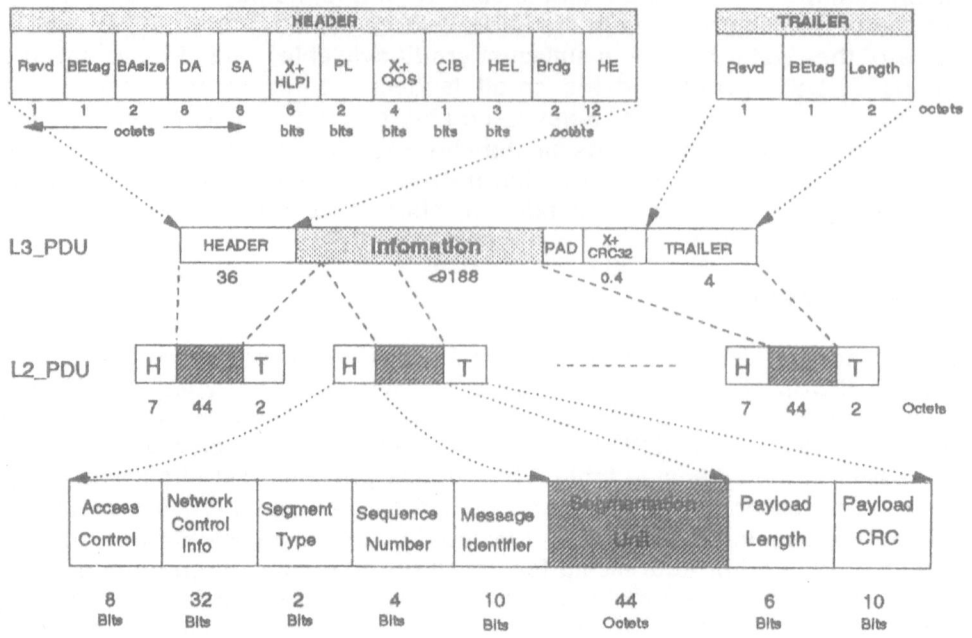

Figure 4. SIP Level 3 Protocol Data Unit.

through 255. These tags are used to form an association between the first and the last segments of L3PDU.

- **BAsize** - This two octet field provides the length in octets for the part of the L3PDU which begins at the Destination Address field and ends at and including CRC32 field.

- **DA** - Destination Address - This 8 octet field is divided into 4 bits of address type and 60 bits of address. The address type is used to distinguish between an individual address (1100) and group address (1110). The field has enough space to contain an international E.164 addressing of the destination. Initial deployment of SMDS will require a prefix "1" followed by digit address followed by 1s to fill up the unused digits.

- **Source Address** - Similar to Destination Address, this field is also an 8 octet long subdivided into 4 bits of address type and 60 bits of address. The address type will always contain "1100" to indicate individual address. Similarly, the address field will contain E.164 address for the source.

- **Higher Layer Protocol Identifier** - This is a 6 bit field which aligns SIP format with the DQDB format.

- **PAD Length** - This 2 bit field indicates the number of octets in the PAD field to make the L3PDU 32 bit aligned.

- **Quality of Service** - This 4 bit field is meant to align SIP format with the DQDB format.

- **CRC32 Indication Bit** - This 1 bit indicates whether the CRC32 field is used. "1" indicates that CRC32 is present and "0" indicates CRC32 was not generated.

- **Header Extension Length** - This 3 bit field indicates the number of 32 bit words in the Header Extension field. For SMDS the Header Extension is always 12 octets. So, the Header Extension Length field always contains the value "011".

- **Bridging** - This 2 octet field is intended to align the SIP and DQDB formats.

- **Header Extension** - This is a 12 octet field which is broken up into subfields (Element Length, Element Type and Element Value) which can be repeated. It is used to indicate Carrier Selection etc. The unused area is PAD which starts with at least one octet of all zeros.

- **Information** - This variable length field is used to contain up to 9188 octets of user data.

- **PAD** - This field can be anywhere from 0 octet to 3 octets in length and contains all zeros. It enables L3PDU to be 32 bit aligned. This is related to PAD Length field described earlier.

- **CRC32** - As described in CRC32 Indication Bit earlier, CRC32 may be present or absent. The CRC32 can perform error detection for the L3PDU fields beginning from the Destination Address field and ending in CRC32. This field is used optionally by the user.

- **Length** - This 2 byte field contains the same value as BAsize field described earlier.

Level 2 Protocol Data Unit (L2PDU)

L3PDU is segmented and carried into 53 octet L2PDUs. A new header and trailer including Payload CRC are generated for each L2PDU. Each L2PDU has 44 octet payload (Segmentation Unit) field to carry the segmented L3PDU information. The last L2PDU may not be completely filled with 44 octet payload and may need to be padded. The highlights of L2PDU format are as follows.

- **Access Control Field (ACF)** - This 8 bit field is divided into the Busy subfield and a 7-bit subfield for cells passing between CPE and SS. The Busy subfield contains "1" or "0" to indicate whether the L2PDU contains information or is empty.

- **Network Control Information** - This 4 octet field contains a value of FFFFF022 for L2PDUs containing information and zeros for empty L2PDUs.

- **Segment Type** - This 2 bit field indicates the position of this L2PDU with respect to the entire L3PDU. A value of "10" is used to indicate the Beginning of Message (BOM), "00" for Continuation of Message (COM), "01" for the End of Message (EOM) and "11" for a Single Segment Message (SSM). For an empty L2PDU, the Switching System will put "00" in this field.

- **Sequence Number** This 4 bit field contains the sequence numbers of L2PDUs within a L3PDU. It is used to reassemble L3PDU and ensure that all the L2PDUs have arrived in correct sequence.

- **Message Identifier (MID)** - This 10 bit field indicates that the L2PDUs are associated with same L3PDU. For a Single Segment Message the MID value is zero.

- **Segmentation Unit** - This 44 octet field is the information field obtained by segmenting L3PDU.

- **Payload Length** - This 6 bit field is used to indicate how many of the 44 octets of the Segmentation Units contain actual data. For BOM and COM, this field indicates 44 octets. For EOM, this field may indicate any value from 28 to 44 in multiples of 4 octets. For empty L2PDUs the switching system set this field to zero.

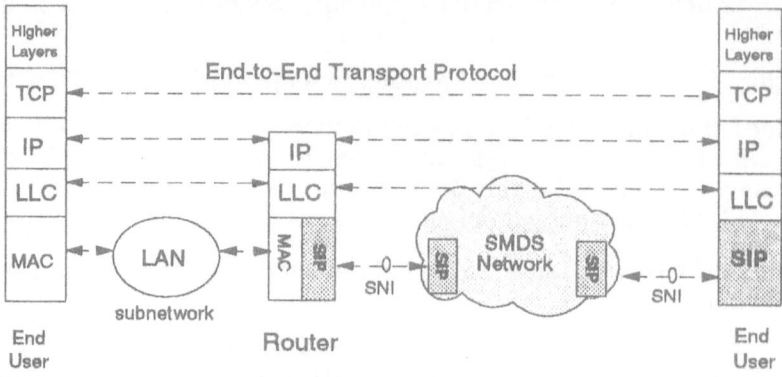

TCP : Transmission Control Protocol
IP : Internet Protocol
LLC : Logical Link Control
MAC : Medium Access Control
SIP : SMDS Interface Protocol
SMDS : Switched Multi-megabit Data Service

Figure 5. Protocol Stack of a Routing Example

- **Payload CRC** - This 10 bit CRC performs error detection for the Sequence Number, Segment Type, Message Identifier, Segmentation Unit, Payload Length and Payload CRC. However, it does not cover the Access Control Field and Network Control Information field.

Multiple L3PDUs in Transit

SMDS allows multiple L3PDUs to travel across the network concurrently. The subscriber may specify at subscription time whether only one L3PDU or upto 16 L3PDUs should be allowed to travel across the network concurrently.

Data eXchange Interface (DXI)

SMDS Interface Protocol (SIP) defined in Bellcore TR-TSV-000772 is required by Data Terminal Equipments (DTEs) to access SMDS Network (see Figure 5).

There are three levels of SIP according to the function performed. DXI is the interface between a DTE and a Data Service Unit (DSU) and the implementation of the three SIP levels is split between the DTE and the DSU [12]. Level 3 of the SIP is performed by the DTE. Level 1 and 2 of the SIP are implemented by the DSU. The DXI interface described here is based on Data eXchange Interface revision 3.2 according to SMDS Interest Group function specification, SIG-TS-001/1991. As shown in Figure 6, the DXI physical layer defines some common physical interfaces, such as V.35, EIA 449, and HSSI, across the link between the DSU and the DTE. The DXI link layer is the HDLC based protocol which is used to exchange SIP 3 protocol data unit (L3_PDU) and local management information between the DTE and the DSU.

The frame structure is described in Figure 7. The flag is 8 bits long with a binary value of 01111110. Bit stuffing scheme is applied when the DTE or DSU is transmitting data, that is, a "zero" bit is stuffed into the data frame after five contiguous "one" bits. The stuffed bits shall be removed when the DTE or DSU receives the frame. The least significant bit is the right most bit and shall be

320

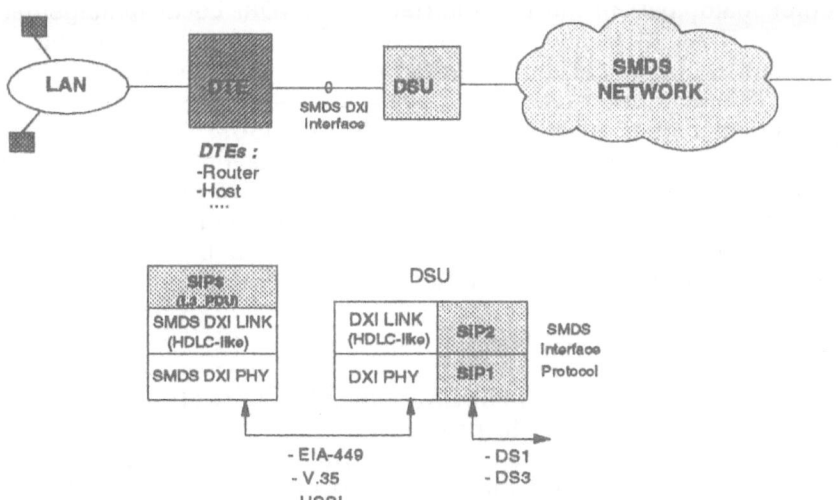

Figure 6. DXI Protocol Stack

transmitted first. The address and the control fields are also described in Figure 7.

The Address field is used to identify whether the frame is a command or response frame, the station address for the frame, and the logical link type contained within the frame. The Control field is a one octet field to identify the frame type. There are two types of frames, UI frame or TEST frame. The UI frame contains L3_PDU or management information.

(a) *Frame Format*

Flag	Addr	Ctrl	(INFORMATION)	FCS	Flag	
1	1	1	0=< N =< 9234	2/4	1	Octets

(b) *Address Field Encoding*

TYPE	DIRECTION	ADDRESS							
		L 8	L 7	L 6	L 5	L 4	S 3	CR 2	AE 1
Command	To DSU	L	L	L	L	L	1	0	1
Response	To DSU	L	L	L	L	L	1	1	1
Command	From DSU	L	L	L	L	L	0	1	1
Response	From DSU	L	L	L	L	L	0	0	1

AE : Address Extension, reserved, set to 1   S : Station Address

CR : Command Response   L : Logical Link

(c) *Control Field Encoding*

COMMAND/RESPONSE	CONTROL							
	8	7	6	5	4	3	2	1
UI (un-numbered Information)	0	0	0	0	0	0	1	1
TEST	1	1	1	P/F	0	0	1	1

Figure 7. DXI Frame Format, Address Field, and Control Field.

The other major part of the DXI interface is the DXI Local Management Interface (LMI). The LMI is based on the Simple Network Management Protocol (SNMP) and in particular the SNMP's Management Information Base (MIB). The DTE (Router/Host) is a proxy agent for SNMP requests that relate to network management information stored in the DSUs. The DXI LMI defines a procedure to access the information in the DSU using SNMP like commands without complicating the design of the DSU. For more detail information about the DXI interface and the DXI LMI, an interested reader may refer to SMDS Interest Group Technical Specification, "SMDS Data Exchange Interface Protocol Revision 3.2", SIG-TA-001/1991.

DXI/Subscriber Network Interface

A frame based interface protocol, DXI/Subscriber Network Interface, is agreed by the SMDS Interest Group to provide an alternate user access interface to networks supporting SMDS [17]. As described before, Bellcore TR-TSV-000772 defines SMDS Interface Protocol (SIP) level 2 and level 3 protocols and procedures to access SMDS. The DXI/SNI defines an alternate level 2 protocol to be used in place of the level 2 protocol. The level 3 protocol and procedures are unchanged from the TR-TSV-000772.

The new alternate level 2 protocol is based on the SMDS Data eXchange Interface Protocol, Revision 3.2, which defines an HDLC based Layer 2 protocol between an SMDS DSU/CSU (DCE) and a Router/Host (DCE). Thus, the DXI/SNI provides the HDLC based interface protocol between a network supporting SMDS and the Customer Premises Equipment at 56, 64, N*56 and N*64 kbps speeds. Based on this interface, the DQDB based interface as well as the segmentation and reassembly functions are not required in the CPE, that is, a special SMDS DSU/CSU is not required in the CPE. However, the standard DSU/CSU (T1 DSU/DSU, for example) is still needed.

Encapsulation of Protocols

An easy scheme to identify the protocol of a packet transmitted through a network is very important for a bridge, router, or a host since a single station could support multiple protocols under today's networking environment. Given that most of the protocols are designed independently, it's very difficult to find a common way to parse a packet to find out the associated protocol of the packet. There are two commonly used encapsulation methods for carrying

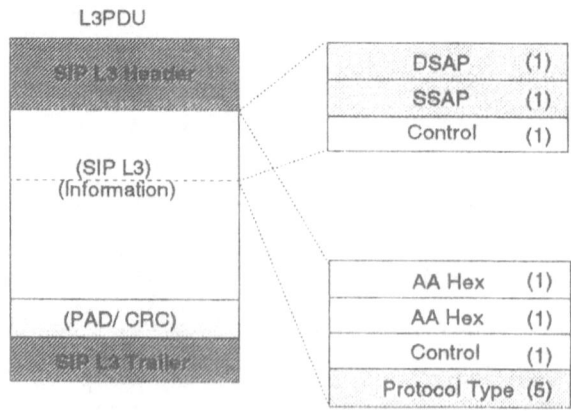

Figure 8. Encapsulation Scheme.

interconnect traffic over a network [14,15,16]. The first scheme is to offer a "protocol type" field in the network layer header, such as Network Level Protocol IDentification (NLPID) field. As it is in the Frame Relay, the NLPID field is administered by ISO and CCITT. It contains values for many different protocols, such as IP, CLNP, and IEEE Subnetwork Access Protocol (SNAP). This NLPID field tells the receiver what encapsulation or what protocol follows. The other method is to use IEEE 802.2 LLC encoding approach. The concept is to use "service access point (SAP)", or so called "socket", which provides independent numbering in the source (SSAP) and the destination (DSAP). [13].

There are two contributions related to the encapsulation of interconnect traffic over SMDS. One is RFC-1209 "IP over SMDS" and the other is the contribution from DEC, "Specification for Implementation of Connectionless OSI over SMDS". Even though there is a protocol type field in the SMDS header, both contributions are based on LLC encoding or SAPs scheme. The reasons are as follows:

- Some protocols have been using 802.2 encoding scheme.
- The concern of running out of values (6 bits protocol type field).
- The problem of administering the protocol field.

In general, the rule of the encapsulation is that protocols that have an assigned SAP value will have the DSAP, SSAP, and CTL fields following the SIP L3PDU header (see Figure 8). Protocols that do not have an assigned SAP value will use SNAP SAP encoding. In this case, following the SIP L3PDU header, there are two octets each with hexadecimal value AA, one octet CTL, and 5 octets protocol type field

The Relation With Other High Speed Networking Alternates

SMDS and Frame Relay

Frame Relay and Switched Multi-megabit Data Service (SMDS) are two of the important emerging high-speed telecommunication services that the carriers and the Regional Bell Operating Companies are currently introducing. Both frame relay and SMDS are packet-switched services and are targeting as LAN interconnect solutions [21]. Users are grappling with the differences between SMDS and frame relay. Some users are sold on frame relay, some are waiting for the wide deployment of SMDS, while others are watching for the progress of asynchronous transfer mode (ATM). In addition, there are users still using the leased lines. From transmission technology point of view, there are some differences between frame relay and SMDS.

Frame relay is a standard that is based on transmission of variable length packets or frames and is defined at the Data Link Layer of the Open System Interconnection model. As a service, frame relay is targeted both to LAN-interconnect and to X.25 improvement. By pushing the error control to the end nodes, frame relay can be considered to be the first protocol that takes advantage of the new communications environment of high transmission speeds and low bit error rates provided by fiber-optic communications. Today, all frame relay services are based on permanent virtual circuits which are pre-established connections. The addressing scheme is to use an ID field called DLCI which has a local meaning only. The speed of frame relay service varies from 56 Kbps to 1.544 Mbps (T1).

SMDS is a switched, high-speed, connectionless data service and is based on the IEEE 802.6 metropolitan area network specification. Instead of variable

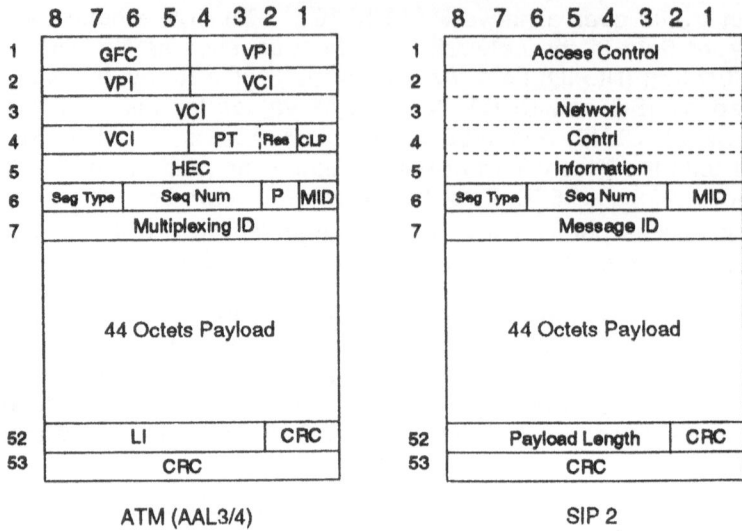

	8 7 6 5 4 3 2 1				8 7 6 5 4 3 2 1		
1	GFC	VPI		1	Access Control		
2	VPI	VCI		2			
3	VCI			3	Network		
4	VCI	PT	Res CLP	4	Contrl		
5	HEC			5	Information		
6	Seg Type	Seq Num	P MID	6	Seg Type	Seq Num	MID
7	Multiplexing ID			7	Message ID		
	44 Octets Payload				44 Octets Payload		
52	LI	CRC		52	Payload Length	CRC	
53	CRC			53	CRC		

ATM (AAL3/4)   SIP 2

Figure 9. Cell Formats.

sized frames, SMDS transmits data in fixed-length of 53 bytes cells. At present, SMDS targets at LAN interconnect within a metropolitan area but future phases should also support wide-area connectivity. SMDS can provide users with the capability of building virtual private networks by using the E.164 addressing scheme. Frame relay is defined at speed up to 2 Megabits per second whereas SMDS is targeted toward higher speeds (T1/T3/OC3). Due to the speed differences between the two services, a migration path from frame relay to SMDS has been suggested towards the broadband service.

SMDS and ATM

When we compare SMDS with ATM, one might argue that SMDS is a carrier service and ATM is a technology that will be used by carriers to carry many services including SMDS. The current SMDS communications interface between the customer and the network uses IEEE 802.6 protocol. Next, we discuss some relations between DQDB and ATM technologies.

ATM has been agreed by CCITT as the technique for transmission, multiplexing, and switching in BISDN [8,9]. Under the CCITT definition, information to be transferred is divided into small, fixed size cells. It is capable of supporting high quality voice, video, and high speed data traffic. ATM is a transfer mode suitable for implementing WAN, MAN, and LAN and could be implemented in different topologies and on top of different physical layers (fiber, coax, or UTP). SMDS deployment is not limited in MAN, however, initial SMDS service is provided in MAN area and the IEEE 802.6 is a MAN standard. Both IEEE 802.6 DQDB and ATM utilize 53 octet cells. The 44 byte payload of a DQDB cell (SIP L2PDU), in Figure 9, can be transparently transported through the ATM network using ATM adaptation layer (AAL) type 3/4 cells. This provides SMDS a natural migration path to ATM. The long term goal is that when a carrier introduces ATM-based switching system, customers need not see any effect on their SMDS services [22]. However, the wide-spread ATM deployment can not be achieved in one day. In the period of deploying ATM-based switching service, a transparent interconnection between SMDS (DQDB based) and the ATM network is required. To get the connection-oriented ATM and connectionless SMDS in line will take considerable effort from the carriers.

Observations

The journey of SMDS has been somewhat turbulent. Even though, SMDS has its roots in 802.6 and QPSX; its creators, Bellcore, chose to make it a strictly data service and not support the pre-arbitrated access mode for the isochronous traffic. This obviously worried a segment of potential users as isochronous traffic could not be carried on SMDS network. The initial thrust for SMDS came mainly from Bellcore and public carriers; as the potential users were nervously sorting out the future choices....frame relay, SMDS, ATM/BISDN. Even the CPE vendors took "wait and see" attitude on SMDS. The real break-through was the availability of DS-1 and DS-3 rate SMDS DSUs which enabled CPE equipments to support SMDS by implementing DXI interface in the microcode/software and no hardware change as the segmentation/reassembly and T1/T3 framing was performed by DSUs. The introduction of SMDS silicons was a positive step, but the CPE vendors have continued to use DXI and avoid hardware development. Of course, the endorsement of DXI by SMDS Interest Group (Technical Work Group) contributed to it. Finally the light at the end of the tunnel came in the form of attractive SMDS tarriffs. Also, the users are beginning to accept that Frame Relay and SMDS services will co-exist and ride on the ATM backbone. In fact, there have been a number of sub-missions in the SMDS Interest Group (Technical Work Group) related to Frame Relay / SMDS Interworking [22,23,24]. At this writing, SMDS is being offered and tariffed by many RBOCs and MCI and stands to benefit from any delays in the introduction of Frame Relay SVC and BISDN.

References

1. Bell Communication Research Technical Advisory, TR-TSV-000772, "Generic System Requirements in Support of Switched Multi-Megabit Data Services".

2. Bell Communication Research Technical Advisory, TR-TSV-000773, "Local Access System Generic Requirements, Objectives, and Interfaces in Support of Switched Multi-megabit Data Service".

3. Bell Communication Research Technical Advisory, TR-TSV-001059, "Inter-Switching Interface Generic Requirements for SMDS Service".

4. IEEE Std 802.6, "IEEE Standards for Local and Metropolitan Area Networks: DQDB Subnetwork of a MAN", 1990.

5. H.R. van As, J.W. Wong, and P. Zafiropulo, "Fairness, Priority and Predictability of the DQDB MAC Protocol under Heavy Load", International Zurich Seminar on Digital Communications, March 1990.

6. M. Spratt, "Allocation of bandwidth in IEEE 802.6 with non-unity ratio bandwidth balancing", Proc., IEEE ICC 91, 1991.

7. R.G. Foldvik, "The evolutionary path to Broadband ISDN", Ninth Annual International Phoenix Conference on Computer and Communications, March 1990.

8. J-Y. Le Boudec, "The Asynchronous Transfer Mode: a tutorial", Computer Networks and ISDN Systems, pp. 279-309, 1992.

9. Draft Recommendation I.363, "B-ISDN AAL Specification", CCITT SG XVIII 1990.

10. C.F. Hemrick, R.W. Klessig, and J.M. McRoberts, "Switched Multi-megabit Data Service and Early Availability Via MAN Technology," IEEE Communications Magazine, April 1988 p9-14.

11. J.F. Mollenauer, "Standards for Metropolitan Area Networks," IEEE Communications Magazine, April 1988 p15-19.

12. SMDS Interest Group Technical Specification, SIG-TS-001 Revision 3.2, "SMDS Data Exchange Interface Protocol," Oct. 1991.

13. R. Perlman, "Specification for Encapsulation of Protocols on SMDS", Draft Technical Specification, SMDS Interest Group, Dec. 1992.

14. D. Piscitello and J. Lawrence, "The Transmission of IP over the SMDS Service", RFC 1209, March 1991.

15. T. Bradley, C. Brown, and A. Malis, "Multiprotocol Interconnect over Frame Relay", RFC 1294, Jan. 1992.

16. R. Perlman, "Specification for Implementation of Connectionless OSI over SMDS", Technical Specification, SMDS Interest Group, Dec. 1992.

17. D. Pierson, "Frame Based Interface Protocol for SMDS Networks - DXI", Technical Specification, SMDS Interest Group, Jan. 1993.

18. C.Hemrick and J. McRoberts, "SMDS: The Beginning of WAN Superhighways", Data Communications, April 1991.

19. D. Minoli, "The New Wide Area Technologies: SMDS and BISDN", Network Computing, Aug. 1991.

20. R. Gareiss and J. Mulqueen, "Frame Relay vs. SMDS - Heading in the Same Direction ?", Communications Week, Oct.19, 1992.

21. F. Gratzer, "ATM and SMDS: A good Match", Communications Week, Feb.1, 1993.

22. P. Sibille and V. Phung, "Specification for implementation of Frame Relay data transfer over SMDS", SMDS Interest Group, March 1993.

23. D. Abensour, M. Joshi and F. Lai, "Frame Relay/SMDS Interworking Proposal", SMDS Interest Group, September 16,1992.

24. J-P Bernoux, "Specification of a Frame Relay Interface over SMDS", SMDS Interest Group, September 25,1991.

INTERWORKING OF B-ISDN AND DQDB—A SIMULATION STUDY

Peter Martini and Markus Rümekasten

University of Paderborn
Dept of Math. & Comp. Science
33095 Paderborn, Germany
e-mail: rueme@uni-paderborn.de

ABSTRACT

In this paper, we present an analysis of two interworking modes for interconnection of Metropolitan Area Networks (MANs) and ATM-based Broadband-ISDN (B-ISDN). We present simulation results of a DQDB network receiving traffic from B-ISDN. This connection is modelled by an Interworking Unit (IWU) within a DQDB network getting aggregated traffic from B-ISDN. The aggregated traffic is a mixture of various constant bitrate (CBR) and variable bitrate (VBR) traffic sources. To meet the bandwidth requirements of connection oriented (CO) traffic in MANs, an enhancement of the DQDB protocol for MANs has been studied: the Guaranteed Bandwidth (GBW) proposal, currently studied by IEEE 802.6 under the Project Authorization Request for Connection Oriented Data Service in the DQDB standard. The two interworking modes considered are the "cell-to-slot interworking" and the "frame interworking". The cell-to-slot mode modifies ATM cells to DQDB slots and vice versa. In contrast to this, the frame mode reassembles cells/slots and modifies B-ISDN frames to DQDB frames and vice versa.

It turns out that the cell-to-slot mode is much better for meeting delay requirements than the frame mode, which causes additional delays due to the reassembly of cells.

1. INTRODUCTION

In late 1990, the IEEE Standards Board accepted a standard entitled "Distributed Queue Dual Bus Network (DQDB) as a Subnetwork of a Metropolitan Area Network (MAN)" [6].

This standard names three different services to be provided:

- Connectionless (CL) service
- Isochronous service and
- Connection Oriented (CO) Data Service.

Up to now, [6] only specifies the first of these three services in sufficient detail, whereas the other services are only mentioned as to be specified in the future. For the interworking of MANs and B-ISDN the two services not yet standardized are very important because B-ISDN basically is a Connection Oriented network. So, for the interworking of MANs and ATM based B-ISDN we are faced with a situation where

1) B-ISDN is based on CO technology, whereas
2) DQDB does not support CO services.

The first problem may be solved by using Connectionless Servers (CLS) in B-ISDN: the aggregate CL traffic will use dedicated ATM connections between Connectionless Servers and MANs, where the "switching" is done in the CLS. The second problem is even harder to solve: the Distributed Queueing Protocol does not support the bandwidth guarantees required by CO traffic. Thus, the protocol itself has to be completed to support CO services.

As far as isochronous services are concerned Pre-Arbitrated (PA) slots are a solution. For Connection Oriented Data Service they are not:

1) The standard [6] clearly points out that this service will be provided by Queued Arbitrated (QA) slots.
2) The usage of PA slots would waste bandwidth because Connection Oriented Data Service will be used by variable bitrate (VBR) traffic.

To solve these problems, a protocol for CO services is needed, which

• uses QA slots to transmit CO data,
• uses high priority transmission in DQDB (CL traffic is limited to the lowest priority)
• guarantees the required bandwidth when needed
• redistributes bandwidth guaranteed but momentarily not needed.

In late 1990, QPSX (who originally proposed the DQDB mechanism) proposed the Guaranteed Bandwidth (GBW) protocol which is compatible to the DQDB protocol at the lowest priority and is able to allocate a required bandwidth for CO Data Service at higher priorities. GBW uses Queued Arbitrated (QA) slots and it gives the flexibility needed for VBR services: there is no pre-allocation of specific slots. Instead, GBW makes low priority stations refrain from transmission whenever CO traffic needs bandwidth. Since March 1992 [1] GBW is officially regarded as the (or at least one) protocol for CO Data Service to be included in the DQDB standard. GBW has all the properties mentioned above.

The future of DQDB strongly depends on the availability of IWUs to interconnect MANs to B-ISDN for all services mentioned above. If efficient service interworking of B-ISDN and DQDB is not possible in the near future, the introduction of broadband networks will be delayed.

The RACE II project COMBINE (Composite Broadband interworking and end-to-end models) is one of the major efforts working on the interconnection of heterogeneous broadband networks in Europe. In COMBINE, a multinational consortium including vendors, network operators, universities and research institutes works on a prototype for a DQDB / B-ISDN Interworking Unit (IWU). This prototype will have two modes of operation: the "cell-to-slot mode" and the "frame mode".

The cell-to-slot mode modifies ATM cells to DQDB slots (and vice versa) on the fly, whereas the frame mode reassembles the ATM cells and modifies the resulting frame to a DQDB Initial MAC PDU (IMPDU) and vice versa. A detailed description of the two modes is included in [11]. The two modes are analyzed in this paper to find out an optimum design for a B-ISDN to DQDB IWU.

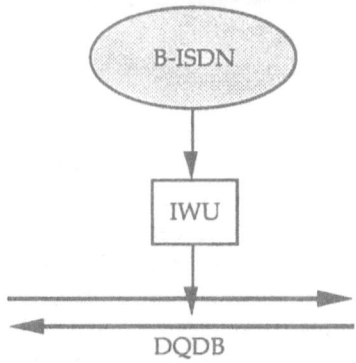

Figure 1: Interworking from B-ISDN and DQDB

Therefore, this paper studies a DQDB MAN receiving traffic from B-ISDN. The B-ISDN is represented by a single DQDB station used as an IWU from B-ISDN and DQDB (see figure 1). This IWU uses the GBW protocol for media access, because the traffic from B-ISDN is assumed to be CO traffic with a mixture of CBR traffic and VBR traffic. The simulation results are dealing with the performance aspects of the two interworking modes that are possible for the B-ISDN/DQDB IWU.

The paper is organized as follows: the next section describes the GBW protocol, section 3 presents the two interworking modes, section 4 describes the simulation model and section 5 presents the simulation results for the IWU using the two different modes of operation.

2. GUARANTEED BANDWIDTH (GBW) IN DQDB

In this paper, the description of DQDB is limited to the GBW protocol. For a description of the basic DQDB protocol the reader is referred to the standard [6] or to papers like [5].

2.1 THE DISTRIBUTED QUEUE

Since GBW uses higher priorities, every station has a specific queue for each of the three DQDB priorities. These queues for each priority include all segments not yet transmitted but "accepted" for immediate transmission, where accepted means that the segment has passed a traffic shaper. The traffic shaping mechanism is outlined in section 2.2.

Each queue can be described as a linked list of 0's and 1's. In this linked list, a "1" represents a segment which has been queued by the station itself and a "0" represents a request by a downstream station. This kind of describing the distributed queue is exactly the same as using state machines like in the DQDB standard [6]. A description of the distributed queue for the GBW protocol by using state machines is also included in [9].

The queue is updated each time

- a segment is transmitted or
- a segment is admitted by the traffic shaper or
- a request is received.

At each station the update of the local queues is done according to the following rules:

1) if a new segment is admitted by the traffic shaper, a "1" is appended to the linked list.

2) if a Request bit of the same priority has been set in a slot on the reverse bus, a "0" is appended to the linked list (this indicates a reservation by a downstream station operating at the same priority).

3) if a Request bit of higher priority has been set in a slot on the reverse bus, a "0" is inserted at the beginning of the linked list (this indicates a reservation by a downstream station operating at a higher priority).

4) if an empty slot arrives, the first element at the head of the linked list is removed. If this element is a "1", then the station is allowed to transmit one segment.

By this kind of distributed queue, GBW allows multiple outstanding requests.

2.2 TRAFFIC SHAPING

GBW shapes the traffic by limiting the rate of enqueueing segments for a specific connection according to the bandwidth accepted at call set-up. This mechanism is similar to the well known Leaky bucket mechanism.

2.2.1 TRAFFIC SHAPING FOR HIGHER PRIORITIES

CO traffic will be transmitted at priority 1 or 2. To shape the traffic, a variable called "credit" and the system parameters "income", "slotcost" and "creditmax" are used for priorities 1 and 2. For each slot passing on the bus to be used for transmission, the parameter income is added to the value of credit, which represents the current credit for this connection.

A connection is only allowed to transmit a segment, if the value of credit exceeds the parameter slotcost. Slotcost is the amount of "money" which has to be spent to transmit one segment. The transmission rights have to be "paid" from the accumulated credit. So, if a segment has been transmitted, the current value of credit is immediately reduced by slotcost.

To avoid accumulation of credit to an "infinite" value in idle times, the value of credit is not allowed to exceed the parameter creditmax. This is to reduce the burstiness by limiting the number of transmission rights at a given time.

As an example: If income is chosen as 30 then the station books 30 for each slot passing on the bus. In this example, slotcost is set to 150 which is equal to the bandwidth in the network. Starting from zero, a prospective sender must wait $150 / 30 = 5$ slots until it has enough money to buy a high priority request resulting in a free slot arriving on the reverse bus. By limiting creditmax to $\lceil slotcost/income \rceil \bullet income = 150$ in this example, the station is allowed to save money for the transmission of exactly one slot. It is obvious that the minimum value for creditmax must be $\lceil slotcost/income \rceil \bullet income$. Otherwise, the station would not be able to transmit segments. Thus, the bandwidth guaranteed to the station is exactly "30 Mbit/s out of 150 Mbit/s". In addition, it may transmit at priority 0.

2.2.2 TRAFFIC SHAPING FOR LOWEST PRIORITY

Since CL traffic has to be transmitted at the lowest priority, the shaping mechanism of priorities 1 and 2 cannot be used for priority 0. So, GBW allows

CL traffic to be transmitted at priority 0 with a mechanism compatible to DQDB. In addition to the DQDB standard it can easily be made to allow multiple outstanding requests.

The shaping mechanism at priority 0 is based on a maximum allowable number of segments in the local queue for priority 0. This number of segments is called "gamma". If gamma is set to 1, multiple outstanding requests are not allowed and the mechanism is fully compatible to the DQDB protocol.

3. INTERWORKING OF B-ISDN AND DQDB

This section presents a model for a part of an IWU for MANs and B-ISDN. This model only takes into account the so-called Receiving Interworking Unit (RIWU). The RIWU terminates a Virtual Path (VP) from B-ISDN for the reception of this traffic from the ATM network and transmits it on the MAN as shown in figure 1. The so-called Sending Interworking Unit (SIWU) receives traffic from a MAN and terminates a VP for the transmission of this traffic over the ATM network. The model and the analysis of the SIWU is outside the scope of this paper.

3.1 THE CELL-TO-SLOT MODE RIWU

In this interworking mode, the protocol conversion can be done for each single cell: a Beginning of Message (BOM) cell contains all the relevant information for protocol conversion like source and destination addresses or the frame length. With this information obtained from the first cell of a frame it is possible to convert the following cells of that frame on the fly.

As well as performing the protocol conversion for each single cell, the RIWU has to decide in which order the DQDB segments obtained from the ATM cells are sent to the DQDB network. This is not a straightforward problem, since the timing relationship between successive cells of an isochronous connection must be restored for connections originating from possibly different MANs or LANs.

It is assumed here that for all CO communication, there is some internetwork connection set-up process and that co-ordination of the timing requirements of different connections is performed at this point.

It is the task of the polling mechanism to decide which buffers to serve first when sending a segment on the DQDB network. The segment buffers effectively act as the DQDB bus access queues. For example, if a PA slot arrives at the RIWU, the appropriate isochronous segment should be transmitted. Otherwise, when a segment can be transmitted (according to the MAC protocol), priority is first given to the CO traffic and then to the CL traffic.

3.2 THE FRAME MODE RIWU

The model of the frame mode RIWU is very similar to the model of the cell-to-slot mode RIWU, except from the fact that the arriving segments are reassembled, whereas the cell-to-slot mode RIWU streams single cells through the IWU to the DQDB MAN. So, the frame mode RIWU needs reassembly buffers, whereas the cell-to-slot mode RIWU only needs segment buffers to store one single cell for the protocol conversion of this cell.

The reassembly buffers contain only one frame per buffer. Upon the arrival of a BOM cell, a free reassembly buffer is allocated to that frame. From this point onwards, it receives all the segments belonging to that frame, but cannot be read by the MAN. Upon the arrival of the End of Message (EOM) cell, the reassembly buffer enters a state where it can be read by the MAN after protocol conversion of the whole frame.

There are two approaches for the implementation of the frame mode RIWU; either the IWU has a fixed number of reassembly buffers, each dimensioned to accommodate the maximum frame size of 9188 bytes, or a "memory pool" is employed which behaves like a heap and allows a fully flexible, more efficient, allocation of memory.

4. THE SIMULATION MODEL

In this section, we present the simulation model we use for our studies. Firstly we describe the traffic load model and the basic scenario and then the model for an IWU.

4.1 THE LOAD MODEL

In our simulations, we use stations sending various kinds of CBR and VBR traffic.

The CBR traffic consists of packets with a length of exactly one DQDB QA segment (44 byte). The data rates are as follows:

- 64 Kbit/s
- 2 Mbit/s
- 34 Mbit/s.

For the VBR traffic, three different traffic sources were used. All the types generate frames with exponentially distributed length. The average lengths of these frames are assumed as follows:

- 200 segments for VBR type A
- 20 segments for VBR type B and
- 2 segments for VBR type C.

The different types have different access rates to the DQDB network due to the different applications they represent. The access rates are chosen as follows:

- 700 Kbit/s for type A (representing bursty CO data) and
- 25 Mbit/s for type B (representing high speed CO data, for example video transmission)
- 1 Mbit/s for type C (representing background CO data).

To enable bandwidth guarantees for those traffic types, a peak rate control is introduced after segmentation of the frames. This peak rate control is as follows:

- 2 Mbit/s for type A and type C traffic and
- 34 Mbit/s for type B traffic.

This peak rate control represents the 2 Mbit/s ISDN connection and a CCITT G.703 connection with 34 Mbit/s connected to the DQDB MAN. Figure 2 shows the VBR traffic model.

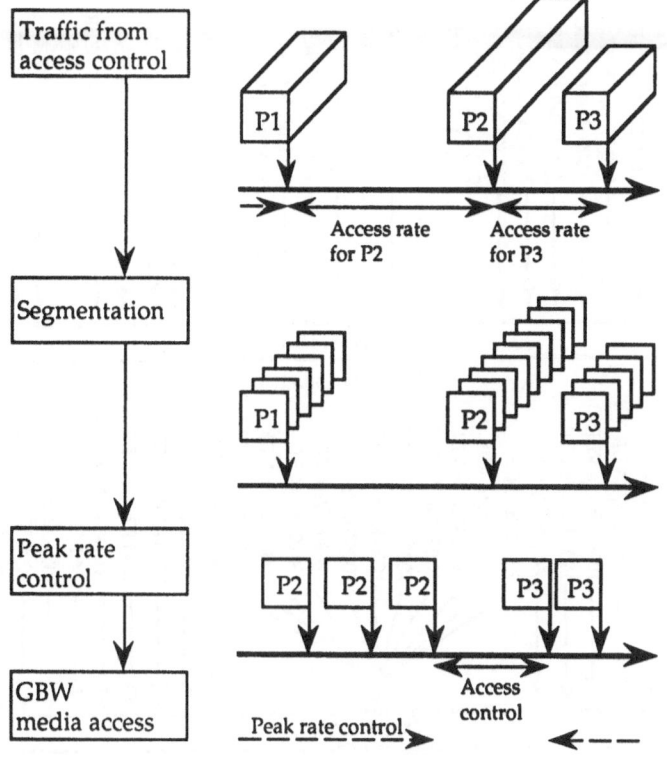

Figure 2: VBR traffic load model

The CBR and VBR traffic sources are originally described in [3], where additional types of CBR and VBR sources can be found.

4.2 THE BASIC SCENARIO

The simulation scenario is based on a 10 km bus with 32 stations equidistantly spaced on the bus. Each station is sending on exactly one of the buses exactly one of the CBR or VBR traffic types. with bandwidth guarantees of the peak rate for VBR sources and access rate for CBR sources. This scenario has been proposed in [3] to compare different media access mechanisms for the ATM Customer Premises Network (CPN). GBW is one of the proposals for this.

The location of the different traffic sources is shown in figure 3 depicting both the scenario with and without an IWU model.

4.3 THE MODEL OF THE IWU

The IWU model shown in figure 3 is a useful approach to describe the RIWU getting traffic from B-ISDN to be forwarded to a DQDB. This kind of RIWU simulation model is actually used in the RACE project COMBINE where we analyze the end-to-end performance in heterogeneous broadband networks.

The idea of the RIWU model is to aggregate traffic of different stations in one single station representing the RIWU. This approach has the advantage to be realistic and flexible.

It is realistic because a real RIWU gets different types of traffic from all over the B-ISDN, which is modelled by the different traffic streams arriving at the

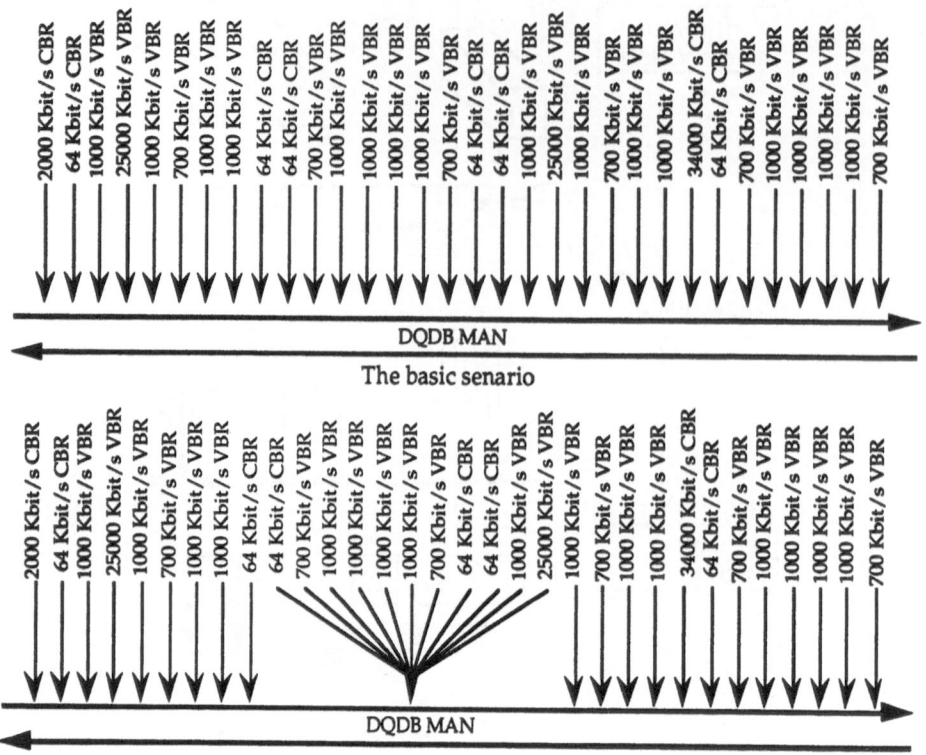

Figure 3: The simulation scenario

RIWU station. It is flexible because the position of the RIWU can be changed easily and the total traffic generated by the RIWU can be modified just by changing the number of stations aggregated as the RIWU.

The bandwidth guarantees for a RIWU station are determined as the sum of the peak rates for the VBR streams and the access rates for the CBR streams. Thus, the whole traffic of the RIWU is guaranteed by the GBW protocol to be transmitted at the highest priority.

To be able to simulate two interworking mode (cell-to-slot mode and frame mode), the RIWU reassembles the arriving segments to frames only in case of frame interworking. For cell-to-slot interworking, the individual ATM cells are transferred to the DQDB.

5. SIMULATION RESULTS

The simulation results presented in this paper have been obtained using the simulation tool ATLAS (Analysis Tool for Local Area Network Simulation). ATLAS has been developed at the Aachen University of Technology, Germany, and is used in COMBINE for performance analysis based on simulations [14].

All simulations use the scenario described above for a 150 Mbit/s bus with a simulation time of 1.5 million slot times (about 4.25 sec. real time). The GBW parameter creditmax is set to save credit for the transmission of one slot to reduce the burstiness in transmission. Income is set to the required bandwidth in Mbit/s. The measurements started after 0.5 million slot times (about 1.41 sec. real time) following the requirements of CCITT [3].

5.1 RESULTS FOR THE BASIC SCENARIO

The simulation of the basic scenario without an IWU clearly shows the excellent behavior of the GBW protocol with segment delays that are very low (less than 30 slot times, which is less than 0.1 msec).

These results are similar to earlier simulation studies testing GBW in comparison to another proposal as a protocol for real time communication in DQDB (see [7] and [8]). Based on these results, the decision for GBW as the protocol for the RIWU getting an aggregated stream of CO traffic is obvious.

Thus, the following simulations of the two interworking modes consider the GBW protocol only. Furthermore, it has been shown that the efficiency of interconnected scenarios is considerably improved by giving high priority to IWU stations [2]. This has been done in our simulations by bandwidth guarantees at highest priority. The bandwidth guarantees for CBR sources and for the RIWU are as follows:

• access rates for CBR sources
• sum of peak rates (VBR) + sum of access rates (CBR) for the RIWU.

All additional traffic, i.e. VBR traffic not generated by the RIWU, is transmitted at priority 0. Obviously, these assumptions give priority for constant bit rate traffic and to internet traffic.

5.2 RESULTS FOR DIFFERENT IWU SCENARIOS

In this section, we present the main results for B-ISDN / DQDB interconnection. Simulation results for cell-to-slot interworking and for frame interworking are presented and discussed using the same scenarios for both modes.

5.2.1 RESULTS FOR THE CELL-TO-SLOT INTERWORKING MODE

The first interesting result can be achieved by modifying the location of the IWU. Changing the position of the IWU in the basic scenario implicitly includes a changing of the traffic mix. Figure 4 shows that those influences have a very low impact on the segment delays. Similar effects have been shown in earlier papers like [7].

The maximum segment delay does not exceed 20 slot times (about 0.05 msec) with a mean segment delay of about 2 slot times in all of the three different simulations. Figure 4 shows slightly increasing percentiles from stations at the head of the bus to stations at the end of the bus. This effect is well-known in DQDB networks (more empty slots at the beginning of the bus), so that this effect is a generic problem and not a problem of cell-to-slot interworking using GBW.

One of the most interesting results of these simulations can be seen when looking closer at the IWU data rate: the IWU representing stations 1 to 11 and the one representing stations 10 to 20 generate traffic at a peak rate of 48 Mbit/s, whereas the IWU representing stations 20 to 30 generates traffic at a peak rate of 86 Mbit/s with the total bandwidth guaranteed in all cases.

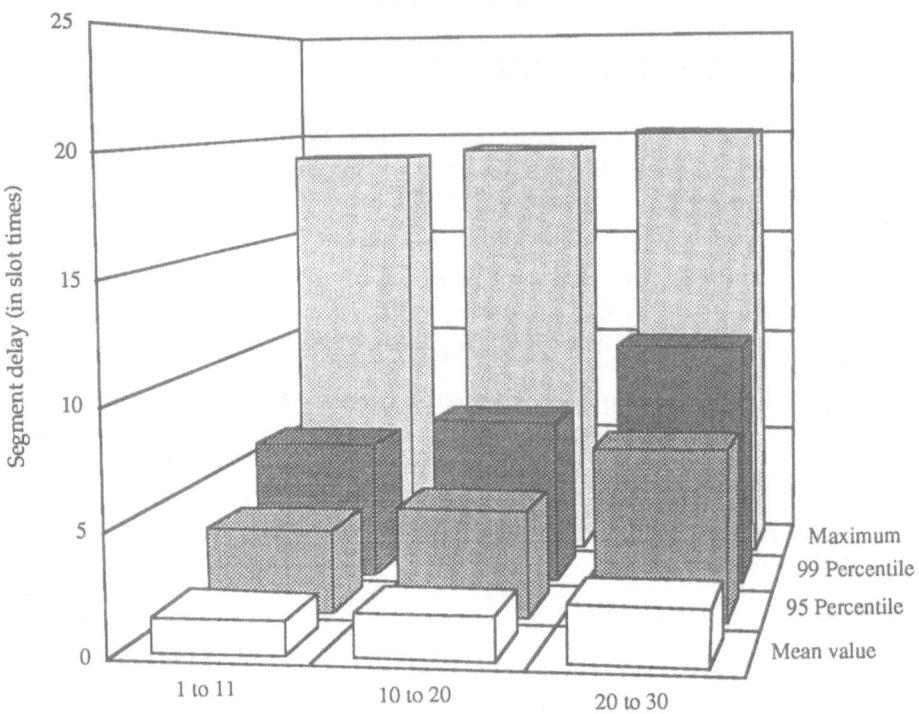

Figure 4: Impact of different IWU locations (cell-to-slot mode)

Since the segment delays do not differ very much, the influence of the different traffic mix and data rates of an IWU in cell-to-slot mode using GBW may be ignored.

Other simulation runs for additional traffic mixes and data rates also showed this very small impact.

Summing up, the main results shown in figure 4 are that the location of the IWU and the variation of traffic patterns in the IWU have a low impact in terms of segment delay.

Another important result can be seen in figure 5. It shows simulation runs using the same scenario with stations 10 to 20 representing the IWU, but with different bus lengths, namely 10 km, 30 km and 60 km. Again, the maximum segment delay is very small (equal to 25 slot times in the worst case, about 0.063 msec).

Figure 5 shows an increasing segment delay from less than 20 slot times for the 10 km bus to 25 slot times for the 60 km bus, i.e. an increase of about 25%. Almost the same behavior was observed for other stations representing the IWU. This increasing delay is again a well-known DQDB feature caused by larger propagation delays of reservations by stations downstream on a longer bus. So, the increase of delays is not due to the cell-to-slot mode. When compared to the results of the same tests for the frame mode described in the next section, the cell-to-slot mode has an excellent performance even for this "unfair" DQDB scenario.

Summing up the results for the cell-to-slot mode, it may be stated that GBW is an excellent protocol for such an IWU. The impacts of different traffic mix, IWU location and bus length can be ignored in the main.

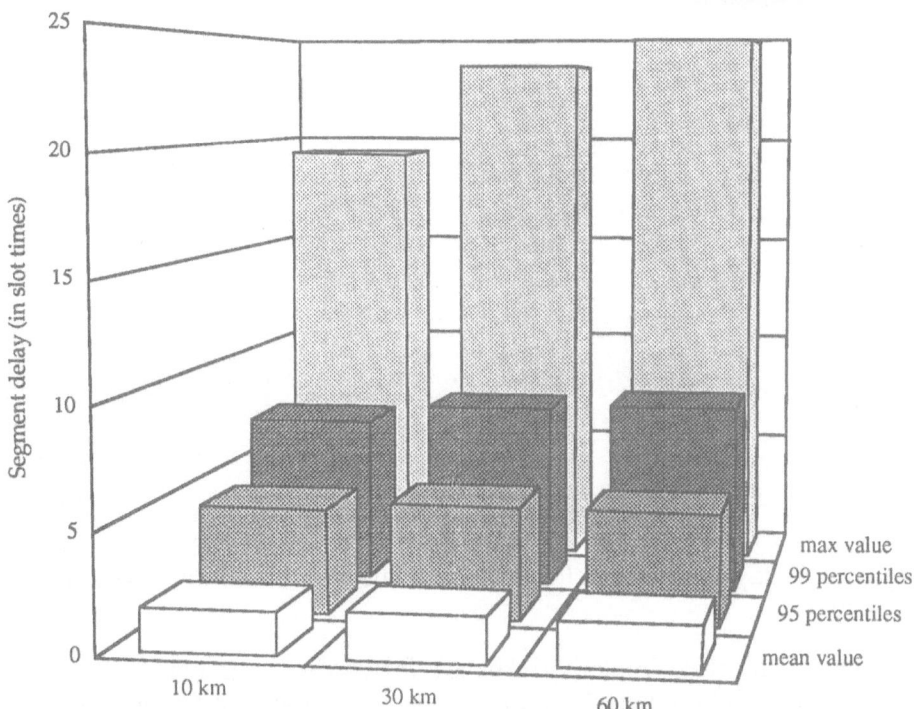

Figure 5: Impact of the variation of bus length (cell-to-slot mode)

5.2.2 RESULTS FOR THE FRAME INTERWORKING MODE

Because of reassembly of the segments to IMPDUs in the case of frame mode, the peak rate control of the segments is "destroyed" by the IWU operating in frame mode. So, the bandwidth guarantees are very difficult to manage, because of the unpredictable peak rates for the more bursty traffic coming from the frame mode IWU.

The results presented in figure 6 show simulations of the scenario with different IWU locations (see figure 4, note the difference in scaling). The bandwidth guarantees for the IWU are either of the access rate or of the peak rate. The measured delays are the 99percentiles of the segment delays.

For station 1 - 11 representing the IWU the access rate is 32 Mbit/s and the peak rate is 48 Mbit/s, for station 10 - 20 they are 31 Mbit/s and 48 Mbit/s and for station 20 - 30 it is 67 Mbit/s and 86 Mbit/s.

Figure 6 shows that the segment delays for guarantees at access rate are much higher than the segment delays for guarantees at peak rate. The delays are much higher (1000 slot times is about 2.8 msec) in both cases than they are for the same scenario for the IWU working with cell-to-slot mode (see figure 4). When compared to figure 4 the delays for bandwidth guarantees at the peak rate are about 40 times higher in the frame mode. For bandwidth guarantees at the access rate the increase of the delays is even worse (about 200 times higher than for cell-to-slot mode).

The high segment delays are caused by the burstiness of the traffic coming from the IWU working in frame mode. The bandwidth for all the segments of a reassembled very long packet (8800 bytes in mean for type A VBR traffic) cannot be guaranteed by any mechanism. This problem arises because all the

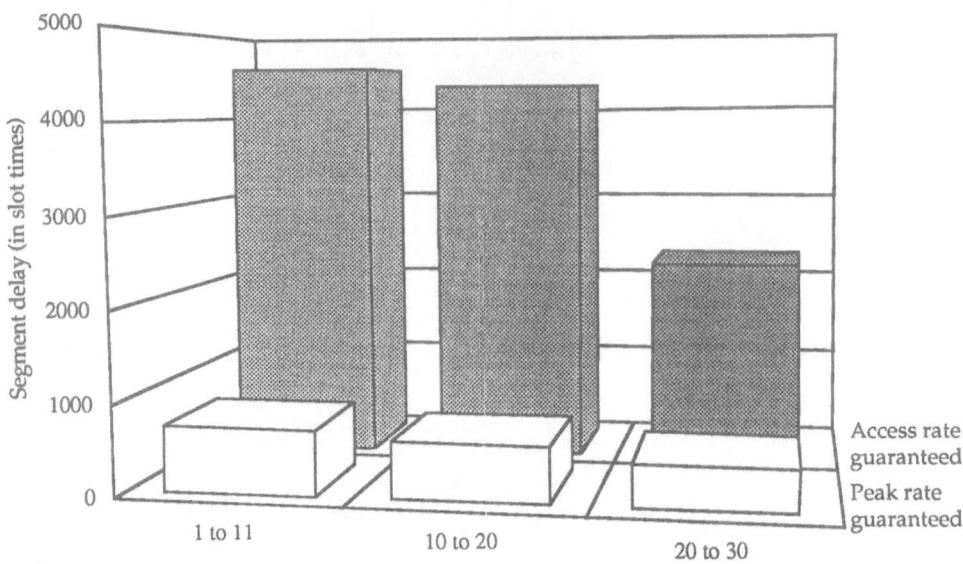

Figure 6: Delay comparison for different bandwidth guarantees (frame mode)

segments of the reassembled packet are forwarded at the same time by the IWU, whereas GBW needs a minimum spacing between arriving segments.

The burstiness of the traffic "generated" by the frame mode RIWU also causes the big differences in the segment delays for the guaranteed access rate in figure 6: since the reassembled frames have an exponentially distributed length, the length of a frame can not be foreseen at all. So, the big differences in figure 6 between the delays for the IWU at the end of the bus (stations 20-30) and the others can be explained by the unpredictable length of the frames waiting for media access. This is of course a big disadvantage of the frame mode, because buffer space for maximum sized frames (9188 bytes) must be foreseen in the frame mode IWU. In opposite to this, the cell-to-slot mode only needs buffers for a few segments.

The difference of the two modes is that the cell-to-slot mode provides single segments with a known peak rate, whereas the frame mode creates new and unknown traffic patterns by the reassembled frames.

The cell-to-slot mode working with GBW has the advantage that enough credit has already been accumulated by the time a new segment arrives at the IWU. Thus, an empty slot for that segment can be "ordered" immediately. The frame mode has the disadvantage that many new segments arrive at the same time. Thus, credit must be accumulated again and again to transfer all these segments. Since the bandwidth guarantees at peak rate are higher than at the access rate, the delays are lower for bandwidth guarantees at peak rate.

When comparing figure 6 to figure 4, it comes clear that the differences in the segment delays for the two interworking modes are considerable large. A very interesting difference may be observed when comparing the values for station 20 to 30 as IWU to the values for the other IWUs. For the cell-to-slot mode there was no significant difference, whereas for the frame mode the differences for the bandwidth guarantees at the access rate are very big. So, the cell-to-slot mode using GBW is less sensitive to different aggregated traffic streams and IWU locations.

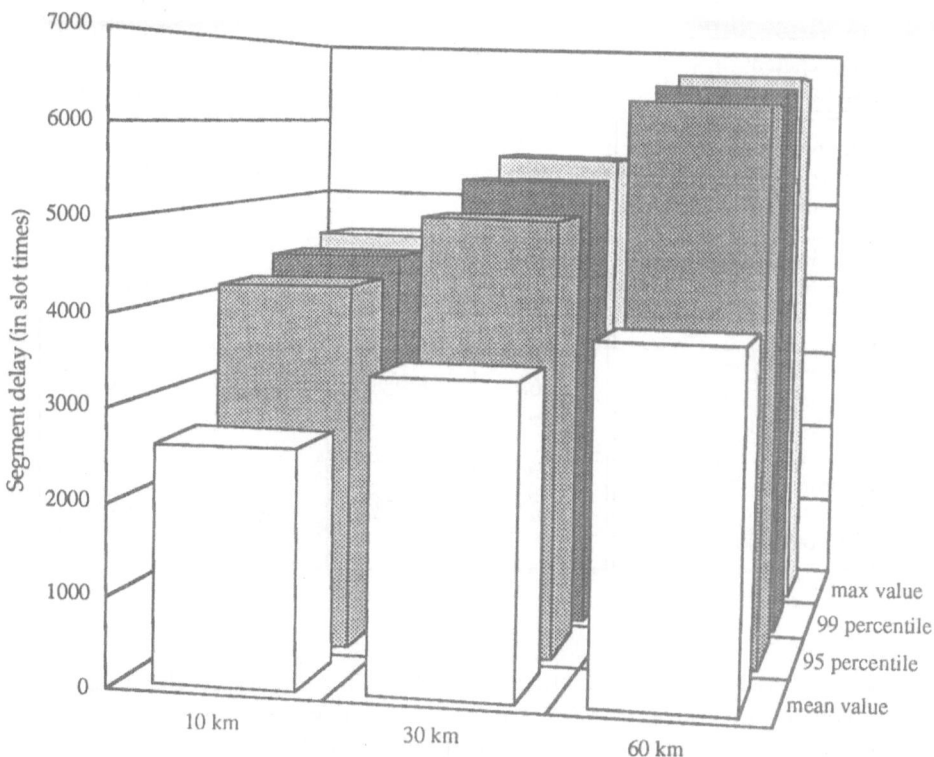

Figure 7: Impact of the variation of bus length (frame mode)

Figure 7 shows results of the simulation scenarios for the frame mode IWU for various bus lengths (see also figure 5).

When comparing the results shown in figure 7 to those in figure 5, the delays for the frame mode are much higher than for the cell-to-slot mode.

The most interesting aspect of figure 7 is the fact that the increase of delays from the simulation using a 10 km bus to the one using a 60 km bus is more than 50% for the frame mode, but only about 25% for the cell-to-slot mode (cf. figure 5). In absolute values this means an increase from about 4000 slot times to 6500 slot times (from about 20 to 25 slot times for the cell-to-slot mode).

This significant difference can be explained by characteristics of the GBW protocol. The more bursty traffic pattern makes the IWU in frame mode more sensitive to propagation delays. For cell-to-slot interworking, waiting times for medium access are delays for individual segments. In most cases delays for one segment only. In fact, for cell-to-slot mode arriving segments and bandwidth allocation are synchronized to a certain extent.

For frame interworking, most segments have a "convoy" following behind. Thus, delays are delays for the whole bunch.

5.2.3 COMPARISON OF THE TWO INTERWORKING MODES

The following table shows simulation results for both modes using the same scenarios. Both IWU alternatives have the same bandwidth guarantees (peak rate guarantees for the cell-to-slot mode) and the measured delays are segment delays in both cases.

Table 1: Comparison of segment delays for both interworking modes

Aggregated stations	Cell-to-Slot mode		Frame mode	
	Mean value	99percentile	Mean value	99percentile
Stations: 1-5 Sum of Access: 28 Mb/s Sum of Peak : 40 Mb/s	1.3	4.76	39.79	965.06
Stations: 1-10 Sum of Access: 32 Mb/s Sum of Peak: 48 Mb/s	1.42	5.95	32.29	716.35
Stations: 20-25 Sum of Access: 3.5 Mb/s Sum of Peak: 10 Mb/s	1.63	6.95	46.21	630.56
Stations: 20-30 Sum of Access: 67 Mb/s Sum of Peak: 86 Mb/s	2.33	10.72	38.37	463.0

The table clearly shows the considerable delay differences of the two modes. For the cell-to-slot mode, the IWU is no bottleneck on a DQDB network and even real-time requirements can be fulfilled using this interworking mode. In contrast to this, the delays for the frame mode are much more unpredictable due to frame reassembly causing a bursty arrival process.

6. CONCLUSIONS AND FUTURE WORK

In this paper, we have studied the performance of two different interworking strategies for B-ISDN and DQDB using the GBW protocol.

The GBW protocol only requires additions to the DQDB standard. It only adds rules which make high priority senders use the service provided by DQDB in a specific way. Since GBW guarantees a required bandwidth for a station at higher priorities it is the optimum protocol for an IWU to prefer long distance traffic (cf. [2]).

It has been shown that the GBW protocol efficiently supports the cell-to-slot interworking mode: the shaping mechanism of GBW corresponds perfectly to the peak rate control in ATM-based B-ISDN.

The frame mode unnecessarily increases the burstiness on the DQDB network by numerous segments arriving at the same time without any spacing. This results in considerable waiting times.

Because of the high delays, the architecture of the frame mode IWU is much more complex than for the cell-to-slot mode IWU. The different services classes (CL traffic, CO VBR traffic and CO CBR traffic) have to be separated according to their delay requirements and shapers have to be introduced to reduce the burstiness of the frame mode IWU. In opposite to this, the cell-to-slot mode IWU can be implemented much more simple: the low delays do not require separation of traffic classes for peak rate bandwidth guarantees and shapers are not needed at all.

Since it is not possible to use the cell-to-slot interworking mode for any other interconnection of standardized networks (only ATM and DQDB have the same slot size), GBW and cell-to-slot interworking mode seem to be the best way to create a worldwide, integrated broadband communication system, providing different kinds of services including new services for multimedia and real-time applications.

It should be noted that GBW is also useful for CL communication in interconnected networks. The protocol may be used to give higher priorities to internet traffic reaching the final destination network (cf. [2]). This reduces the unfairness caused by multihop scenarios. Of course, as an overall goal the user should not be able to tell internetwork transmission from the local transmission.

Further work on CO service in DQDB includes the specification of a draft standard which requires further clarifications to [1]. Additional studies are necessary on the following points:

• guaranteed delays: In its current form, the GBW protocol only guarantees bandwidth. It may easily be shown that for an infinite user population delays may be infinitely large. This problem can be fixed in several ways to be studied.

• mean rate control: In November 1992, IEEE 802.6 accepted a proposal presented by Peter Martini that requires the GBW protocol to include mean rate control (in addition to peak rate control) to allow for statistical multiplexing.

Acknowledgements

This work has been supported by the Deutsche Forschungsgemeinschaft (DFG). The work on the interworking modes and performance analysis of these modes is supported by the Commission of the European Communities under a contract of the RACE II project 2032 COMBINE.

The authors also would like to thank the HP Laboratories, Bristol, England, especially Pat Baker and Michael Spratt for their information about GBW and providing the University of Paderborn with machinery for simulations.

References

[1] Z. L. Budrikis, "Working Document for IEEE 802.6 CO Services Standard", IEEE 802.6 working group March 92
[2] W. Bux, D. Grillo, "Flow Control in Local-Area Networks of Interconnected Token Rings", IEEE Transactions on Communications, Oct. 1985
[3] CCITT Study group XVIII, Temporary document 34(XVIII/8), Geneva, June 1991
[4] M. Gerla et al.,"Interconnecting LANs and MANs to ATM", Proceedings of the 16th Annual Conference on Local Area Networks", IEEE Computer Science Press, Sept. 1991
[5] IBM report RC 17016 "A journey through the DQDB literature" in "Special Issue of Performance Evaluation journal", 1992
[6] IEEE standards for Local and Metropolitan Area Networks: Distributed Queue Dual Bus (DQDB) Subnetwork of a Metropolitan Area Network (MAN), 802.6- 1990
[7] P. Martini, G. Werschmann, "Real-time Communication in DQDB - A Comparison of Different Strategies", Proceedings of the 17th Annual Conference on Local Area Networks", IEEE Computer Science Press, Sept. 1992

[8] P. Martini, G. Werschmann, "Connection Oriented Data Service in DQDB - Simulation Studies of the Guaranteed Bandwidth Protocol", Proceedings of the International Conference on Computers and Communications, March 1993

[9] G. Mercankosk, Z.L. Budrikis, A. Cantoni, "DQDB for Time Constrained Services", 4th IFIP conference on High Performance Networking, 1992

[10] J. F. Mollenauer, "Metropolitan Area Networks and ATM Technology", International Journal of Digital and Cabled Systems, Vol. 1, pp.223-228, 1988

[11] A. K. Noergaard, "Functional Specification of an ATM - DQDB Interworking Unit", CEC, Proceedings of the Interworking'92 - First International Symposium on Interworking, Berne 1992

[12] I. Norros (Editor), "Models for heterogeneous networks and their elements", Deliverable D3 of WP1 of RACE II project COMBINE, Document ID R2032/KTA/WP1/DS/P/003/b1

[13] J. Ottensmeyer et al., "A Concept for Interconnecting DQDB MANs through ATM-based B-ISDN and Related Issues with respect to Simulation", CEC, Proceedings of the Interworking'92 - First International Symposium on Interworking, Berne 1992

[14] M. Rümekasten, "ATLAS reference manual", Working document of RACE II project COMBINE, Document ID COMBINE/UPB/WP1/003/01, March 1992

[15] M. Rümekasten, "Strategies of ATM and DQDB interconnection", Proceedings of the EOS/SPIE Conference "Local and Metropolitan Area Networks", Berlin, April 1993

PERFORMANCE EVALUATION OF A BUFFERED ATM
CONNECTIONLESS SERVER

Kenn Kvols[1] and Enrique Vázquez[2]

[1] KTAS
Address: Norregade 21
DK-1199 Copenhagen K   Denmark
e-mail: kk@ktas.dk

[2] Technical University of Madrid
Address: ETS Ing. Telecomunicacion
E-28040 Madrid   Spain
e-mail : enrique@dit.upm.es

Abstract

In an ATM-based network it is foreseen that connectionless traffic will be handled
by Connectionless Servers (CLSs) interconnected to each other and to the users
via semi-permanent connections, forming a *Connectionless Overlay Network*.
This paper discusses the necessity of frame-level buffering in the CLSs in order
to meet the connectionless service requirements, in particular the maximum
allowed frame loss ratio, with a reasonable utilization of network resources. The
frame loss and delay performance of a buffered CLS are evaluated both by
analysis and by simulation, showing the utilization improvements that can be
achieved with properly dimensioned CLS buffers.

1   Introduction

The concept of a Connectionless Overlay Network (CLON), based on semi-permanent
VPCs between the sources of connectionless traffic and Connectionless Servers
(CLSs) and between different CLSs, is a promising solution for LAN and MAN interconnection over B-ISDN (see figure 1). However, achieving a reasonable utilization
of a CLON is not an easy task. This is due to the unpredictable nature of the connectionless traffic. The traffic on each of the VPCs will be composed of an unpredictable
number of frames, and the individual frame transmission rates will not be known.

If the Connectionless Broadband Data Service (CBDS) is provided by the CLON,
the traffic is controlled by the Access Class Enforcement (ACE) mechanism [CBDS-92].
The ACE ensures that traffic entering the network conforms to one of the CBDS access classes. It, however, does not give any attention to the route that the individual

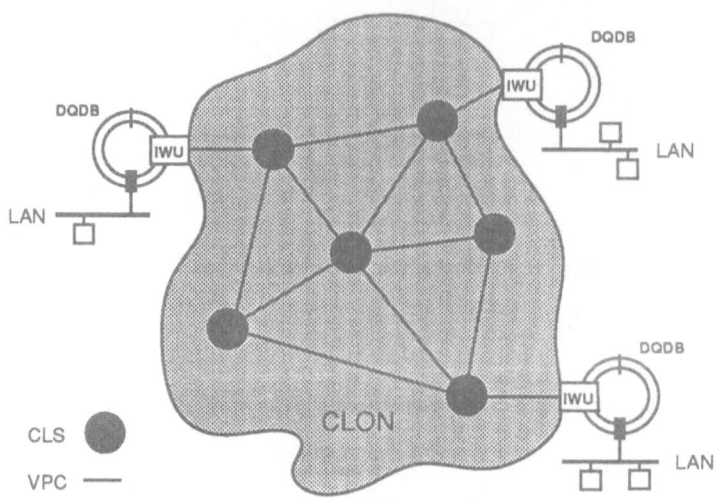

Figure 1: A Connectionless Overlay Network

frames will follow through the CLON, or to their transmission rates. Therefore, the traffic within the network remains largely unknown.

The problem with the unknown characteristics of the connectionless traffic is especially pronounced if the CLSs do not provide buffering for at least a few frames (i.e. several hundreds of cells). Without buffering, or with buffers limited to a few tens of cells, it is very likely that some cells will be lost due to buffer overflow, for example when several frames arriving to a CLS are routed to the same output VPC. In this case, even if some buffer congestion control mechanism is used to decide whether or not an incoming frame can get access to an outgoing VPC, a reasonable throughput can only be achieved if the frame transmission rates are small compared to the rate allocated to the VPCs [Kvo92]. Small frame rates can be achieved, for example, if frame rate shaping is performed at the access to the CLON. However, this may unnecessarily increase the frame transmission delay. A better approach, as discussed below, is to introduce larger buffers in the CLSs, making it possible to interleave several frames on the outgoing VPCs regardless of the frame rates.

The aim of this paper is to investigate how buffering in the CLSs can increase the throughput performance of a CLON, while still meeting the frame loss and delay

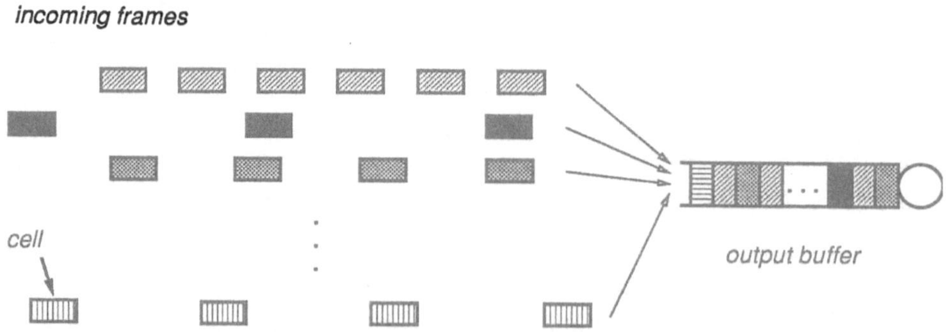

Figure 2: A Model of one of the CLS Outlets

344

requirements of the connectionless service to be provided. A proposal for a service strategy to be used at each outlet of a CLS is given in section 2. Its performance is then evaluated by both analysis and simulation. Section 3 presents an analytical evaluation of the frame loss ratio in a CLS, based on the assumption of Poisson frame arrivals. The analysis has been done for two modes of operation: cell-to-slot mode and frame mode (see section 2). The results make apparent the necessity for frame-level buffering in the CLS. The performance effects of more complex frame arrival patterns, and the introduction of output congestion control in the CLS have been studied with a discrete-event simulation model of the CLS behavior. Section 4 presents the simulation results. Finally, section 5 summarizes the conclusions of the paper.

2 The CLS Model

The service strategy proposed here allows several frames to be interleaved on each outgoing VPC. A buffer with room for M frames makes it possible to interleave at least M frames without losing any of them. The idea behind this approach is similar to the idea of having at least a cell-scale buffer in an ATM switch [Ras91], although the argument is used on the frame level here. The model used in the paper is illustrated in figure 2.

The cells from the incoming frames are interleaved in a FIFO buffer with capacity M times the maximum frame size (210 cells). Cell losses may occur if more than M frames flow through the buffer simultaneously. Therefore, in order to concentrate losses on full frames it is advantageous either to use a push-out mechanism, that automatically removes from the buffer cells belonging to frames that have effectively become useless due to the loss of one or more cells [Aal92] [Man93], or to use an admission control on all arriving frames at the buffer, so that full frames are deleted if the output is overloaded [Vog92] [Tur91].

The maximum value for M depends on the delay constraints. If the VPC rate is 155 Mbps and considering a maximum delay per hop (i.e. per CLS) of 5 ms in line with the SMDS requirements [SMDS-91] [Bel92], the maximum value is M = 8. This value will be reduced if the frames experience a processing and/or routing delay before they get access to the output buffer. On the other hand, many frames are shorter than the maximum length, so actually the output buffer has room for a number of frames exceeding the buffer size divided by the maximum frame size. The value 8 derived above may therefore be increased without violating the delay requirement. M may also be increased because the maximum delay is not required for all frames, but for only 95 % of them.

The proposal given here is based on the assumption that the CLS operates in *cell-to-slot mode*, i.e. the cells of an incoming frame can be transmitted to the output as soon as they arrive to the CLS. However, a similar proposal can be given if the CLS operates in *frame mode*, i.e. the last cell (EOM) of each frame shall arrive before the cells of the frame can be transmitted to the output VPC. In this case, the memory space associated with each CLS outlet must be enough for the reassembly of at least M frames at a time. Frames are deleted if not enough memory space is available upon arrival of their first cell (BOM).

3 Analytical Results

In order to evaluate the proposal presented in the preceding section analytically, let us assume that frames arrive to the CLS outlet according to a Poisson process.

Table 1: Allowed Output Load ρ vs M and N

	N = 1	N = 2	N = 5	N = 10
M = 1	1.0 %	3.7 %	14.1 %	26.9 %
M = 2	4.7 %	8.7 %	19.3 %	31.0 %
M = 5	22.6 %	26.2 %	34.1 %	42.3 %
M = 10	46.1 %	47.7 %	52.1 %	56.4 %
M = 20	68.1 %	68.8 %	70.3 %	72.0 %
M = 30	78.0 %	78.4 %	79.1 %	80.0 %

M = buffer size in frames
N = output VPC rate / frame rate
Cell-to-slot mode; FLR = 10^{-4}

Further it is assumed that the frame length distribution is exponential, and that the VPC rate to frame transmission rate ratio is N.

If the CLS operates in cell-to-slot mode, no frames get lost if the number of frames in the system is less than $N + M$. (The VPC can handle N frames and there is room for M frames in the output buffer.) Therefore, a conservative expression for the frame loss ratio (FLR) can be obtained by assuming that frames are lost whenever the number of frames in the system exceeds $N + M$. Let A be N times the load ρ of the output buffer. Then:

$$\text{FLR}_{\text{cell-to-slot}} = \frac{\left(\frac{A}{N}\right)^M \frac{A^N}{N!}}{\sum\limits_{i=0}^{N} \frac{A^i}{i!} + \frac{A^N}{N!}\sum\limits_{i=1}^{M}\left(\frac{A}{N}\right)^i} \quad \text{with } A = N\rho$$

If, alternatively, the CLS operates in frame mode, and the frames are read out one by one in the order their EOM arrived (FIFO discipline), then the output system can be approximated by a M/M/1 queue. Here let A denote the load of the system. Then the frame loss ratio is equal to:

$$\text{FLR}_{\text{frame}} = \frac{(1 - A)A^M}{1 - A^{M+1}}$$

In order to evaluate the performance of the proposed service strategy, some figures for the allowed load with frame loss probability 10^{-4} are given in table 1 for the cell-to-slot mode. The results for the frame mode are similar to the first column in the table (N = 1).

It can be seen that in order to have a reasonable utilization irrespective of the frame transmission rate, i.e. irrespective of the value N, then there must be room for several frames in the output buffer. Increasing the buffer size above 20 frames gives only a little gain in utilization.

The analysis presented here has not taken into account the multi-timescale variations exhibited by connectionless traffic [Lel91]. Therefore, the figures given in the table are only indicative, and further analysis is needed to accurately determine the loss behavior of the CLS. However, there is no doubt that buffering in the CLS is needed, unless the frame transmission rates are small compared to the rate allocated to the VPCs. This conclusion is confirmed by the simulation study described in the next section.

4 Simulation Results

This section presents simulation results for the frame loss ratio in the CLS, which provide further insight into the critical aspects identified and analyzed in the preceding sections. The simulation results have been obtained with a discrete-event simulation model of the CLS, developed in the C programming language with the tool ATLAS [Rüm91]. The CLS simulation model is described in detail in [Vaz93a].

In the simulation experiments, the CLS model has been loaded with two types of traffic. Firstly, a superposition of frame sources with exponentially distributed on and off periods. The model allowed us to simulate several input VPCs, each of them supporting interleaved frames from several sources, and with selectable average frame length, average traffic per source, etc. Secondly, a model of mixed file transfer and interactive traffic generated by Ethernet LANs attached to a DQDB network. The DQDB network was modeled as a process sharing queue, and was connected to the CLS model via a DQDB-to-ATM interworking unit operating in cell-to-slot mode [Aal92], [Man93] as illustrated in figure 1. The results reported below correspond to the first traffic type only. The results obtained with the other traffic type are very similar.

4.1 CLS Design Considerations

The CLS simulation model represents a particular connectionless server, manufactured by Alcatel [Del92], which will be used in the test-bed of the COMBINE project. From the point of view of frame loss performance, one of the main characteristics of this CLS is the use of a mechanism called Predictive Congestion Control (PCC), designed to reduce the number of frames corrupted by cell losses. This mechanism tries to avoid the overflow of the CLS output buffers by preventively rejecting (full) incoming frames when the total cell input rate, predicted by the PCC, is above a certain threshold. The idea is that the rejection of a few selected frames will avoid the overflow of the output buffers, which would cause an uncontrolled loss of cells potentially corrupting many of the frames in transit through the CLS.

One key point of the PCC algorithm is the estimation of the aggregated input rate of all the cells going to a given CLS output buffer at a certain time. This input cell rate, which results from all the frames being processed by the CLS which are addressed to the considered output, must be kept below a certain level to avoid the overflow of the output buffer. The algorithm, however, does not completely eliminate the buffer overflows for several reasons. Firstly, the cell rate of each incoming frame is considered constant, but in fact it may change significantly throughout the frame duration, for example in case of frame interleaving in the CLS inputs. Secondly, it has been found in practice that the threshold for the total traffic allowed to pass to a given CLS output must be higher than the capacity allocated for the VPC serving this output. Otherwise the PCC algorithm would be too pessimistic, i.e. it would reject many frames unnecessarily. This means that during some periods the total cell input rate may exceed the capacity of a CLS output, so its buffer may overflow.

The ratio between the maximum aggregate input rate allowed on a CLS output and the capacity of the corresponding VPC is called the factor F in the simulation results presented later. This factor should be dimensioned in order to minimize the total number of lost frames, that is, frames rejected by the PCC plus frames corrupted due to cell overflows in the output buffers. (In general, the criterion for dimensioning

F could weight differently the importance of rejected frames and that of corrupted ones. The cells of rejected frames are discarded immediately. Corrupted frames, on the contrary, are partially transmitted to the next CLS or to the user, causing a waste of network resources. Therefore, their impact on the overall network performance is worse.)

In principle, the number of rejected plus corrupted frames as a function of F will have a minimum for some value of F. If F is too low, too many frames will be rejected. If it is too high, not enough frames are rejected and, therefore, the number of corrupted frames will increase. In practice, however, the optimization of F is not easy, because the optimum value will depend on the characteristics of the traffic loading the CLS, and also on internal CLS characteristics (e.g. the output buffer size). The simulation results presented in this paper have been obtained with a value F = 2, which has been chosen based on the results of preliminary simulation runs not reported here.

Additional simulation results about the PCC algorithm can be found in documents of the COMBINE project [Vaz93b]. Anyway, a detailed study of congestion control algorithms such as the PCC or others reported in the literature, e.g. [Boi93], [Man93], [Vog92], is out of the scope of this paper, which focuses on the overall performance of the CLON. A careful tuning of the parameter F in the case of the PCC (possibly using a dynamic algorithm in order to adapt F to changing traffic conditions), or the use of more sophisticated output control mechanisms may reduce the buffer size required to achieve a given FLR. Nevertheless, if the required FLR is in the order of 10^{-4}, the CLS buffer capacity will have to be in the order of *frames* (the buffer size expressed in frames is the parameter M of section 3), not just a few tens of cells. As discussed in section 3, if the CLS buffers are small, an alternative way of dealing with several concurrent frames directed to the same CLS outlet is to increase the ratio between the rates of the outgoing VPCs and the frame transmission rate (that is, the parameter N). From the point of view of the allowable network load, see table 1, the former alternative is better.

4.2 Frame Loss Performance

The curves in figure 3 show the FLR in the CLS versus the total load in the output VPC for different input rates. In all cases, the output VPC rate is 100 Mbps and the buffer capacity is approximately 1 frame. (Actually the buffer size was Q = 64 cells for an average frame length of B = 50 cells.) Other relevant parameter values are shown below the graph.

As predicted by the analysis, the comparison of the different curves shows that for a given traffic load, the higher the input rate, the larger the number of lost frames. In other words, for a given maximum FLR, a high output VPC utilization can only be achieved if the input VPC rate is reduced. For example, it can be seen that with this small buffer capacity, a FLR below 10^{-4} can only be obtained for reasonable load values if the input VPC rate is at least 5 times smaller than the output rate. Of course, if the amount of traffic generated by the users is fixed, the reduced VPC input rate is paid in terms of longer transmission and queueing delays, because the input VPCs will be more loaded.

In this experiment the frame transmission rate is equal to the (peak) rate allocated to the input VPCs, because there is only one frame source using each input VPC. Therefore, the factor N introduced in the analysis is here equal to the ratio between

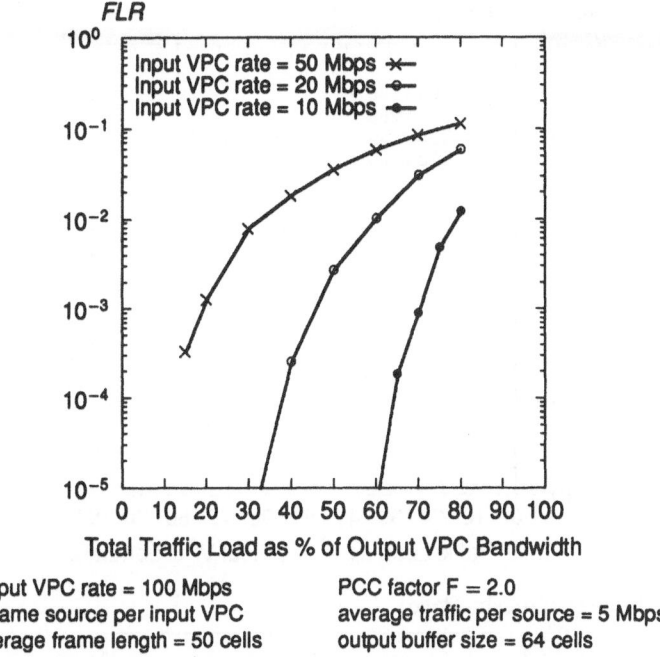

output VPC rate = 100 Mbps   PCC factor F = 2.0
1 frame source per input VPC   average traffic per source = 5 Mbps
Average frame length = 50 cells   output buffer size = 64 cells

Figure 3: Frame Loss Ratio vs Input/Output VPC Rates

the output VPC rate and the input VPC rate. It can be seen that the results in table 1, based on Poisson arrivals, provide a conservative estimation of the FLR shown in figure 3. Note that the simulation includes the effect of the PCC algorithm, which was not considered in the analysis.

Figure 4 shows the FLR that can be achieved by using buffers between 1 and 10 times the average frame size (Q from 64 to 512 cells for B = 50 cells). The curves

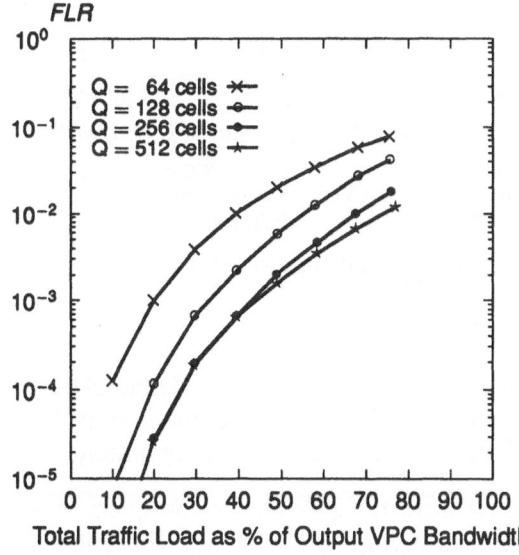

input VPC rate = 10 Mbps   output VPC rate = 25 Mbps
5 input VPCs x 4 frame sources per VPC   average traffic per source = 0.125 ... 1 Mbps
Average frame length = 50 cells   PCC factor F = 2.0

Figure 4: Frame Loss Ratio vs CLS Buffer Size

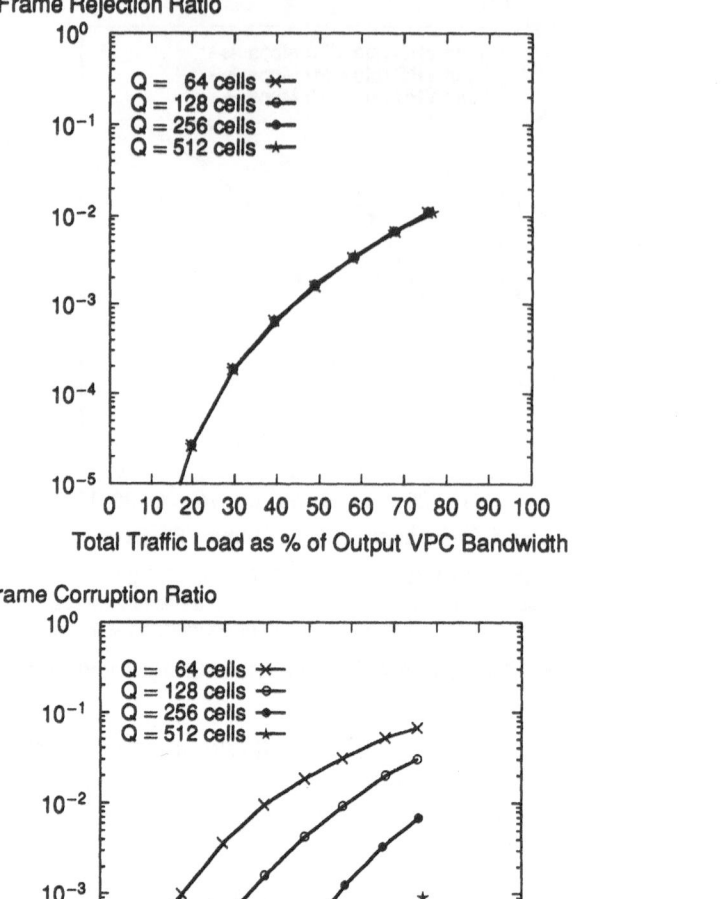

Figure 5: Rejected and Corrupted Frames vs CLS Buffer Size

show that with small buffer sizes an acceptable loss ratio can only be obtained if the traffic is very low, approximately 10 % or 20 % of the output VPC capacity. Figure 5 shows the rejected and the corrupted frame ratio separately.

For the smallest buffer size, the amount of corrupted frames is the dominant component of the FLR. However, note that with the value F = 2, the amount of rejected frames is itself above the 10^{-4} limit, except for small traffic values. Therefore, if the PCC algorithm was made more strict, i.e. smaller F, in order to reduce the frame corruptions caused by the frequent overflow of the small output buffer, the rejected frame ratio alone would exceed the total loss ratio allowed. In order to have an acceptably low corruption ratio for medium VPC utilization, it is necessary to have larger buffers. For example in this case, where the average frame length is 50 cells, the curves for buffer sizes of 256 or 512 cells, that is, about 5 to 10 times the average frame length, exhibit a much better loss performance. Note that if the frames were larger (the maximum frame length is over 200 cells), the buffer size should be increased accordingly.

When the buffer is large, the situation of the two FLR components is the opposite as before. For example, with buffers of 512 cells the total loss rate is dominated by the frames rejected in the PCC, which in fact becomes the limiting factor for the overall CLS frame loss performance. This can be seen by comparing the graphs in figures 4 and 5. For a buffer size of 512 cells, the number of corrupted frames, shown in the second graph of figure 5, is very small. However, the total frame loss ratio, shown in figure 4, is dominated by frame rejections in the PCC, so the potential FLR improvement in passing from a buffer of 256 cells to 512 cells is almost completely spoiled (except for high traffic loads).

The cause of this problem is that the PCC algorithm as currently simulated does not take into account any information about the maximum size or the current occupancy of the buffer it controls. (That is why all the curves of rejected frames in the first graph of figure 5 are the same, regardless of the value of Q.) If the buffer is very short, this information probably makes little difference, but if the buffer is large, the additional knowledge about its occupancy can be exploited to improve the control algorithm performance. In general, this would mean varying the factor F dynamically as a function of the traffic.

A particular case of this is simply to allow frame rejections only when the buffer occupancy is above a certain threshold. A more sophisticate extension of the PCC could be to predict the future evolution of the buffer occupancy, taken as input data the current level of occupancy, the capacity of the output VPC serving the buffer (which is known), and a prediction of the future cell input rate (already included in the PCC). Frames would be rejected only when the predicted buffer occupancy would exceed the buffer size.

Besides the reasons mentioned above, an additional cause of the relatively bad performance of the PCC algorithm in this experiment is the presence of frame interleaving in the input VPCs (note that figures 4 and 5 correspond to a traffic configuration where each input VPC is used by four frame sources). As discussed in section 4.1, frame interleaving introduces errors in the estimation algorithm used by the PCC, which may cause unnecessary frame rejections.

4.3 Delay Performance

As mentioned in section 2, the delay of a frame going through the CLS is limited to a few milliseconds, which imposes a limitation on the maximum queueing delay in the CLS, and hence on the maximum buffer space in the CLS outputs. Figure 6 shows the maximum delay for 95 % of the frames in the same scenario of figures 4 and 5.

The delay in the CLS is defined as the time since the first cell of the frame enters the CLS until it is transmitted to the output VPC. At low loads, the delay curves tend to a value of 2 ms, because the CLS simulated in this paper introduces a constant processing delay of 2 ms for all cells, which is added to the queueing delay in the output buffers [Del92]. At high loads, the delay curves show the impact of increasing the CLS buffer size. For example, for the highest traffic load (80 %) the 95 percentile

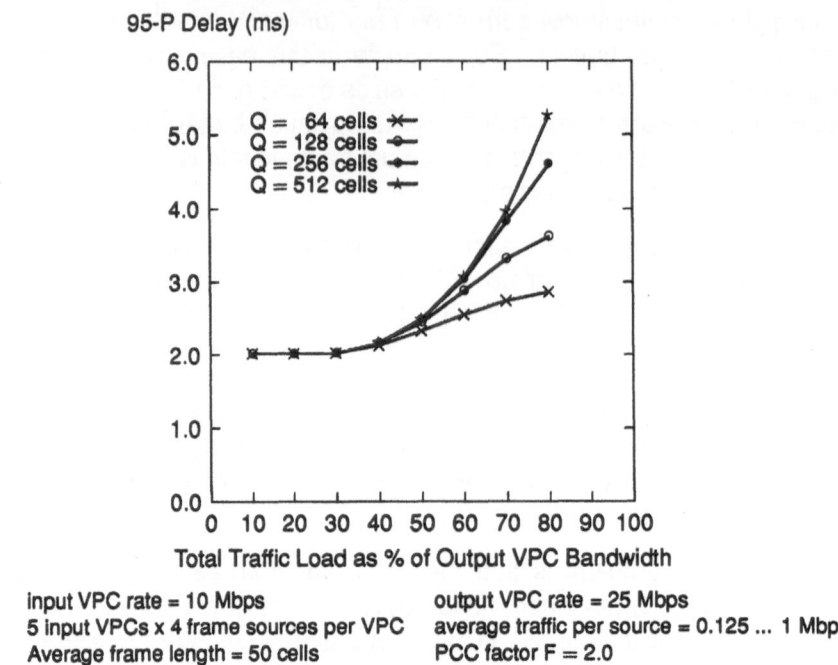

95-P Delay (ms)

Q = 64 cells
Q = 128 cells
Q = 256 cells
Q = 512 cells

Total Traffic Load as % of Output VPC Bandwidth

input VPC rate = 10 Mbps
5 input VPCs x 4 frame sources per VPC
Average frame length = 50 cells

output VPC rate = 25 Mbps
average traffic per source = 0.125 ... 1 Mbps
PCC factor F = 2.0

Figure 6: 95-Percentile Delay vs CLS Buffer Size

delay with buffers of 512 cells is almost the double of the same delay with buffers of only 64 cells.

The impact of the buffer size on the average delay is less noticeable. For the same load as before, when the buffer size increases from 64 to 512 cells, the increment of the average delay is only 20 % approximately (from 2.17 ms to 2.65 ms). Finally, note that the curves in figure 6 correspond to a relatively small output VPC capacity. If more capacity were allocated to the output VPCs, the delay would be reduced accordingly.

The delay curves can be used to determine the maximum buffer size that can be used in the CLSs for a given CLON configuration and maximum delay requirements. Then, the FLR curves can be used to determine the maximum load that can be reached without exceeding the required maximum FLR.

5 Conclusions

This paper has studied how buffering in the CLSs influences the achievable through-put of a connectionless overlay network for given maximum frame loss rate and delay requirements. A service strategy based on frame-level buffering, which allows the outlets of the CLS to handle several concurrent frames, has been proposed and evaluated. A conservative upper limit for the required buffer size has been derived by analysis assuming Poisson frame arrivals. Then, simulation results have been presented for a more detailed traffic model and different combinations of parameter values, such as output buffer size, average frame length, input and output VPC rates, traffic load, etc. In particular, the simulation included the effect of a preventive congestion control mechanism designed to reduce the frame loss ratio in the CLS.

The results obtained lead to the conclusion that buffering for at least a few frames (e.g. 10 to 20) is needed in the CLSs if the frame transmission rates are comparable to the rates allocated to the VPCs of the CLON. Increasing the buffer size above 20 frames gives only a little gain in utilization. In practice, the maximum buffer size is limited by the delay requirements. If the average frame length is small, it is possible to buffer the proposed number of frames, e.g. 20, without violating the delay requirements. In other case, the number of buffered frames will be limited by the maximum delay allowed. For example, the delay results presented for an average frame length of 50 cells give a maximum buffer of about 10 frames only. An additional consideration is that the PCC algorithm studied in this paper should include information about the buffer occupancy, in order to dynamically adapt itself to traffic variations.

In summary, without enough CLS buffering space, the frame loss performance of the CLS is severely degraded, and the only alternatives to meet the typical FLR requirements (e.g. 10^{-4}) are to accept either a reduced transmission rate from the user side, or a low traffic load, i.e. poor utilization of resources, from the network side.

Acknowledgements

The work described in this paper has been done in the framework of the RACE project COMBINE (R2032). The authors want to acknowledge the members of Work Package 1 (performance evaluation) of this project for fruitful discussions.

References

[Aal92] S. Aalto, et al. *Performance aspects of streaming and message modes of operation*. First International Symposium on Interworking, Interworking '92, Bern, November 1992.

[Bel92] Bellcore. *Broadband ISDN Switching System Generic Requirements*. Technical Advisory TA-NWT-001110, issue 1, August 1992.

[Boi93] G. Boiocchi, et al. *ATM Connectionless Server: Performance Evaluation*. IFIP Workshop TC 6: Modeling and Performance Evaluation of ATM Technology, La Martinique, January 1993.

[CBDS-92] ETSI. *Network Aspects (NA); CBDS over ATM*. Doc. ref. DE/NA-53203, Draft, November 1992.

[Del92] D. Deloddere, P. Reynders, P. Verbeeck. *Architecture and Implementation of a Connectionless Server for B-ISDN*. International Switching Symposium ISS'92, Yokohama, October 1992.

[Kvo92] K. Kvols (leader), I. Norros (editor), RACE 2032 COMBINE deliverable: *Models for Heterogeneous Networks and Their Elements*. Delivered to CEC December 11, 1992.

[Lel91] W. Leland, D. Wilson. *High Time-Resolution Measurement and Analysis of LAN Traffic: Implications for LAN Interconnection*. IEEE INFOCOM'91, 1991.

[Man93] S. Manthorpe. *Buffering and Packet Loss in the DQDB to ATM Interworking Unit*. Integrated Broadband Communication Networks and Services, IBCN&S, Copenhagen, April 1993.

[Ras91] C. Rasmussen, et al. *Source Independent Call Acceptance Procedures in ATM Networks*. IEEE JSAC, 9(3), 1991.

[Rüm91] M. Rümekasten. *ATLAS Reference Manual*. Dep. of Mathematics and Computer Science, University of Paderborn, December 1991.

[SMDS-91] Bellcore. *Generic System Requirements in support of Switched Multi-Megabit Data Service*. Doc. TR-TSV-0007772, issue 1, May 1991.

[Tur91] J.S. Turner. *A Proposed Bandwidth and Congestion Control Scheme for Multicast ATM Networks*. Washington University Technical Report WUCCRC-91-1, May 1991.

[Vaz93a] E. Vázquez. *CLS Simulation Model*. Tech. Report COMBINE/DIT/WP1/-003/02, RACE R2032 (COMBINE), March 1993.

[Vaz93b] E. Vázquez. *CLS Packet Loss Performance*. Tech. Report COMBINE/DIT/-WP1/007/01, RACE R2032 (COMBINE), April 1993.

[Vog92] R. Vogt, et al. *A Concept for Interconnecting DQDB MANs through ATM-based B-ISDN and Related Issues with respect to Simulation*. First International Symposium on Interworking, Interworking '92, Bern, November 1992.

INDEX

Token ring, 146, 295-298, 314
TP-4, 1, 125
Traffic
    characteristics, 66, 77
    locality, 233-234
Transient analysis, 109
Transport connection, 23

Uniformization, 109, 111, 115-118, 120-123
Universal interface, 67
Urn-model, 84
Utilization factor, 52

Virtual
    access point, 29, 32
    capacity, 51
    database, 28
    image, 29
    link capacities, 51
    networks, 68, 187
    paths, 50, 341
Visa protocol, 210, 216, 218, 226
Visible interface, 31

Wide Area Networks (WAN), 48, 79, 312, 322

X.25, 23-24, 36,146, 321
XNS, 212